Murat Özyavuz (ed.)

Sustainable Landscape Planning and Design

PETER LANG

Bibliographic Information published by the Deutsche Nationalbibliothek
The Deutsche Nationalbibliothek lists this publication in the Deutsche Nationalbibliografie; detailed bibliographic data is available in the internet at http://dnb.d-nb.de.
Library of Congress Cataloging-in-Publication Data
A CIP catalog record for this book has been applied for at the Library of Congress

ISBN 978-3-631-73439-1 (Print)
E-ISBN 978-3-631-73440-7 (E-PDF)
E-ISBN 978-3-631-73441-4 (EPUB)
E-ISBN 978-3-631-73442-1 (MOBI)
DOI 10.3726/b11768

Peter Lang – Frankfurt am Main · Bern · Bruxelles · New York · Oxford · Warszawa · Wien

This publication has been peer reviewed.

www.peterlang.com

Sustainable Landscape Planning and Design

Contents

Assoc. Prof. Dr. Seda H. Bostanci[1]
Assist. Prof. Dr. Pınar Öztürk Demirtaş[2]

Influence of Urban Furniture Design on Environmental Behavior

[1]Namık Kemal University, Faculty of Economics and Administrative Sciences, Political Science and Public Administration Department

[2]Okan University, Faculty of Art, Design and Architecture, Industrial Product Design Department

1. Introduction

Today, at different scales, design products have become the main research area for designers in urban to neighborhood scale design, allowing them to communicate with the user and to meet the user's behaviors, tendencies, and anticipations. This approach in the field of environmental psychology emerged from a combination of experimental and behavioral issues, which are aesthetic dimensions in the formation of innovative research. Developments in the field of environmental psychology have been demonstrated by various pieces of social, cultural, cognitive and experimental research by experts in this field, where positive urban experiences are seen to cause a change of behavior beyond a change in people's taste and a level of happiness with aesthetic, cognitive, semantic, functional and perceptual variables [1; 2; 3; 4; 5; 6].

Sustainability is a framework concept that includes economic, social and ecological dimensions to form what we consider to be a human-focused environmental approach, and one that seeks to find an equilibrium between the resources available today and a rational and decent usage of these by future generations, as well as striking a balance between urban conservation and development. The work on sustainability means that the community can fulfill its responsibilities in this field by developing a community structure which is itself sustainable [7]. Sustainable communities are those where there is harmony between nature and the environment. Their lifestyle inspires thoughts about sustainable development and sustainable governance [8]. We can see that the cultural, sociological and economic strata that build diversity and difference within the development of the environmental consciousness, as well as the environmental values of societies among themselves are as decisive in this area as they are elsewhere. It may be thought that some societies need to be directed in the field of environmentalism whereas others set a precedent for their traditions in this area. The design of urban furniture can be seen as an innovative tool to design and guide the environmental awareness of social life for the inhabitants of urban open spaces.

Academic studies intended for fostering environmentally friendly behavior are increasing in the fields of product design and marketing [9; 10; 11]. " 'Design for Sustainable Behavior' (DfSB) is a new field of enquiry exploring how design can influence user behavior to reduce negative social or environmental use impacts" [12].

Urban furniture has the main function of improving the quality of life in public spaces such as lighting at night, comfortable elements in parks, garbage collection, recycling of waste, navigation and wayfinding. This study discusses the effects of urban furniture on environmental sustainability. By studying sustainable design concepts with different aspects, the design processes of urban furniture are discussed considering the relation between environmental psychology and design. The study brings forward suggestions about the future of this field by giving examples of local governments, non-governmental organizations and university studies on the achievements of urban furniture design which aims at sustainability.

2. Sustainable Design Approaches

The responsibility of the designer is influential in the effectiveness of the design in terms of sustainability. Design can contribute to sustainability in many ways, such as the use of materials, the form of production, environmental awareness, and the orientation of user behavior. It is important to develop the life cycle of the products [13]. Within the framework of sustainable design, ecosystem design for the environment is included in relevant literature and research. The definitions regarding where the concept of "environment" in terms of sustainability appears in the study can be presented with the titles of design for environment – DfE, eco-design, eco-effective design, environmentally responsible manufacturing, green design and life cycle design. Tingstrom (2007), basing his work on a broad set of references, explained the definitions of these concepts as follows: "Design for environment-DfE: Systematic consideration of environmental, human health and safety impacts in design products and processes. Eco-design: Multi-dimensional features of environmental values to be considered before the cost in product design. Eco-effective design: Design function, quality, product and process targeted at the efficiency and productivity of environmental performance. Environmentally responsible manufacturing: Promotion of the adaptive approach, as in the design process of resource use. Green design: An innovative design concept that is seen as an opportunity in the design process, in the direction of sustainable goals of protecting environmental values. Life cycle design: A design concept that evaluates recycled materials by taking the design product as a life cycle" [14].

Considering the above concepts, it is apparent that there are environmental sustainability approaches such as function, cost, quality, minimizing environmental effects, preventing waste formation and environmental sensitivity. Accordingly, the design itself must bear environmental responsibility in terms of structure and function – it should be able to create an environmental point of view for the user and be responsible for creating social behavior.

Recycling waste packaging such as plastic bottles, batteries and beverage cans, as well as receiving a payment or an award in return, is another example of this approach. These recycling vehicles, which are located in public areas, either open or closed, have widespread use in many countries [15].

Urban furniture, open to public use and designed for the city, is important in terms of environmental responsibility. From this point of view, there are many design approaches to achieve sustainable behavioral habits. For example, garbage separation and the development of people-oriented design is an approach to direct behavior towards waste separation; and today, as seen in Figure 1, it is used as a social responsibility and policy in many countries.

Figure 1: Developing behavioral patterns for environmental sustainability–Waste disposal–Düsseldorf / Germany (Photo by Pınar Öztürk Demirtaş).

Cultural differences among societies have brought with them different behavioral habits. These differences must be taken into account when creating design approaches to develop environmental responsibility in the community. For example; in some societies, seed sales and consumption have become a habit in open spaces, in parks and in coastal areas, and as a result, seed shells create an environmental problem. As shown in Figure 2, it may be a matter of design to develop a behavioral model for the user to solve this problem.

Figure 2: Student design project on seed sales and garbage collection unit in areas such as park, coastal region, etc. (Okan University – Department of Industrial Product Design).

"It is aimed to make people do the things – that they are not willing – voluntarily by performing a positive action. It seems that the designed product is a 'seed sales unit' which gives back some of the money as a "deposit" when it is given the garbage. It is planned that the user will be rewarded when the refuse provided during the sale is filled and returned" [16].

Figure 3. Urban road separator areas, the use of waste tires to create an environmental layout in green islands–Düsseldorf–Germany (Photo by Pınar Öztürk Demirtaş).

One of the methods for sustainability is using a recyclable material for the design of the chosen material, whereby the expired products can be reused for different purposes is another solution,– as we can see in the example in Figure 3. The design area produces creative solutions with environmental values and sustainable and user-friendly product development interface.

3. Urban Furniture in the Context of the Environmental Behavior and Design Relation

Being able to relate the causal relationships of human behaviors to the field of environmental psychology using observation has the qualities from the disciplines of psychology, environmental sciences, and design. According to Skinner (1953), "behavior is a process, rather than a thing; it cannot easily be held still for observation. Education emphasizes the acquisition of behavior rather than its maintenance" [17].

Sustainable behavior training comes in from the outset as one of the main determinants in the mold development process. Urban open spaces provide significant information on observation related to the quality of life. These areas including squares, parking areas, central business districts, and coastal area design are among the main objectives of urban design. Urban open spaces that enable people to move, perform various activities, and spend time can create areas which meet basic needs such as design features and safety. The design features of various urban furniture such as lighting elements, seating groups, garbage cans, recycling boxes, stops, water items, and panels are influential in making these areas more comfortable and attractive.

The vitality of the urban environment can be determined by examining the activities of people in the outdoor environment. “Some activities within the building may benefit from being able to extend outwards into adjacent public outdoor space. When this occurs, they will contribute to the activity in the public space itself” [18]. According to Lynch (1984) the view from inside the person shows that “the environment is to be valued by the way in which it is individually experienced and for such qualities as openness, legibility, meaning, educativeness, and sensory pleasure” [19]. “The visitor experiences not only a sequence of sights but the constant gradual transformation created by perspective and lighting in every wall or constellation of elements. The architect has helped to translate the visitors’ purely physical movement into a corresponding visual event” [20]. Urban experience constitutes the mental memory of the user’s visual and other sensory impressions.

Urban life, morphology, design and the result of human-specific patterns of behavior and expectations are gaining authenticity and meaning. “One of the basic reasons for creating or redeveloping cities, constructing buildings, making plazas, gardens, and new infrastructure systems is to provide for some existing or potential set of human activities” [2]. “Human brain constitutes cognitive models for various conditions and events in order to adapt to the urban environment and lead a life based on experiences. Based on multidimensional sensory experiences, people create an internal model of a city” [21]. “The pre-perceptual environment is described as ordered, self-regulated, and composed of stable patterns and nesting subsystems. The unit of analysis is the behavior setting and the behavior episode which are natural molar units. Within a behavior setting a person contributes to the setting by which he, himself, is constrained. The life space is the means by which the setting secures the behavior appropriate to it” [22]. Considering these aspects, environmental psychology should have imagination about how users will behave and what they will feel in designed spaces beyond the ability of designers to realize formal features and visual effect. Today, such a complex and difficult task is not addressed by a designer, who undertakes it on its own, but design comes together with experts and urbanists from different disciplines in a participatory process. Building blocks and roads have features that are difficult to change for many years after having been built. Besides, urban furniture can be replaced more flexibly and easily. Thus, the expectation of innovation in the city can be influenced by the attractive, aesthetic value of urban furniture. With another approach, this negative image can be reduced with urban furniture in a region where the city is difficult to renovate and is not liked by its users. Thus, urban furniture can mediate the breakdown between the user and the space with innovative designs in settlements that are inadequate in terms of environmental psychology.

4. Urban Furniture Design for Promoting Environmental Awareness

All over the world, professional institutions from different disciplines and local governments continue to work on environmental pollution, the disappearance of natural resources, production and waste problems. Various studies address urban furniture designs as being among the elements of public space design which can ensure wellbeing and sustainability [23].

Foundations, companies and educational institutions also work in Turkey to develop environmental protection and reduce environmental pollution in society. Claim Your Waste Foundation is an institution that cooperates with media, non-governmental organizations and municipalities to raise environmental awareness. For example, this institution used cigarette butt boxes with questions and answers on them in order to raise the awareness about throwing the butts in the garbage. This worked via a "game" and monitored the reaction of society in the region of Üsküdar in İstanbul [24].

In 2015, in the Claim Your Waste Foundation workshops collaborating with Okan University (Istanbul), products from waste PET bottles were produced within the festival of "Treasury Came Out" [25], serving different purposes to students.

Nevertheless, municipalities organize projects, events and competitions to raise awareness of this issue. For example, in the context of Istanbul Kadikoy Municipality Environment Festival, the Kadikoy Design Workshop and Sustainable Urban Furniture Workshop are organized to encourage designers, media and students [26].

Considering the decisive influence of design work on design in industrial design departments in universities, it is seen that students' projects are carried on especially through sustainable product designs [27; 28].

Different experimental work is being carried out into developing the habit of throwing garbage in garbage cans in society. The 'Talking Garbage Cans' project is one of them. As an approach to encourage behavior to encourage the recycling of glass and plastic containers, the plan here is that the garbage cans communicate with people and set up verbal incentives for the garbage to be thrown into them. For two days the garbage cans called out to people who passed by them and asked them to throw plastic and glass bottles to 'them'. At the end of these two days, 275kg of glass and plastic packaging were collected, and it was determined that this total was 3.5 times more that of the regular collection [29].

5. Conclusion

This study involves the interpretation of urban furniture and sustainable design relation in the context of environmental psychology. The role of urban furniture designers in the developmentof user relationships and sustainable attitudes and behaviors are of common interest in the fields of behavioral sciences and environmental psychology.

Progressive technology and social progress towards excessive consumption, such as the consumption of natural resources and waste production, result in pollution of the environment. Design plays an important role in fighting this situation, because it is part of the design of all the products that people get and prefer to use. Designers and products that are concerned about sustainability can create new forms of behavior for their users that can develop responsibility for the environment. It applies to all design categories from personal use products to public use products. City furniture is a structure that is open to the common use of society, can follow others' behaviors and exemplify them, and can create a series of habits. These products, which are located in the external environment of people, should be a positive and responsible contribution to the environment. For this reason, sustainable urban furniture is now explored in many areas such as municipalities, foundations, and educational institutions. This study deals with the increasing importance of this area through limited examples. At the same time, the potential impact of changes in design on behavior, the sustainable society structure, the contribution to the goal of co-design and co-operation are discussed. Findings have thus been made that sustainable communities can mediate design on different scales in the process of transformation.

References

1. I. ALTMAN, M. M. CHEMERS: *Culture and Environment. 2, CUP Archive* (1984).
2. J. T. LANG: *Creating Architectural Theory: The Role of the Behavioral Sciences in Environmental Design. New York: Van Nostrand Reinhold* (1987).
3. A. RAPOPORT: *Human Aspects of Urban Form. 3, Oxford: Pergamon* (1977).
4. J. L. NASAR: *The Evaluative Image of the City. Thousand Oaks, CA: Sage Publications* (1998).
5. S. KAPLAN: *New Ways to Promote Proenvironmental Behavior: Human Nature and Environmentally Responsible Behavior. Journal of Social Issues*, 56 (3), 491–508 (2000).

6. R. GIFFORD: *Environmental Psychology: Principles and Practice. Colville, WA: Optimal Books* (2007).
7. M. ROSELAND: *Toward Sustainable Communities: Solutions for Citizens and Their Governments. 6, New Society Publishers* (2012).
8. S. H. BOSTANCI, A. N. ALBAYRAK: *The Role of Eco-Municipalities in Climate Change for a Sustainable Future. In Environmental Sustainability and Climate Change Adaptation Strategies (Ed. Ganpat, Wayne, Isaac, Wendy-Ann). IGI Global*, 213–231 (2017).
9. D. LOCKTON, D. HARRISON, N. STANTON: *Making the User More Efficient: Design for Sustainable Behaviour. International Journal of Sustainable Engineering*, 1(1), 3–8 (2008).
10. D. LILLEY: *Design for Sustainable Behaviour: Strategies and Perceptions. Design Studies*. 30(6), 704–720 (2009).
11. K. LIDMAN, S. RENSTROM: *How to Design for Sustainable Behavior? Chalmers University of Technology, Sweden* (2011).
12. T. BHAMRA, D. LILLEY, T. TANG: *Design for Sustainable Behaviour: Using Products to Change Consumer Behaviour. The Design Journal*, 14(4), 427–445 (2011).
13. N. BEYAZIT: *Endüstri Tasarımı Temel Kavramları. İdeal Kültür Yayıncılık* (2011).
14. J. TINGSTROM: *Product Development with a Focus on Integration of Environmental Aspects. PhD. Thesis, Royal Institute of Technology, Stockholm* (2007).
15. *Greece: An Advertising Based Recycling Program*. Retrieved on 4 27, 2017 from http://www.envipco.com/installations/greece.php
16. P. Ö. DEMİRTAŞ, S. H. BOSTANCI: *Analysing of Design Approaches of Urban Furniture at City Centres Over Students Projects. Balkan Journal of Social Sciences* (2017).
17. B. F. SKINNER: *Science and Human Behavior. Simon and Schuster. The Free Press* (1953).
18. I. BENTLEY: *Responsive Environments: A Manual For Designers. Routledge* (1985).
19. K. LYNCH: *Good City Form. MIT Press* (1984).
20. R. ARNHEIM: *The Dynamics of Architectural Form. (Vol. 376). University of California Press* (1977).
21. S. H. BOSTANCI, M. ORAL: *Experimental Approach on the Cognitive Perception of Historic Urban Skyline. ICONARCH III*, 192–201 (2017).
22. R. G. BARKER: *On the Nature of the Environment. Journal of Social Issues* (1963).

23. K. W. M. SIU, P. H. WAN: *Sustainable Urban Living Environment: Street Furniture Design for an Urban Fishing Village. International Journal of Environmental, Cultural, Economic and Social Sustainability* (2011).
24. *Atma Oyla Projesi,* Retrieved on April 27, 2017 from http://www.copunesahipcik.org/faaliyetlerimiz/saha-calismalari/
25. *Çöpten Hazine Çıktı,* Retrieved on April 25 *2017* from https://www.okan.edu.tr/arastirma/sayfa/5185/copten-hazine-cikti/
26. *Kadıköy'den Kendi Elektriğini Üreten Kent Mobilyası Tasarım Çağrısı,* Retrieved on April 25 2017 from http://bigumigu.com/haber/kadikoyden-kendi-elektrigini-uretenkent-mobilyasi-tasarim-cagrisi/
27. O. ER: *Endüstri Ürünleri Tasarımı Bölümlerinde Sürdürülebilir Tasarım Eğitimine Yönelik Öğretim Yaklaşımları ve Proje Dersleri İçin Uygulama Önerileri. Anadolu Üniversitesi Sosyal Bilimler Dergisi,* 15(4) (2015).
28. M. RAMIREZ: *Promoting Sustainability Through Industrial Design Studio Projects, International Conference on Design Education. University of New South Wales, Sydney, Australia* (2007).
29. *Talking Garbage Cans,* Retrieved on May 2 2017 from https://www.youtube.com/watch?v=blulbrClL90

Fatma Aşilioğlu[1]

Permeable Pavements for Pedestrian Use

[1]*Landscape and Ornamental Plants Program, Kalecik Vocational School, Ankara University, Kalecik 06870, Ankara, Turkey, e-mail: fatma.asilioglu@ankara.edu.tr*

1. Introduction

The majority of structural environments in urban areas are pavements. The area covered by pavements is almost twice the area of that by buildings, and this has a significant impact on the urban environment [1, 2]. Approximately 50% of the world's population live in urban environments covered by impermeable surfaces and environmental problems related to this are increasing day by day [3]. Many researchers regard permeable pavement (PP) systems as a solution, and they commonly focus on interlocking concrete pavers, pervious concrete and pervious asphalt [2, 4–9]. Some researchers add open-celled paving grids and plastic geocells to them [3, 10, 11]. Moreover, unbound aggregate, turf, soft pervious pavements, decks and stone pavers, and bricks designed with open joints are also found on this list [1, 12]. PPs are a valuable alternative that can be used in the place of conventional impermeable surfaces in areas such as roads, carparks, pedestrian areas, walkways, playgrounds and parks [1, 3, 4, 7]. Open-celled paving grids and plastic geocells are not preferred for pedestrian use because they do not provide a suitable surface to walk on. PPs provide a direct contribution to sustainability and the environment [2] through benefits such as stormwater infiltration, improved water quality, reduction in runoff and urban heat island effect, recharging of groundwater, and improving conditions for trees [1, 4, 7–10, 13]. Moreover, they provide safety in wet conditions and are aesthetically pleasing. It is also an advantage to obtain some PP materials from natural and recycled sources [1]. However, there are disadvantages such as not allowing surface water harvesting, clogging, disputes about the geotextile layer [1, 3, 14], high cost, moisture damage and groundwater contamination [9]. In this chapter, PP types suitable for pedestrian use are examined and their functions and benefits, limitations and disadvantages are discussed.

2. Types of Permeable Pavements

PPs are a new technology with different purposes and design to conventional pavements [3, 4]. There are various types of PP, and the main differences between the types are the amount of void, the design of layers and structural strength.

Almost all PPs have a similar general structure [5, 8–11] (see Fig. 1):

Top layer (surface layer, cover or permeable paver units): This is the top layer seen by humans and the pavement type is defined according to this layer.

Bedding layer (bedding course): This is the layer which fill the voids of interlocking concrete and stone paver and is laid under the pave material. Usually sand is preferred.

Gravel base (base reservoir, base course, small stone granular base): In the majority of permeable pavements, a gravel support layer is needed to increase strength and load- bearing capacity. This layer also allows the stormwater to be removed quickly.

Sub-base (sub-base reservoir, rock sub-base, large stone granular base): This is the actual layer where the water is stored, which contains a larger aggregate than the gravel base.

Underdrains: These are drainage lines consisting of small-diameter plastic pipes and are located just below or near the sub-base to collect the water and convey it to the sewer network. They are frequently used in clay containing soils and high water tables.

Geotextile: This is laid under the sub-base to separate the subgrade from the aggregate so that the soil does not fill the gravel pores.

Subgrade (sub-base, in situ soil, underlying soil): This is the bottom natural ground and is compressed at different levels depending on the purpose of the pavement.

Fig. 1. General structure of PPs

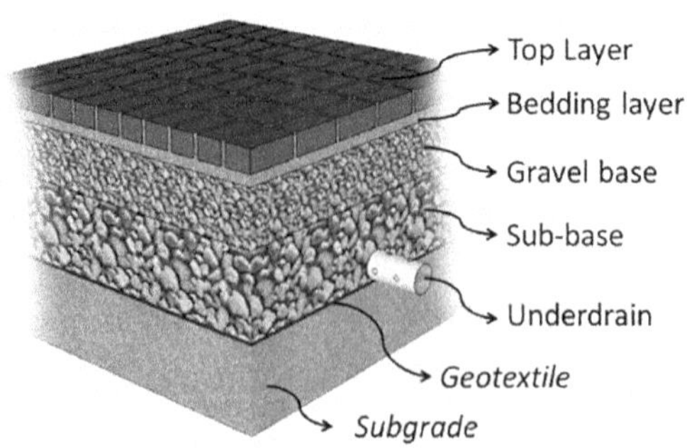

2.1. Open-jointed paver units

Paving blocks are concrete, brick or stone units and aggregate or turf in the joints enables these to be porous [1] (see Fig. 2). They are used on driveways, walkways and low-use parking lots. They can be found in many different colors, forms, sizes and thicknesses. Thicker pavers are suitable for heavier use. It is visually impressive [12]. However, paver blocks are more expensive, deformation can occur in the lower layers and require coating to prevent swelling [1]. Retaining is required to fix the edges. It is not suitable for areas with more than a 5% slope. Occasionally, joint material needs to be renewed and weeded [12].

Fig 2. Types of open–jointed paver units

2.2. Pervious concrete

Pervious concrete is a mixture of cement, coarse aggregate, water and admixtures. It does not contain sand, and has a void content between 15–30% [15], see Fig. 3 [16].

Fig. 3. Pervious concrete

If necessary, the addition of fine aggregate reduces the void ratio but increases the strength [17]. The single-size stone layer under pervious concrete allows the water to be stored before it seeps into the soil. This removes the need for surface drainage structures that are expensive and cause loss of land. Pervious concrete drains the water quickly and prevents ponding, so the risk of slipping and falling is reduced. It is weaker than conventional concrete due to its low mortar and high void content. But it is sufficient for pedestrian use. It requires little maintenance [18]. The cost of pervious concrete can be higher but it is a long-lasting pavement. Cracks can occur due to soil movement [1].

2.3. Pervious asphalt

Pervious asphalt is a mixture of fine and coarse aggregate and bituminous asphalt binder [1, 12]. It contains void between 15–20%. Depending on user load, the thickness ranges from 7.5 to 18cm. A base course is required to increase strength and storage capacity [10]. Pervious asphalt, parking lots, walkways and roads are suitable. [12]. Asphalt is a common and cheap paving material. In addition, pervious asphalt is permeable [1]. Pervious asphalt is paved quickly and easily. The cost is similar to conventional asphalt. With it, uninterrupted and flexible surfaces can be obtained. Its maintenance requirement is low and it should be cleaned with low pressure water or mechanical sweepers [12]. In some applications the asphalt binder may cause clogging. Or, when the binder is too liquid, it may flow down the pores and release the aggregate [1]. Recycling is limited and the surface is rougher [12].

2.4. Unbound aggregate

Aggregate types are gravel, crushed stone, crushed brick, decomposed granite etc, see Fig. 4. A single-size aggregate mass contains void between 30–40% and has high permeability. Aggregate can consist of rounded or angular granules and can be in dense-graded or open-graded form. The granule size affects the porosity, stability and walking comfort [1].

Fig. 4. Unbound aggregate samples

Suitable aggregate size for walking and parking areas is between 0.074 mm and 1.27cm [19]. Unbound aggregate can be obtained from limestone, dolomite, granite, traprock, sandstone, marble and quartzite. When strong and durable aggregate is obtained from granite, hard sandstone, hard limestone and traprock, weak aggregate is obtained from soft limestone. Weak aggregate is sensitive to pressure and water effect [1]. The surfaces of compressed fine aggregate and clay are also used as unbound PPs, which offer natural, cheap and easy to build options for pedestrian use, especially in large parks [20]. Unbound aggregate is suitable for residential applications, low-use parking lots and pedestrian walkways with light traffic and pedestrian use [1, 12]. For most areas, aggregate offers a cheap and durable surface. In addition, the single-size aggregate is a good choice with high permeability for the tree root zones [1]. Depending on the main material used, it can be obtained in very different colors and it is a natural, non-invasive and recyclable material [12]. Plowing can deform the surface. Raking can be required to fix appearance and squeeze. Runoff may cause splitting and crusting. When the surface layer is deformed, the underlying coarse gravel particles release. In pavement and non-retaining edges weeds grow and work is required with herbicides [1]. Unbound aggregate requires a retaining edge not to scatter. In some cases, a geotextile can be laid to prevent the aggregate from filling in the sub-base [1, 12]. Taking precautions against loss of aggregate, replacing lost material and weeding are necessary. It is not suitable for areas with a slope of more than 3%. It is not suitable for frequent use by roller skates, pushchairs, wheelchairs, bicycles and skateboards [12].

2.5. Turf

Turf areas cope easily with low pedestrian traffic. Roots maintain the permeability of the soil. In such areas, it is necessary to use a seed mixture suitable for being trodden on, as well as climate and soil. Maintenance work such as mowing, irrigation, fertilization and weeding is required as for all green areas.

If the soil is clay, heels and wheels damage the surface in wet weather. Such surfaces can be reinforced for heavier traffic using geocells, grids or blocks [1].

2.6. Soft pervious pavements

Soft paving materials are rubber granules, crushed shells or bark mulch from a natural or recycled source [5], see Fig. 5. These pavements cannot carry loads heavier than use by pedestrians and lose their permeability due to compression. Excessive use and loads cause granules to crush and move. Pavements containing light material can be dispersed by wind effect. In pedestrian areas, it can be preferred for its natural appearance, cheapness, the availability of recycling resources and ease of application. [1].

Fig. 5. Bark mulch

3. Functions and Benefits

Storm water management: A significant part of the rainfall in urban areas ends up on impermeable surfaces [3]. Impervious surfaces cause runoff during even small storms, because they are impermeable and they have limited storage [21]. Impermeable pavements can cause flooding and erosion of stormwater drainage systems. PPs reduce the amount of runoff by leaking some of the rainfall to the ground thanks to the voids. At this point, groundwater is charged and the need for a storm water drainage system is removed or the size of the necessary system is minimized [1–3, 5, 6, 10]. Compared to conventional drainage systems, storm water retention and infiltration is a more sustainable and cheaper process for urban areas [10].

Pollutant control and improving water quality: Pollutants such as heavy metals, total phosphorous, total nitrogen, oils, hydrocarbons, nutrients, bacteria, and organic matters are accumulated on impermeable surfaces and transported by runoff to reach water resources directly without being cleaned [3, 8, 21]. PPs are the technologies that can be installed to control and retain these pollutants [8]. They reduce the contamination of water resources by reducing the amount of runoff, filtering pollutants during underground infiltration, and helping to biodegrade hydrocarbon pollutants [2, 3, 5, 21]. They also charge aquifers and groundwater [2].

Improving conditions for urban trees: In order to take full advantage of urban trees, it is necessary to prepare the conditions for their normal growth. The tree's life and growth depends on the sufficiency of air, water and nutrients in the root zone. In a root zone which is buried under an impervious surface, they are quickly consumed and renewal is not possible. As a result, the tree remains small for its age, and becomes drought-stressed and susceptible to disease and insects [1]. Besides, the root of the drought-stressed tree seeks water near the surface and causes damage to the pavements, water pipes and buildings [3]. PPs allow optimal growth of trees by allowing air and moisture exchange in the root zone [1, 3], (see Fig. 6).

Fig. 6. PPs for urban trees

Reducing urban heat island effect: Approximately 90% of the increase in temperature in cities is due to building surfaces that absorb and store solar heat. These surfaces cause the temperature to increase even after sunrise by re-radiating the heat they have stored [1]. This is called the urban heat island effect, and impermeable pavements enhance it [1, 5, 22]. PPs, also referred to by the EPA as cool pavements, have been accepted by the United States Environmental Protection Agency (EPA) as a strategy to reduce the urban heat island effect [23]. The increase in temperature causes more energy to be used for cooling and more carbon dioxide emissions from more energy-producing plants [1, 7, 24]. High temperatures accelerate chemical reactions that convert emissions into ozone and cause air pollution. At the same time, runoff from the hot surfaces increases the temperature of the nearby waters, reduces the oxygen holding capacity and affects vitality [1] and sensitive biological communities [6] in the waters. High temperatures also bring about problems related to human comfort and health [1, 21]. The voided structure of PPs cools the surface with natural evapotranspiration. It reduces the urban heat island effect by allowing long-lived shade trees to grow [1].

Safety: PPs are more successful than impermeable pavements in terms of safety of pedestrians and vehicles. Water and oil that accumulate on the pavements leak down from the pores and surface accumulation is prevented. This reduces the risk of slipping. For example, the coefficient of friction of the impermeable asphalt drops to one in four when it is wet, whereas the permeable asphalt keeps its friction value in dry weather conditions [1].

Using recycled and regional sources: It is also an advantage to obtain some PP materials from natural and recycled sources [1, 25]. For example, aggregate can be a material such as gravel, crushed stone, crushed recycled brick, or decomposed granite [1]. Some aggregate types reduce the consumption of natural resources.

Plants that produce bricks and vitrified-clay products crush their faulty and unwanted products and leave them as aggregate [1]. Recycled aggregate can be used to make pervious concrete without compromising strength and performance [15]. Almost all ready concretes contain recycled materials such as fly ash, slag, or silica fume [18]. Aggregate is obtained from regional sources such as natural stones and other local material, and does not require transport [15, 18, 25]. Soft paving materials include any granular material from an organic or recycled source such as bark mulch, crushed shells, or rubber granules [5].

Aesthetic and naturalness: Open-jointed pavements made from natural stones, unbound aggregate, turf and soft pervious pavements provide a natural image and a satisfactory effect in pedestrian areas. PPs soften the character of paved areas and make them compatible with residential communities. On one hand they restore the perceptual link between the environment and society; on the other hand they improve natural functions. They also maintain the quality of the environment and human life as a whole [1].

4. Fulfilling ASLA and LEED sustainable design criteria

The Sustainable Sites Initiative, constituted by American Society Landscape Architects (ASLA), has developed a rating system for sustainable land design and development [26]. PPs can contribute to some of the credits in this system, see Table 1.

Table 1. ASLA's sustainability credits which are fulfilled by PPs

Credits	Description of the credits	How the PPs fulfill the credit?
Water	Credit 3.1 Manage precipitation on site	PPs retain precipitation volume through on-site infiltration, evapotranspiration and reuse and implement runoff-reduction strategies that also improve water quality.
	Credit 3.3 Manage precipitation beyond baseline	PPs prevent pollutants from reaching underground waters.

Credits	**Description of the credits**	**How the PPs fulfill the credit?**
Soil + Vegetation	Credit 4.6 Conserve and use native plants	PPs allow air and water exchange in the root zone of existing urban trees.
	Credit 4.9 Reduce urban heat island effects	PPs cool the surface thanks to natural evapotranspiration provided by their porous structure and allow the growth of long-lived shade trees.
Materials Selection	Credit 5.5 Use recycled content materials	Unbound aggregate and soft pervious pavements are constructed of recyclable materials.
	Credit 5.6 Use regional materials	Materials used in the construction of some PPs, such as stone, aggregate and crushed bricks, are obtained from the natural resources of the region or from unwanted products of nearby plants.
Human Health + Well-Being	Credit 6.11 Support local economy	

Leadership in Energy and Environmental Design (LEED) is a rating system developed by the United States Green Building Council (USGBC) to evaluate the buildings in terms of sustainability and being green [27]. Pervious concrete can contribute to some of the credits in this system, see Table 2.

Table 2. LEED's credits which are fulfilled by pervious concrete (adapted from [15, 18, 25])

Credits	**Description of the credits**	**How the PPs fulfill the credit?**
Sustainable Sites	Credit 6.1 and 6.2 Stormwater management	Pervious concrete reduces stormwater flow by allowing water to soak through and infiltrate to the ground below. It can also reduce the pollutant loads by filtering pollutants as the water is transferred through the pavement.
	Credit 7.1 Heat island effect	Pervious concrete absorbs less heat than dark-colored pavements. It reflects the heat due to its porous structure and light color. It provides a suitable environment for the trees, which provide shade and cool pavements.

Credits	Description of the credits	How the PPs fulfill the credit?
Water Efficiency	Credit 1.1 and 1.2 Water efficient landscaping	The gravel sub-base of pervious concrete can store stormwater for irrigation. Trees and other plants grow optimally due to easy air and moisture exchange. Using pervious concrete reduces the consumption of potable water for irrigation.
Materials & Resources	Credit 4.1 and 4.2 Recycled content	Pervious concrete contains recycled aggregate and other recycled materials such as fly ash, slag, or silica fume.
	Credit 5.1 and 5.2 Local/Regional Materials	The aggregate used in pervious concrete construction can be obtained from natural and local stones, thus transport needs can be reduced.

5. Limitations and Disadvantages

In some special areas impermeable pavements are required for the protection of resources and the environment. For example, it is not desirable to infiltrate stormwater under ground where there will be an effective rain harvest [1]. Pollutants leaking from PPs in intensive industrial areas may contaminate groundwater [1, 6]. As a result of overuse PPs may become compressible and impermeable [1, 3]. The urban rainwater flow contains significant amounts of sediment. This leads to a rapid clogging of the PPs. As a result, high maintenance and replacement costs arise [4]. The research reveals different findings about how long the PPs are clogged, and this time varies as widely as 5–35 years. Some researchers suggest that geotextiles should be applied between layers to prolong the permeability life of PPs. But others consider that geotextiles reduce permeability and destroy the structural integrity of the pavement [3]. There are also studies showing that an impermeable membrane must be installed at an appropriate level to prevent pollutants from infiltrating into groundwater [10].

The fact that some PPs are more expensive than conventional pavements is related to the need for a water storage layer. In addition, the surface must be constantly cleaned of dust and particles to prevent clogging. In permeable asphalt and concrete, moisture damage can be seen, resulting in loss of adhesion and loss of cohesion.. The loss of adhesion is the stripping away of asphalt film due to water getting between the asphalt and aggregate. The loss of cohesion is the weakening of the bond between granules in the asphalt and concrete through the effect of water [6].

6. Conclusion

Many studies show that PPs are part of a solution and sustainability for urban environmental problems arising from impermeable pavements. PPs meet some of the sustainable design criteria of ASLA and LEED and they considerably mitigate the negative effects of paved surfaces in cities by reducing runoff, controlling pollutants, improving water quality, improving conditions for urban trees, reducing the urban heat island effect, providing walking safety, using recycled and regional sources and being aesthetic and natural. Most of the paved surfaces in urban areas consist of roads with heavy traffic and parking lots. However, the areas used by pedestrians are large enough not to be underestimated. The use of PPs in these areas may be a good example and may pioneer the widespread use of PPs. Moreover, the strength that can be provided by conventional pavements alone is not necessary for pedestrian use and PPs can be applied easily. In cities, impermeable pavements are persistently used for sidewalks, pedestrian areas, squares, walkways and even in parks that aim to offer a natural environment to people. The appropriate means of giving up this persistence is to be aware that these small and dispersed areas cover large surfaces in total, and then focus on the remarkable benefits we can see from the studies undertaken on this issue.

References

1. Ferguson, B.K., *Porous Pavements*. Integrative Studies in Water Management and Land Development, ed. R.L. France. 2005, Florida: CRC Press. 577.
2. Shackel, B., *The Design, Construction and Evaluation of Permeable Pavements in Australia*, in *24th ARRB Conference – Building on 50 Years of Road and Transport Research*. 2010: Melbourne, Australia.
3. Mullaney, J. and T. Lucke, *Practical Review of Pervious Pavement Designs*. CLEAN – Soil, Air, Water, 2014. **42**(2): p. 111–124.
4. Beecham, S., T. Lucke, and B. Myers, *Designing Porous and Permeable Pavements for Storm water Harvesting and Reuse*, in *1st European Congress of the International Association for Hydro-Environment Engineering and Research (IAHR)*. 2010: Edinburgh, United Kingdom
5. Drake, J.A.P., A. Bradford, and J. Marsalek, *Review of Environmental Performance of Permeable Pavement Systems: State of the Knowledge*. Water Quality Research Journal of Canada, 2013. **48**(3): p. 203–222.
6. Kayhanian, M., et al., *The Application of Permeable Pavement with Emphasis on Successful Design, Water Quality Benefits and Identification of Knowledge*

and Data Gaps, A summary report from the National Center for Sustainable Transportation. 2015, National Center for Sustainable Transportation.
7. Li, H., et al., *Corrigendum: The use of reflective and permeable pavements as a potential practice for heat island mitigation and storm water management*. Environmental Research Letters, 2013. **8**(4): p. 049501.
8. Scholz, M. and P. Grabowiecki, *Review of permeable pavement systems*. Building and Environment, 2007. **42**(11): p. 3830–3836.
9. Weiss, P.T., et al., *Permeable pavement in northern North American urban areas: research review and knowledge gaps*. International Journal of Pavement Engineering, 2017: p. 1–20.
10. Abustan, I., M.O. Hamzah, and M.A. Rashid, *Review of Permeable Pavement Systems in Malaysia Conditions*. OIDA International Journal of Sustainable Development, 2012. **4**(2): p. 27–36.
11. Hunt, W.F. and K.A. Collins, *Permeable Pavement: Research Update and Design Implications*. 2008, North Carolina Cooperative Extension Service.
12. Dunn, S. and K. Bailey, *Permeable Surfaces*, in *BRANZ Build Magazine*. 2010, BRANZ: New Zealand. p. 28–29.
13. Doygun, N. and Ş. Kısakürek, *Kahramanmaraş'da Bazı Kent Parklarının Geçirimli Zemin Yeterliği Bakımından İncelenmesi*. İnönü Üniversitesi Sanat ve Tasarım Dergisi, 2013. **3**(7): p. 23–29.
14. Mullaney, J., T. Lucke, and S.J. Trueman, *A review of benefits and challenges in growing street trees in paved urban environments*. Landscape and Urban Planning, 2015. **134**: p. 157–166.
15. Topličić-Ćurčić, G., et al., *Pervİous Concrete in Sustainable Pavement Design*, in *International Conference Contemporary Achievements in Civil Engineering, 24 April 2015*. 2015: Subotica, SERBIA. p. 47–53.
16. Association, S.C.P. *Pervious Concrete*. [cited 2017 25 April]; Available from: http://secement.org/products/pervious-concrete/.
17. NRMCA. *CIP-38 Pervious Concrete, Concrete in Practice* 2004 [cited 2017 10 April]; Available from: http://www.nrmca.org/GreenConcrete/CIP%2038p.pdf.
18. Ashley, E. *Using Pervious Concrete to Achieve LEED Points*. 2008 [cited 2017 07.04.2017]; Available from: http://www.perviouspavement.org/downloads/LEED_PointsAshley.pdf.
19. Thompson, M.R., *Aggregate as a Structural Product*, in *The Aggregate Handbook*, R.D. Barksdale, Editor. 1991, National Stone Association: Washington.

20. Hildebrand, G., P. Kristoffersen, and T. Dam. *Slotsgrus®: A new concept for footpaths with unbound surfaces.* in *1st International Conference on Transportation Geotechnics.* 2008. Nottingham, UK: CRC Press.
21. Barnes, K.B., J.M. Morgan, and M.C. Roberge. *Impervious Surfaces and the Quality of Natural and Built Environments.* 2002 [cited 2017 07.04.207]; Available from: http://citeseerx.ist.psu.edu/viewdoc/download?doi=10.1.1.477.2834&rep=rep1&type=pdf.
22. Wang, D.-C., et al., *Benefit Analysis of Permeable Pavement on Sideealks.* International Journal of Pavement Research and Technology, 2010. **3**(4): p. 207–215.
23. EPA, *Reducing Urban Heat Islands: Compendium of Strategies, Cool Pavements.* 2005.
24. Akbari, H., M. Pomerantz, and H. Taha, *Cool Surfaces and Shade Tress to Reduce Energy Use and Improve Air Quality in Urban Areas.* Solar Energy 2001. **70**(3): p. 295–310.
25. Tennis, P.D., M.L. Leming, and D.J. Akers, *Pervious Concrete Pavements.* 2 ed. 2004, USD: Portland Cement Association. 28.
26. ASLA, *Sustainable Sites Initiative, SITES v2 Rating System For Sustainable Land Design and Development.* 2014.
27. LEED, *Green Building Rating System.* 2002.

Habibe Acar[1], Demet Ülkü Gülpinar Sekban[2], Cengiz Acar[3]

Sustainable Design Approaches in Children's Playgrounds

[1]*Assoc. Prof. Dr. Karadeniz Technical University, Faculty of Forestry, Department of Landscape Architecture, 61080, Trabzon, TURKEY habibeacar@hotmail.com*

[2]*Res. Assist. Karadeniz Technical University, Instute of Natural Sciences, Landscape Architecture Program, 61080 TRABZON demetgulpinar@hotmail.com*

[3]*Prof. Dr. Karadeniz Technical University, Faculty of Forestry, Department of Landscape Architecture, 61080, Trabzon, TURKEY cengiz@ktu.edu.tr*

1. Introduction

Along with the growing population, the speed with which individuals consume and destroy natural resources has gone beyond that of nature's self-renewal. The destruction of natural resources is hard to repair, and if the destruction continues with the present speed and violence, the possibility that it might not be possible to restore resources to their original status has introduced certain concepts. New approaches that emphasize the restructuring of cities, along with ecological plans to control the use of resources and increase the awareness of individuals, have begun to be discussed. These discussions and their aims have been and are based on the concept of sustainability and the principle of "development that meets the needs of the present without compromising the ability of future generations to meet their own needs", as defined in the Brundtland Report [1] for the first time in 1987.

While only 2% of the world's population lived in urban areas in the 1800s, this figure was 14% in 1900, 30% in 1950, and according to the United Nations World Urbanization Possibilities Report, the same figure exceeded 54% in 2014. However, by 2050, the same report predicted that 66% of the global population would live in urban areas [2]. The above figures demonstrate that urban areas are increasingly becoming significant habitats for humans. Urban areas account for 80% of carbon emissions, 60% of water use and 80% of wood use, despite covering a small percentage of the earth's surface. [3]. Urban areas possess economic, social, environmental and ecological factors, especially for the future of the planet, which are significant for the concept of sustainability. As the United Nations Secretary-General Kofi Annan said, "The future of mankind is in the cities". Thus, urban areas, where human use and human vital needs are so intense, are the areas where sustainability should be truly implemented [4].

1.1. The Concept of Sustainability

The concept of "sustainability" was introduced for the first time at the Environmental Development Conference held in Rio in 1992 [5]. Sustainable development is based on a balanced and harmonious relationship between social needs, economic activities and the environment. It is known by definition that sustainable development is "the development that meets the needs of the present without compromising the ability of future generations to meet their own needs" [6].

In urban areas, the main objective of the sustainability principle is to ensure that the urban landscape is transferred to future generations within a system as a whole. A sustainable urban area should establish balance between environmental protection/utilization, economic development, social welfare, equality and accessibility. In doing so, it is necessary to minimize the consumption of resources, protect ecosystems and human health, design a healthy and viable urban environment or spaces that serve users, provide comfortable and equal access to resources, spaces and services, and protect cultural and social diversity. Thus, all disciplines that produce and implement decisions in urban areas need to consider this approach.

1.2. Sustainability in Landscape Design

Sustainable landscapes are responsive to the environment, regenerative, and can actively contribute to the development of healthy communities. Sustainable landscapes sequester carbon, clean the air and water, increase energy efficiency, restore habitats and create value through significant economic, social and, environmental benefits [7].

As demonstrated above, sustainability is a combination that includes several components and truths. Hence, if we want a defined space to reflect the concept of sustainability, all elements that constitute this space must support the concept. The concept of "sustainable landscape" focuses on three important issues. The first is the design and preservation of natural assets and resources; the second is the emphasis, protection and use of the dynamics that make up the ecosystem and multifunctional connections; and the third is the welfare of people. These three significant points that the sustainable landscape concept focuses on enable us to understand that sustainability is a social responsibility that everyone should share [8]. But this is not as simple as one might think. It is necessary to develop an original and up-to-date understanding of how a particular space would work in an urban area and how that particular space would interact with the dynamics that form the ecosystem on both a local and global scale. This is because every space has its own metabolism and the parameters of this metabolism define the sustainability criteria [9].

Various certification systems were established around the world to design sustainable spaces in different areas. In the field of Landscape Architecture, the American Society of Landscape Architects (ASLA) developed the "Sustainable Sites Initiative" certification system. This system grades, classifies and certifies designed areas [10]. As we could observe in the abovementioned system, the area that was planned to be designed as a sustainable space should have been considered in terms of ecology, economy, social aspects, administration, human health, and environmental and aesthetic principles. Thus, sustainable spaces are defined areas that are designed based on environmental factors, with low resource use for their application, construction and sustainability, with an infrastructure that enables recycling of the waste produced during application, and which are also self-sustaining [11].

1.3. Sustainable Design Criteria for Children's Playgrounds

Playgrounds are significant spaces for children that fulfill their playing needs, while supporting their healthy development in the outdoor environment. These areas provide positive contributions to the physical, cognitive, social and psychological development of children based on the opportunities provided by their location, quality and efficiency. When designers make choices for children's playgrounds, they should both contribute to their playing needs and development and learning in different areas. This is because the best learning tool is play. The form, quality and diversity of the areas where children play would influence the form, quality and diversity of their play. Thus, quality spaces are needed for quality play. A quality playground is a space that meets children's expectations, keeps children in the area for extended periods of time, and provides opportunities to encourage and develop their imagination and creativity, having a spatial construct suitable for exploration and observation, in which all these factors are sustainable. Therefore, creating healthy and sustainable environments is also necessary for children's playgrounds, which contribute significantly to future generations. The adoption of a sustainable philosophy in the design and implementation of these areas would also contribute to urban sustainability.

Literature on the design of children's playgrounds demonstrates that sustainable design criteria include ecological, economic, equality, circulation, indigenous and environmentally adequate material use, accessibility, human health, biological diversity and security concepts. In the present study, the aim was to determine sustainable design approaches for children's playgrounds. For this purpose, design parameters were grouped under more general headings based on the literature review conducted.

2. Material and Method

2.1. Objectives of the Study

Today, sustainability emerges as a concept that is a requirement for the continuity of existing resources and emphasized by different professional disciplines. Within the scope of the present study, children's playgrounds are assessed specifically on the basis of landscape design sustainability. Accordingly, the objectives of the present study were:

- To determine sustainable design approaches for children's playgrounds and to design an assessment scale accordingly
- To create a guide for designers to establish future playground sustainability
- To discuss sustainability of constructed playgrounds based on the designed assessment scale and provide recommendations for their shortcomings.

As a result, an attempt was made to answer the question "How could sustainable design approaches be evaluated in children's playground designs?"

2.2. Method

The study consists of two phases to reach the determined objectives. The first stage is the establishment of sustainable design parameters (SDP) assessment scale; the second stage is the selection of examples of children's playgrounds and the determination of sustainability categories within these playgrounds.

2.2.1. Sustainable Design Parameters (SDP) Assessment Scale

At this stage, an attempt was made to answer the question "What are sustainable design parameters in landscape architecture?" Sustainability criteria were determined based on a literature review. In this context, the criteria were based on a study by Mahayudin et al. [6] and the "Sustainable Sites Initiative" [7] certification system developed by ASLA in 2005, which was used to design sustainable areas in landscape architecture. The probable parameters of sustainability were assessed based on the abovementioned resources and a new list was formed. This list, which included a total of 25 items, was categorized under the four groups of Material, Ecology, Management / Maintenance / Economy and Planning / Design (Table 1) based on the criteria. Subheadings were expressed as questions to identify their presence in the project areas.

To create a numerical scale, 4 points were assigned to each criterion expressed in the table, and an assessment scale with a total score of 100 points was designed. To assess the constructed examples in the study, existing parameters would be marked to determine the rate at which the particular area possessed sustainable design criteria. Furthermore, percentages were grouped in four different categories, namely Gold (100% -80%), Silver (79% -50%), Approved (49% -30%) and Not Sustainable (29% -0%), similar to the criteria depicted by Mahayudin et al. [6] and ASLA [7].

The constructed table could also be used to evaluate the sustainability of application examples that contained different topics. It could also serve as a guide for new landscape design projects that would be designed and implemented, in addition to the examples that had already been implemented.

2.2.2. Selection of Children's Playground Examples and Their Sustainability Criteria

In this stage, a total of 10 different children's playgrounds were designed and implemented in different regions in the world, and examined in order to answer the question "How can sustainable design parameters be used in children's playground designs?" The children's play areas examined were selected from among those that were scrutinized in the literature. Nine of these playgrounds were chosen from among those that emphasised the project content, from the beginning through to implementation, and were designed based on eco-friendly and sustainable approaches. These were the Rainwater-Capturing Children's Playground, Kilburn Grange Adventure Play Park, Greensburg Sustainable Playground, Recycled Wind Turbine Playground, Energy Playground, Shipping Container Playground, Nishi Rokugo Park, Terra Nova Play Experience, and Avic Park. One of these was the first children's playground that received the Sustainable Sites Initiative certificate in the world. This project is The Woodland Discovery Playground. General views of the children's playgrounds examined are presented in Figure 1.

Based on the designed assessment scale, the categories of the sample areas based on the sustainable design criteria were determined. For this purpose, the knowledge found in the literature that explained the project details was used. Based on this information, according to the questions in Table 1, the existing criteria in the examined playground project were marked (Table 2). As a result, the sustainability categories and the presence of each parameter in children's playground implementations were determined.

Figure 1. General views of the children's playgrounds examined [12, 13, 14, 15, 16, 17, 18, 19].

Table 1. Sustainable landscape design parameters

Sustainable Design Parameters (SDP)
Material (20%)
1. Are natural elements used in design and implementation?
2. Are recycled material used in design and implementation?
3. Are the materials used mainly natural?
4. Is the playground pavement / covering material permeable?
5. Is the playground pavement resistant to the pressure of continuous use?

Ecology (28%)			
6. Is plant species diversity adequate in the project area?			
7. Are the utilized plant species mostly natural species?			
8. Was attention paid to lighting elements, energy use and light pollution in the project?			
9. Does the area include various designs that utilize renewable energy resources such as wind, solar energy?			
10. Is the project area capable of providing animal diversity, fulfilling their nutritional, protection and housing needs?			
11. Is there any planning or application about water management?			
12. Are there any facilities where rainwater and excessive water could be utilized?			
Management/Maintenance/Economy (16%)			
13. Are there any units that could provide for the needs of individuals?			
14. Is the area low-maintenance?			
15. Are there any measures taken to decrease water use in the irrigation system?			
16. Are there any recycling applications (such as gathering fallen leaves and using these as compost material)?			
Planning/Design (36%)			
17. Does the design support sustainability awareness and training?			
18. Are there any plans and designs to facilitate the use of the area by individuals with disabilities?			
19. Is the design suitable for the topography of the area? Does the design support topographical use?			
20. Is accessibility considered as a criterion when selecting the location of the playground?			
21. Is it possible to use the area in four seasons?			
22. Is the playground attractive for individuals of all ages?			
23. Does the playground provide activity and furniture diversity?			
24. Are there spatial solutions that support the socialization of the users?			
25. Did the users participate in the design of the playground?			
TOTAL (100%)			
Sustainability Categories			
100–80%	Gold Category	Approved Category	49–30%
79–50%	Silver Category	Non-Sustainable	29–0%

Table 2. Sustainable design categories of the examined children's playgrounds

SDP	Playgrounds										
	1	2	3	4	5	6	7	8	9	10	
Material											**%**
1	x	x	X	–	–	x	–	x	x	x	**70**
2	x	x	X	x	–	x	x	x	–	–	**70**
3	–	x	–	–	–	–	–	x	x	–	**30**
4	x	x	X	x	x	x	x	x	x	–	**90**
5	x	x	–	x	–	x	x	x	x	x	**80**
Ecology											
6	x	–	–	x	–	–	–	x	x	x	**50**
7	–	x	X	–	x	–	–	x	x	x	**60**
8	x	x	X	–	x	x	–	x	x	x	**80**
9	–	–	–	–	x	–	–	–	–	x	**20**
10	–	–	–	–	–	–	–	–	x	–	**10**
11	x	–	X	–	–	–	–	x	x	x	**50**
12	x	–	X	–	–	x	–	x	–	x	**50**
Management/Maintenance/Economy											
13	–	–	–	x	x	x	x	x	–	x	**60**
14	x	x	X	x	–	x	x	x	x	–	**80**
15	x	–	X	–	–	–	–	x	x	x	**50**
16	–	–	–	–	–	–	–	–	–	–	**0**
Planning/Design											
17	x	x	X	x	x	x	–	x	–	–	**70**
18	x	–	–	–	–	–	–	x	–	–	**20**
19	x	x	–	x	–	–	x	x	x	x	**70**
20	x	x	–	x	x	–	x	x	x	x	**80**
21	x	x	–	x	–	x	–	x	–	x	**60**
22	x	x	–	–	x	x	x	x	x		**70**
23	–	x	–	x	–	x	–	x	–	x	**50**
24	–	x	–	x	x	x	–	x	–	x	**60**
25	x	–	X	–	x	–	–	x	x	–	**50**
%	**68**	**60**	**44**	**48**	**40**	**52**	**32**	**88**	**60**	**60**	

SDP	Playgrounds										
	1	2	3	4	5	6	7	8	9	10	

1. Rainwater-Capturing Playground	6. Shipping Container Playground
2. Kilburn Grange Adventure Play Park	7. Nishi Rokugo Park
3. Greensburg Sustainable Playground	8. The Woodland Discovery Playground
4. Recycled Wind Turbine Playground	9. Terra Nova Play Experience
5. Energy Playground	10. Avic Park

It is possible to increase the number of the sample areas. The objective of the present study was not to discuss the children's playground projects that were evaluated. It was only to demonstrate the method and extent of the application of sustainable design parameters specifically in examples of children's playground spaces. In this respect, the aim was to contribute to future design and application studies.

3. Conclusion and Recommendations

The assessment scale for sustainable design parameters that was presented in the current study could be utilized not only in children's playgrounds but also in landscape planning and design studies in different areas. Subtitles in the table could be increased when deemed necessary based on the resources available in the project area. It is also possible to calculate percentages within total values by assigning a numerical value to the added titles.

The data obtained for the 10 children's playground examples (Table 2) demonstrated that SDP rates varied between 88% and 32%. One of the projects (Nishi Rokugo Park – 32%) appeared to be borderline in the approved category based on the designated categories. However, the approval rate would increase if further ecological work were to be conducted in this area.

Furthermore, The Woodland Discovery Playground, which received a sustainability certificate from ASLA, received a score of 88%. In the project, the items (SDP-9, 10, 16) that were missing or neglected based on the table, were in fact missing or neglected or utilized only a little in other examples as well. Therefore, it is possible to design sustainable children's playgrounds when applications similar to this playground are implemented. Based on SDP parameters, it was observed that 9 (90%) of the examined playgrounds had ground material permeability. The second highest rate (80%) was observed in Material – playground pavement is resistant to the pressure of continuous use, Ecology – the majority of the plant species used were natural species, Management/ Maintenance/ Economics – low maintenance playground, and Planning/Design – accessibility criteria were considered in the selection of the playground location categories. Thus, it seems that these criteria were considered in the design of theplaygrounds examined. However, it was observed that there were no recycling plans and practices (such as the collection and composting of fallen leaves). It was also determined that there was only one project that provided animal diversity and fulfilled the nutritional, protection and housing needs of the animals.

The ability to implement the sustainability criteria set out in the present study is related to the potential of the project area. For this purpose, initially, it is necessary to analyze the area well, to identify its strengths and weaknesses, and to discuss its feasibility for design, because it is important for sustainability to protect existing resources, use them effectively and sustain these resources for future use. Only then caneconomic, ecological and "sustainable" practices that can be adopted by all be implemented. Also, when the ecological criteria are adopted in the design of playgrounds, children will find an opportunity to be exposed to nature. This will also contribute to their healthy development.

4. References

1. World Commission on Environment, Development (Wced): *Our Common Future (1987),* http://www.un-documents.net/our-common-future.pdf
2. *World Urbanization Prospects The 2014 Revision Highlights,* https://www.compassion.com/multimedia/world-urbanization-prospects.pdf
3. N. B. GRIMM, S. H. FAETH, N. E. GOLUBIEWSKI, C. L. REDMAN, J. WU, X. BAI, J. M. BRIGGS: *Global change and the ecology of cities. Science,* 319, 756–760, (2008).
4. J. WU: *Urban sustainability: an inevitable goal of landscape research. Landscape Ecology,* 25, 1–4, (2010).
5. B. ERDOĞAN ONUR: *Peyzaj Tasarım ve Yönetiminde Ekolojik Yaklaşım ve Sürdürülebilir Kent Hedefine Katkıları. Inonu University Journal of Art and Design.* Vol. 2, No.5, 245–252, (2012).
6. R. M. MAHAYUDIN, M. Y. M. YUNOS, M. A. O. MYDIN, O. M. TAHIR: *Developing a sustainable campus landscape criteria: an evaluation of Universiti Pendidikan Sultan Idris as a green campus. Advances in Environmental Biology,* 9(4), 201–204, (2015).
7. *Sustainable Sites Initiative (SITES),* https://www.asla.org/sites/
8. P. SELMAN: *What do we mean by sustainable landscape. Sustainability: Science, Practice & Policy,* 4, 23–28, (2008).

9. V. MACLAREN: *Developing indicators of urban sustainability: A focus on the Canadian experience. Intergovernmental Committee on Urban and Regional Research (ICURR) Press,* (1996).
10. R. GÜRBÜZ, L. ARIDAĞ, *Sürdürülebilir peyzaj tasarımı için ASLA ve LEED kriterlerinin karşılaştırılması. Beykent Üniversitesi Fen ve Mühendislik Bilimleri Dergisi,* 6(2), 77–92, (2013).
11. C. VAROL, N. A. SAT, Z. A. GUREL UCEL, S. B. YENIGUL: *Metropoliten alanlarda çok merkezli mekansal gelişmelerin sürdürülebilirlik üzerinden değerlendirilmesi: Avrupa Birliği'ndeki uygulamalar. TUBAV Bilim,* 10 (1), 61–74, (2017).
12. http://housevariety.blogspot.com.tr/
13. https://www.flickr.com/photos/exploringadventure/5041722048/
14. http://www.bnim.com/project/greensburg-sustainable-playground
15. http://www.landezine.com/
16. http://www.campaignbriefasia.com/2012/07/cannes-highlight-ace-saatchi-s.html
17. http://inhabitat.com/skinners-playground-kidscape-made-from-shipping-containers/attachment/12383/
18. http://www.bellissimakids.com/nishi-rokugo-park-tokyo/
19. http://www.sustainablesites.org/woodland-discovery-playground-shelby-farms

Asist.Prof.Dr. Okşan Tandoğan
Asist.Prof.Dr. Deniz Yetkin Aker

Urbanization in Tekırdag and a Case Study about Children's Independent Mobility

Introduction

Running alongside their physical, perceptual-cognitive and social-emotional development, the physical environment for children may refer to a whole building or nearby neighborhoods such as schools and playgrounds. Apart from the availability of such physical spaces, children's independent mobility without adult accompaniment is a fundamental obligation for their development and is also a right for children (Huttenmoser and Degen- Zimmerman, 1995; Zomervrucht, 2005).

Research has proved that children's opportunities for independent mobility contribute to their interaction with physical and social environments (Norinder, 1996), as the experience they gain in this interaction helps children to develop their spatial, motor, personal, cognitive, emotional and social skills and to acquire a sense of identity as individuals (Zomervrucht, 2005; Hillman et al., 1990; Hillman, 1999; Malone, 2007; Ross 2007; Skelton 2009; Zubrick et al., 2010; Nansen et al., 2015). The time that children spend alone or with their friends and the promotion of physical activity are important means of combating some childhood health problems such as asthma, allergies and cognitive problems (Zubrick et al., 2010) and ofpreventing childhood obesity (Tudor-Locke et al., 2001).

The concept of children's independent mobility is defined as the freedom to move around the neighborhood – or similar – without adult accompaniment (Tranter and Whitelegg, 1994; Zubrick et al., 2010). Independent mobility refers to children's travelling alone or travelling with friends from their neighborhoods or cities to outdoor environments such as school or playgrounds, as well as to children being outside their homes (Shaw et al., 2013).

However, nowadays, children's independent mobility in developing and developed countries has decreased due to urbanization and social and environmental factors (Kytta 2004; Hillman et al., 1990; O'Brien et al., 2000; Tranter, 1993; Tandy, 1999; Gaster, 1992). These factors can be summarized as a decrease in open areas through rapid urbanization and dense settlement; an increment in traffic loads following high rates of car ownership; and parental fear (Little and

Wyver, 2008; Norinder, 1996). Recently, parents increasingly see urban spaces as dangerous places for children due to the dangers of traffic and their fear of strangers. Consequently, children have lost their access to their physical environment and have been deprived of independent mobility (Hillman et al., 1990; O'Brien et al., 2000; Tranter, 1993; Tandy, 1999; Malone, 2007; Gaster, 1992; Shaw et al., 2013).

In the relevant literature, variables influencing children's independent mobility are accepted as being the type of living settlements (rural or urban), their size and density (Norinder, 1996; Jones, 2000; Kytta, 1997; Van der Spek & Noyon, 1995; O'Brien et al., 2000), the danger of traffic (Bjorklid, 2002; Prezza et al., 2001) and gender differences (Hilmann et al., 1990; Kytta, 1997; Castelli et al., 2007; Valentine and McKendrick, 1997). It is also possible to include parents' fear of strangers.

Focusing on those variables, as a result of various empirical studies, it can be seen that children in rural areas or low-density settlements are freer than children in high-density settlements and have higher levels of outdoor activity (Norinder, 1996; Jones, 2000; Kytta, 1997; Van der Spek & Noyon, 1995; O'Brien et al., 2000). The road safety of the residential surroundings is another factor affecting children's independent mobility (Bjorklid, 2002; Prezza et al., 2001).

In terms of the "gender" variable, male children are freer than female children in urban and rural settlements. Male children often travel independently. Female children have greater parental restriction (Hilmann et al., 1990; Kytta, 1997; Castelli et al., 2007; Valentine and McKendrick, 1997).

The objective of the study is, by focusing on children's points of view and their statements, to examine how children's independent mobility differs in relation to urban and rural settlements and gender; which factors influence children's independent mobility, and whether these factors vary according to rural or urban neighborhoods. The study focuses on Tekirdag, one of the top ten immigrant-receiving cities across Turkey, which is surrounded by rural and urban settlements and faces rapid urbanization.

Lastly, the study presents the following shortcomings: to begin with, it is a single case study. Besides, it might be better if the study could include additional neighborhoods and thus reach a higher confidence level with a wider sampling, but this could not be realized due to time restraints and cost.

Research Methodology

Tekirdag is the 23rd most crowded city in Turkey and one of the 10 cities which receive the highest number of immigrants due to its closeness to Istanbul (2013, TUİK). For this study, Tekirdag was chosen as a case study since it has both urban and rural areas within the center of the city and is becoming crowded and urbanized mainly because of such a migration flow. It is one of the most developed 15 cities of Turkey as well (Karasu, 2008).

For the study, two different settlement areas, namely Hürriyet and Karadeniz, which are close to the city center of Tekirdag, were chosen to compare urban and rural settlement areas. With rapid urbanization and its dense population, the Hürriyet neighborhood is located in the province of Suleymanpasa in Tekirdag. The Hürriyet eighborhood has a population of 34 000 people living in 12 000 households (face to face meeting with Municipal Headman, 2016). It has an area of 302 53 hectares. The population density per one hectare of land area is 112.4 people (Süleymanpaşa Municipality, Department of Urban Planning, 2017). The Hürriyet neighborhood hosts 123 residential building complexes, in which upper middle class households and higher educated households live. There are two primary schools, two secondary schools and 1 private school in the Hürriyet neighborhood. There are 17 playgrounds in the neighborhood as well (face to face meeting with Municipal Headman, 2016). On the other hand, the Karadeniz neighborhood is a low-populated settlement area (4500 people) and is located in the province of Suleymanpasa (interview with Municipal Headman, 2016). It has an area of 1893.39 hectares. (Süleymanpaşa Municipality, Department of Urban Planning). The population density per hectare of land area is 2.4 people. The Karadeniz neighborhood is home to middle income families with middle education levels and the income of the neighborhood is based on economic activities in agriculture, breeding and the construction sector. 60 percent of the neighborhood areas are allocated for agriculture. There are two primary schools and two secondary schools in the Karadeniz neighborhood. There are two playgrounds in the neighborhood as well, one of which is situated in the schoolyard of the 100. Yıl Mehmet Sadık Özyüksel Primary School (face to face meeting with Municipal Headman, 2016).

This study focuses mainly on the following research questions:

1) Is the independent mobility of children residing in Hürriyet different from that of the Karadeniz neighborhood?
2) Which factors impact on the independent mobility of children residing in the Hürriyet or Karadeniz neighborhoods in Tekirdag?
3) Do these factors declared by children differ between the settlement areas?

The urban space/settlement variable was operationalized as being those residing in the Hürriyet Neighborhood and the rural space/settlement variable was operationalized as being those residing in the Karadeniz neighborhood. Children's independent mobility without adult accompaniment was operationalized as the "mode of travel to school and mode of travel to playground". Travelling on their own or with their friends (on foot or cycling etc.) were accepted as forms of independent mobility without adult accompaniment. Additionally, children were asked about neighborhood safety (it was classified as "safe" and "very safe" in the measurement). Lastly, the distance from school or playground to home, and gender, were taken as variables which might influence children's independent mobility.

The research population consists of all public schools in the Hürriyet and Karadeniz neighborhoods. Each neighborhood has one public secondary school and primary school. A survey was conducted in these schools with the participation of literate children (age 7–14). However, only those children living in the Hürriyet and Karadeniz neighborhoods were included in analysis of these questionnaires.

Since the Karadeniz High School has a low number of students, the aim was to include every child (50 students) from this school, yet 49 students participated in the survey. Similarly, all 86 students from 100. Yıl Mehmet Sadık Primary School were envisaged as being included in the study, yet only 60 students were surveyed.

As the schools in the Hürriyet neighborhood are very crowded, a stratified sampling method was used for the sampling. The total number was 1436 students (only the ones going to the schools in the Hürriyet neighbourhood and living in the Hürriyet or Karadeniz neighbourhoods). Each grade was chosen as a stratum that was proportional to that of the total population. A sample group of about 305 students was selected, with a 95% confidence interval and and ± 5% precision level. From the public school in the Hürriyet neighborhood, 49 students (the 5th grade has 228 students in total, accounting for 16% of the school), 49 students from the 6th grade (232 students in total, 16% of the school), 43 students from the 7th grade (199 students in total, 14% of the school) and 40 students from the 8th grade (185 students in total, 13% of the school) were included. And again, 36 students from the 2nd grade in this neighborhood (176 students in the 2nd grade, 12% of the school), 36 students from the 3rd grade (168 students in the 3rd grade, 12% of the school) and 52 students from the 4th grade (248 students in the 4th grade, 17% of the school) were included in the study. The 1st grade students were excluded from the study as they are illiterate.

The number of children surveyed and included in the analysis was 337 (Table 1). All findings were analyzed using SPSS.

Table 1. Distribution of the Sample by School

Neighborhood	School Name	Questionnaire Number	(%) Percentage
Hürriyet Neighborhood	Marmara Secondary School	81	24
	Ticaret ve Sanayi Odası Primary School	147	43.6
Karadeniz Neighborhood	Karadeniz Secondary School	49	14.5
	100.Yıl Mehmet Sadık Özyüksel Primary School	60	17.8
Total		337	100

Whether urbanization and gender variables in urban and rural settlements were related to children's independent mobility was analyzed through descriptive analysis and nonparametric tests. Similarly, whether the "urbanization" variable in urban and rural spaces was related to the reasons for restrictions imposed by parents was examined through descriptive analysis and nonparametric tests. The results were discussed and analyzed through comparisons with the literature.

Research Results and a Comparative Analysis

51.9% of the 337 children surveyed in total is female and 48.1% of the total sample is male. In terms of age groups, respondents aged 9 comprise 22% of the total.

The great majority of the children surveyed (67.1%) dwell in the Hürriyet neighborhood, a densely populated and urbanized space, while 32.9% of the children surveyed reside in the Karadeniz Neighborhood, a low-populated and ruralized space.

1. Safety in Neighborhood from the Point of View of the Child: Is it safe or not?

Regarding safety in the neighborhood, the proportion of the children who thought their neighborhood was safe (the total of "very safe" and "safe" responses) was over 90%. Only 3.3% of the children stated their neighborhood was not safe.

The percentage of the children who think their neighborhood is unsafe is 5.4% in the Karadeniz neighborhood and 2.2% in the Hürriyet neighborhood. However, as a result of the chi-squared test, it can be concluded that the difference between "neighborhood safety" and "settlement area" variables is not statistically significant.

2. Mode of Travel to School

The study results show that nearly half of the children (42.6%) travel to school by school bus, while 29.2% of them walk to school under adult supervision, or have parents who drive them to school. Only 26.5% of the children travel to school on their own or by walking with friends, or cycling to school. According to the children's responses, it was revealed that more than 70% of the children do not have independent mobility when travelling to school.

In terms of the 'gender' variable, there is no striking difference between female and male children and the percentage of male children who come to school alone or with friends is 29%, while the percentage of female children is 24.1%. These collected findings are compatible with the results of studies realized by Hilmann et al. (1990), Kytta (1997), Castelli et al. (2007), Valentine and McKendrick (1997). However, the results are not statistically significant.

In terms of the 'urban settlement' variable, the percentage of pupils who travel to school alone or with friends or by cycling is 27.1% in the Hürriyet Neighborhood and 25.2% in the Karadeniz Neighborhood. The percentage of the children who travel to school on foot with adult supervision or by their parents' car is 35% in the Hürriyet Neighborhood and 17.1% in the Karadeniz Neighborhood. Children in the Karadeniz neighborhood travel to school by school bus rather than by private car or walking accompanied by parents. We can observe that the school bus is used as a mode of travel at the rate of 36.1% in the Hürriyet neighborhood while it is 55.9% in the Karadeniz neighborhood.

According to chi-squared test results, there is no statistically significant difference in the proportion of children's independent mobility dependent on their settlement areas. In the light of the results obtained, the present study does not support the argument presented in the literature as follows: children who live in rural and low density spaces are much freer (Norinder, 1996; Jones, 2000; Kytta, 1997; Van der Spek&Noyon, 1997; O'Brien et al., 2000). In general, the children travelling to school did not have independent mobility in these two neighborhoods.

3. The Main Reasons for not travelling to School Independently

The pupils in these two neighborhoods were asked to tell us the reason why they were not allowed to travel to school alone without adult accompaniment or with their friends.

Children state the underlying reasons for adult supervision as follows: the first, the school is far away from home (30.6%), the second, danger of traffic (25.9%) and the third, fear of strangers (25.9%). As argued in the literature, the study also supports the argument that the 'neighborhood safety' (Bjorklid, 2002; Prezza et al., 2001) and 'fear of strangers' variables (Little and Wyver, 2008; Norinder, 1996) impact on children's independent mobility. According to the study, the 'home-school distance' should be added to the variables related to children's independent mobility.

As a result of nonparametric tests (chi-squared test), the difference of the obstacles for independent mobility between the groups (children living in urban and rural spaces) set by the "neighborhood" variable is statistically significant (chi-squared value 20.896, P<.005). Thus, while the most critical reason for the children in the Hürriyet neighborhood who were not allowed to travel to school independently was danger of traffic (30.5%), it was the distance issue in Karadeniz neighborhood (39.7%).

4. Mode of Travel to Playground

The findings demonstrate that a great majority of the children go to playground areas (68. 6%). In terms of settlement areas, 72.6% of the children in the Hürriyet neighborhood go to playgrounds while 60.4% of the children in the Karadeniz neighborhood go to playgrounds. The percentage of children travelling to school alone or by walking with friends or cycling to school is 55.6% in the Hürriyet neighborhood and 74.6% in the Karadeniz neighborhood. However, the chi-squared test shows that there is no statistically significant relationship between the "settlement" variable and the "mode of travel to school" variable.

Regarding the gender variable, we can see that the percentage of male children who come to school alone or with friends is 66.4%, while the percentage of female children who do the same is 55.8%. As suggested in the literature, this result is not compatible with the findings arguing that gender variable (Hilmann et al.,1990; Kytta, 1997; Castelli et al., 2007; Valentine and McKendrick, 1997) is among the factors related to children's independent mobility. Although it seems that male children are freer than female children in terms of transportation to playgrounds, this difference is not statistically significant according to nonparametric tests.

5. The main reasons for not travelling to playgrounds independently

Children stress that they are not allowed to travel to playgrounds alone or with their friends due to danger of traffic (41.4%). This is followed by fear of strangers or fear of crime (21%) and the home-playground distance (21%).

When the results are compared by settlement areas, 44.5% of the children in the Hürriyet neighborhood state that they are not allowed to travel to playground areas due to danger of traffic, On the other hand, only 27.6% of the children in the Karadeniz neighborhood pointed out the danger of traffic. The most significant factor from the children's point of view in the Karadeniz neighborhood is the distance of the playgrounds from the school (55.2%). The percentage of children who are not allowed to go to playgrounds alone due to parents' fear of strangers is 22.7% in the Hürriyet neighbourhood while this Figure is lower in the Karadeniz neighborhood (13.8%).

As presented in the literature, the "danger of traffic" variable (Bjorklid, 2002; Prezza et al., 2001) emerges as the most significant variable restricting children's independent mobility. According to this study, another variable related to children's independent mobility is "home-school distance".

According to chi-square test results, the difference between settlement areas and the reasons declared by children about travelling to playgrounds alone are statistically significant (chi-square value 25.956, P<.000). In other words, although children do not often have the freedom to travel to playgrounds on their own, the reasons restricting their independent mobility significantly differ between the settlement areas. While the major reason for children who are not allowed to travel to playgrounds on their own is the home-playground distance, the major reason in the Hürriyet neighborhood is heavy road congestion, which is naturally to be expected as a result of urbanization.

Conclusions

This present study focuses on children's responses and how children's independent mobility differs between urban and rural areas due to rapid urbanization in Turkey. This is what led to a case study being conducted in Tekirdag, where one of the fastest paces of urbanization has been experienced. The aim was to explore how children's independent mobility in Tekirdağ differentiated between urban and rural areas; which factors impacted on children's independent mobility, and whether these factors differed between dwelling areas. Children who were pupils in public schools and resided in two different neighborhoods, namely Hürriyet and Karadeniz, were surveyed.

According to the children's responses, children travelling to school do not have freedom in either rural or urban areas in terms of independent mobility. In both areas, the great majority of children arrive at school by school bus or with their parents (by walking or by parents' car). On the other hand, non-parametric tests indicate that the differentiation regarding the children's independent mobility is statistically significant in terms of settlement areas.

Another striking result of the present study is that children in the Hürriyet neighborhood are freer in travelling to school than children in the Karadeniz neighborhood.

However, children who do not have the right to travel to school alone in either settlement are freer when it comes to travelling to playgrounds. In particular, children in the Karadeniz neighborhood, a low-populated and ruralized space, are freer than children living in the Hürriyet neighborhood, a densely populated and urbanized space. Although we cannot see any similar results in terms of travelling to school, this result is compatible with the findings indicating that children living in low-populated and rural spaces are freer.

In terms of the "gender" variable, there is no striking difference between female and male children. In the two settlement areas, male children are freer than female children in terms of travelling to school and playgrounds. This finding supports the argument that gender is one of the factors related to the independent mobility of children as presented in the literature (Hilmann et al., 1990; Kytta, 1997; Castelli et al., 2007; Valentine, McKendrick, 1997). However, in terms of travelling to school, the gender variable shows no statistically significant difference in this study.

On the other hand, the reasons restricting children's independent mobility when travelling to school without adult supervision or with their friends indicate statistically significant differences between settlement areas. Children in the Hürriyet Nneighborhood declared that they were not allowed to travel alone or with their friends due to danger of traffic or fear of strangers, while the children in the Karadeniz neighborhood pointed out the reason that the school was far away from home, along with the fear of strangers. Similarly, the differentiation by settlement area was statistically significant, given the fact that children were restricted with respect to travelling to playgrounds on their own. While the children in the Hürriyet neighborhood do not move around without adult supervision due to danger of traffic and fear of strangers, the children in the Karadeniz neighborhood do not have independent mobility because of the distance between their playgrounds and homes. In the light of the findings collected from the study, we can see that the main barriers to children's independent mobility in the urban area are the danger of traffic, the fear of strangers or, in other words, a fear of crime. As a result, the vast majority of children travel to their physical envirionments such as school or playgrounds only as long as they are accompanied by their parents.

Children's independent mobility is a fundamental obligation for their development and is also a right for children (Huttenmoser andDegen-Zimmerman, 1995; Zomervrucht, 2005). It is not important whether there are available spaces for children's activities or whether there is a wide range of play spaces for children if children do not have the opportunity for independent mobility in urban areas. Lastly, it is believed that the present study will contribute to the current literature in Turkey with respect to the issue of children's independent mobility. However, it should be noted that it is possible to conduct qualitative and quantative studies in various cities and countries in order to draw comparisions.

References

Castelli, D.,M., Hillman, C.,H., Buck, S.,M., and Erwin H.,E., (2007). Physical fitness and academic achievement in third and fifth grade students. Journal of Sport and Exercise Psychology, Vol. 29, pp 239–252.

Gaster, S. (1992). Historical changes in children's access to US cities: A critical review. Children's Environments, 9(2), 23–36.

Gürelli, N. (1973). Şehirleşme ve Suç, İ.Ü.H.F. Mecmuası cilt: 32, sayı:2–4, İstanbul.

Hillman, M., Adams, J., Whitelegg, J. (1990). One false move, a study of children's in dependent mobility. London: Policy Studies Institute.

Hillman, M. (1999). The impact of Transport Policy on Children's Development. Paper presented at the Canterbury Safe Routes To Schools Project Seminar, Canterbury Christ Church University College, 29 May 1999. http://www.spokeseastkent.org.uk/mayer.php.

Huttenmoser M. and Degen-Zimmerman, D. (1995). Lebensräume fur Kinder, Empirische Untersuchungen zur Bedeutung des Wohnumfeldes fur den Alltag und die Entwicklung der Kinder, Bericht 70 des National Forschungsprogram Stadt und Verkehr (Room for Children: Empirical Research into the Influence of the Daily Environment for Children's Development, National Research Program for City and Transport), Zurich.

Jones, O. (2000). Melting geography: purity, disorder, childhood and space. In S. L. Holloway, & G. Valentine (Eds.), Children's geographies. Playing, living learning (pp. 29–47). London: Routledge.

Kytta, M. (1997). Children's independent mobility in urban, small town, and rural environments, growing up in a changing urban landscape, Royal, Van Gorcum, Assen, pp. 41–52.

Kytta, M. (2004). The extent of children's independent mobility and the number of actualized affordances as criteria for child-friendly environments, Journal of Environmental Psychology 24 (2004) 179–198.

Little, H., & Wyver, S. (2008). Outdoor play: does avoiding the risks reduce the benefits? Australian Journal of Early Childhood, 33(2), 33–40.

Malone, K. (2007). The bubble-wrap generation: children growing up in walled gardens. Environmental Education Research 13 (4): 513–527.

Nansen, B., Gibbs, L., MacDougall, C., Vetere, F., Ross, N. J. & J McKendrick, J. (2015) Children's interdependent mobility: compositions, collaborations and compromises, Children's Geographies, 13:4, pp. 467–481, DOI: 10.1080/14733285.2014.887813.

Norinder, M. H. (1996). Children, Environment and Independent Mobility. In Evolving Environmental Ideals – Changing Way of Life, Values and Design Practices: IAPS 14 Conference Proceedings, 314–323. IAPS. Stockholm, Sweden: Royal Institute of Technology.

O'Brien, M., Jones, D., & Sloan, D. (2000). Children's in dependents patial mobility in the urban public realm. Childhood, 7(3), 257–277.

Prezza, M., Stefania, P., Morabito, C., Cinzia, S., Alparone, F. R., &Guiliani, M. V. (2001). The influence of psycho-social and urban factors on children's independent mobility and relationship to peer frequentation. Journal of Community and Applied Social Psychology, 11(Number 6), 435–450.

Ross, N. J. (2007). My journey to school…': foregrounding the meaning of school journeys and children's engagements and interactions in their everyday localities. Children's Geographies 5(4): 373–391.

Shaw, B., Watson, B., Frauendienst, B., Andreas Redecker, A., Tim, J.,Hillman, M. (2013). Children's independent mobility: a comparative study in England and Germany (1971-2010), The Policy Studies Institute, London, 2013.

Skelton, T. (2009) Children's geographies/geographies of children: play, work, mobilities and migration, Geography Compass, 3, 4:1430–1448.

Tandy, C. A. (1999). Children's diminishing playspace: A study of intergenerational change in children's use of their neighborhoods. Australian Geographical Studies, 37(2), 154–164.

Tranter, P. (1993). Children's mobility in Canberra: Confinement or independence? Monograph series No. 7, Department of Geography and Oceanography. University College, Australian Defence Force Academy, Canberra.

Tranter, P., and Whitelegg, J. (1994). Children's travel behaviours in Canberra: car-dependent lifestyles in a low-density city. Journal of Transport Geography, 2(4), 265-27.

Tudor-Locke, C., Ainsworth, B., & Popkin, B. (2001). Active commuting to school: an overlooked source of children's physical activity? Sports Medicine, 31(5), 309–313.

Tuik (2013). MEDAS Merkezi Dağıtım Sistemi. Türkiye, Erişim tarihi: 15 Ekim 2016, http://biruni.tuik.gov.tr/medas/?kn=98&locale=tr.

Valentine, G. and J Mckendrick, J. (1997). Children's outdoor play: exploring parental concerns about children's safety and the changing nature of childhood, Geoforum Volume 28, Issue 2, May 1997, Pages 219–235.

Van der Spek, M. & Noyon, R. (1995). Children's Freedom of Movement in The Streets. Paper presented at the International Conference on Building Identities. In Gender Perspective on Children and Urban Space. Amsterdam.

Zomervrucht, J. (2005). Inviting Streets For Children, Some lessons and results of the Child Street 2005 conference in Delft, Huizen, The Netherlands, Veilig Verkeer Nederland. Retrieved July 08, 2007, from http://www.urban.nl/childstreet2005/downloads/Stutgart/zomwevrucht_paper_stuttgart.pdf

Zubrick, S., Wood, L., Villanueva, K., Wood, G., Giles-Corti, B., Christian, H. (2010). Nothing but fear itself, Parental fear as a determinant impacting on child physical activity and independent mobility, Victorian Health Promotion Foundation

Assist. Prof. Dr. Suzan Girginkaya Akdağ[1]

Bim for Sustainable Landscape Design

[1]*Bahçeşehir University, Faculty of Architecture and Design, İstanbul, Turkey*

1. Introduction

The implementation of BIM has lately been reinforced by mandatory government policies due to its significant benefits: as a tool for enhancing collaboration with other professionals and clients; streamlining information processes; improving decision-making and delivering in-detail design solutions on schedule and on budget [1]. Although BIM has been criticized for its building-centric nature, its standards, applications and information technologies have also been adopted by landscape architects, who prefer to call it as LIM (Landscape Information Modelling) and SIM (Site Information Modelling) [2]. What makes BIM even more critical for landscape design is its potential for testing sustainable/green/eco-friendly design criteria. Its efficiency and accuracy in the analysis and evaluation of design decisions and access to information makes BIM promising for sustainable landscape design, although still there are gaps between theory and practice [3]. This chapter aims to highlight the potential of BIM for sustainable landscape design.

2. Sustainability and BIM

Towards the end of the 20th century, sustainability became a crucial issue in design due to environmental concerns and the rising cost of energy. Each phase within the overall building lifecycle requires the consumption of energy, raw materials and fresh water. The generation of construction and demolition waste as well as the emission of greenhouse gases pose big threats. Hence, the design of sustainable buildings and environments entails an interdisciplinary and integrated workflow. BIM is efficient in enabling this workflow for the landscape designer who is in need of collaboration with the client, architect, engineers, quantity surveyors, energy consultants, facility manager, contractor and design facilitator, etc. [4]. On the other hand, sustainable design requires the efficient use of materials, resources and energy. Therefore designers have to make decisions based on a lifecycle assessment including new construction, refurbishment and in-use stages for master planning projects, infrastructure and buildings [5]. In terms of sustainable land design and development, ecosystem services such as generation, restoration, conservation and management of water, soil, vegetation, material use and access to nature have been evaluated [6]. During each, assessments take into account environmental impacts such as embodied energy, greenhouse gas emissions, solar/shading performance, lifecycle cost and lifespan.

Figure 1: Ecosystem Services in Sustainable Land Design&Development [6]

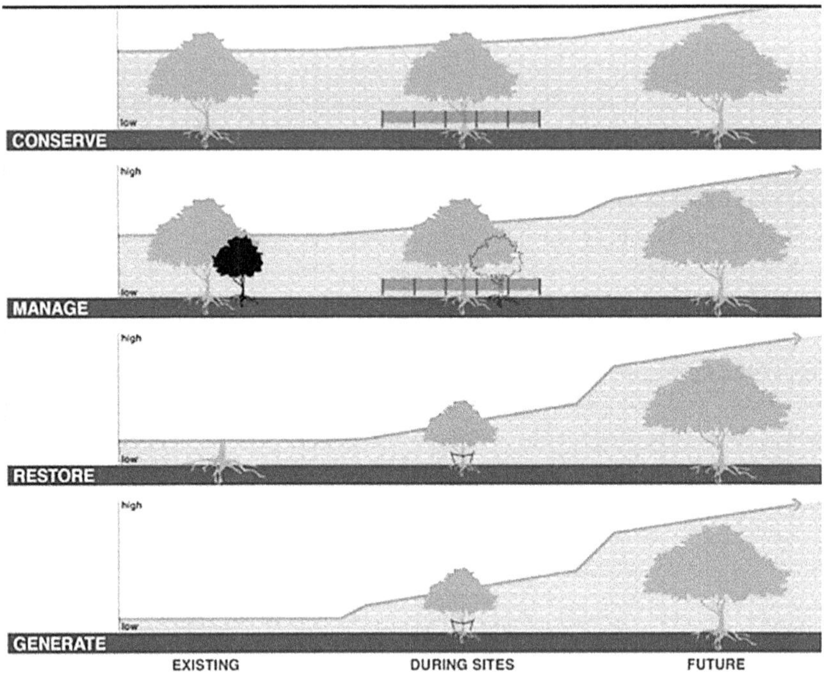

Sustainable design research and practice has been triggered by BIM software owing to their smart built-in objects, site information modeling capabilities, databases, analysis tools, project sharing, flexible documentation and open data standards and interoperability features [7]. For landscape design "BIM is well suited for tracking compliance with rating guidelines. The intelligent objects directly report this data to built-in worksheets where calculations are made to complete the energy efficiency of the documentation process. In many cases, objects perform multiple analysis tasks. For example, when conducting solar studies, a landscape architect may position trees to minimize solar gain and at the same time use the trees to compute the amount of carbon to be sequestered over the number of years that those trees will benefit the project. You can also analyze water efficiency in irrigation and ensure the use of the right equipment and the right type of plants. The user can add other relevant information to these smart objects themselves, like solar reflectors on pavements and performance standards for site furniture [8]".

3. Green Building Certification Systems and BIM

Due to the critical need for a more careful and sensitive approach to land design and development, green building certification systems have become the leading and most widely used environmental assessment methods for buildings and communities. They have changed the way buildings and communities are planned, constructed, maintained and operated. In particular, U.S. LEED (Leadership in Energy and Environmental Design), U.K. BREEAM (Building Research Environment Assessment Method) and German DGNB (German Sustainable Building Council) Certification Systems have widely been accepted as national and international standards.

The comparison of green building certificate systems reveals that each have common criteria concerning landscape design, such as integrative thinking, water management, location and transportation and sustainable sites [5; 9; 10]. However, each system also has its own challenges such as LEED evaluating 'Global, Regional and Local Practices' or DGNB assessing 'Integrated Public Art' within 'Sociocultural and Functional Quality'. Other complementary and supplementary rating systems are available such as: Sustainable Sites evaluating ecosystem services or Park Smart for high-performing, sustainable garages [6; 11].

Table 1: Criteria of Green Building Certification Systems

LEED	BREEAM	DGNB
Integrative thinking	Management	Process Quality
Water	Water	Environmental
Waste	Waste	
Materials	Materials	
Location&Transportation	Transport	Functional
Sustainable Sites	Land Use	Site

LEED	BREEAM	DGNB
Health &Human Experience	Health &Well Being	Sociocultural
Regional impacts	Pollution	
Innovation	Innovation	Economic Economic
Global, Regional, Local		

As the checklists of certification systems, many credits require a huge amount of data and drawings to be submitted. The calculation and the visualization of data can be achieved very efficiently with BIM programs, reducing the time and cost involved in the compliance process [12] . Furthermore those calculations can be embedded in a BIM schedule and refreshed dynamically throughout the entire building life cycle. In the following parts the most specific BIM tools for landscape design are explained: Sustainable Site and Visualization, Energy Efficiency and Simulations, Sustainable Location and Transportation, Water and Rain Water Management.

Figure 2: Landscape Design Software [12]

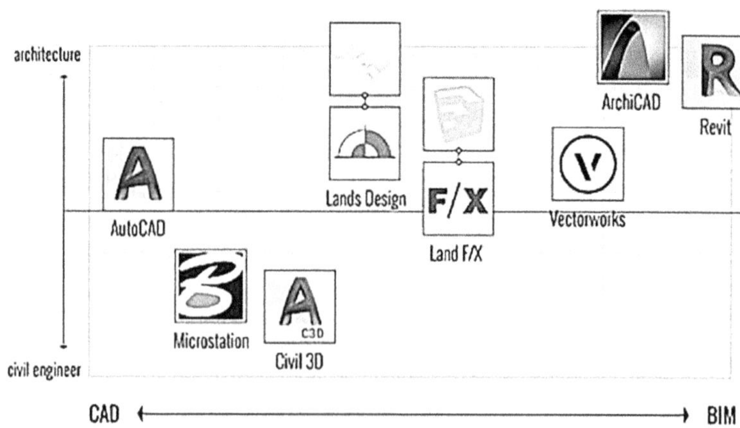

3.1. Sustainable Site and Visualization with BIM

Probably, one of the most well-known BIM tools for landscape design is topography modeling required by sun and shadow studies.

On the topography model, the site can be analyzed for all earthwork based on its boundaries, area, volume, slope, cut and fill, and other location-based data, such as water flow, sun and shade. Complex areas of level changes can be resolved in the early design process. Planting plans can be imported into the 3D project model to provide a visual understanding of massing. Hardscape materials such as site furniture, fencing and walls can be added [13].

Figure 3: Modeling Sites and Landscapes in BIM [13]

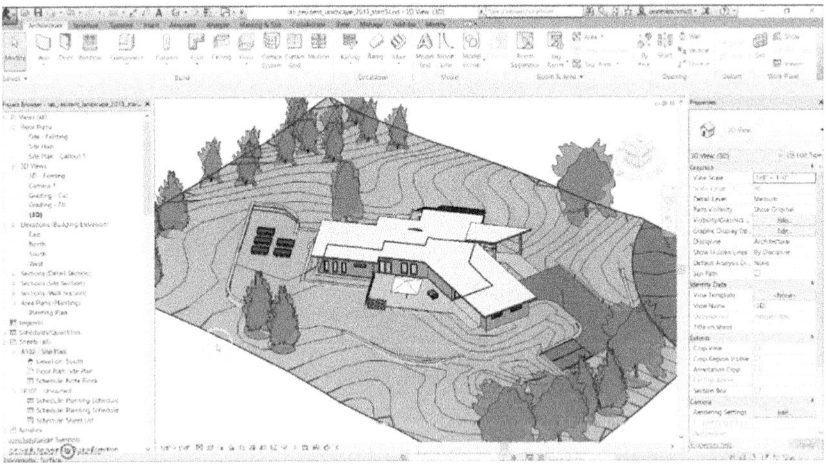

However, BIM lacks landscape-oriented workflows. Topography, which is the base for projects, is created by the triangulation and contouring of input points. Once created, this TINN (Triangulated Irregular Network) surface cannot be delineated to multiple soil types and depths; it is limited to a single material with an infinite depth. And within that single material, there is no way to create a tunnel or underground space below the surface [14].

In addition to dealing with imprecise and difficult topography, Revit does not have any built-in tools for modeling water or hardscape elements (such as roads, sidewalks, curbs, or any other pavement). Though there are other elements, such as floors and in-place components for modelling hardscape and curbs, they do not interact with TINN and display hardscape patterns incorrectly [14]. Hence there is need to improve BIM content libraries for site elements since landscape design necessitates the 3D site reconstruction of detailed topographic land surveys. For example, developing BIM standards that would enable the designer to analyze trees in an urban plaza and determine how they impact wind and sun hitting a building would greatly enhance the site/structure relationship [15].

Figure 4: BIM Site Modelling with Topographical Survey Data [16]

On the other hand, BIM tools can depict 3D models of several sustainable design features associated with built structures such as terraces, courtyards, walkways, green roofs, and roof gardens. For example, the green roof demonstration project at the ASLA (American Society of Landscape Architects) headquarters highlights the potential of green roofs. Using BIM modelling for roof gardens over the layout of structural beams, utilities, and spaces below means that the design phase is integrated with other disciplines. Strategies such as planting wells required by the trees to be planted over the roof garden could be considered, and placed within the structural, electrical, and mechanical parts of the building [17].

Figure 5: National BIM Library Roof Systems for Green Architecture [18]

Figure 6: ASLA Headquarters Green Roof [17]

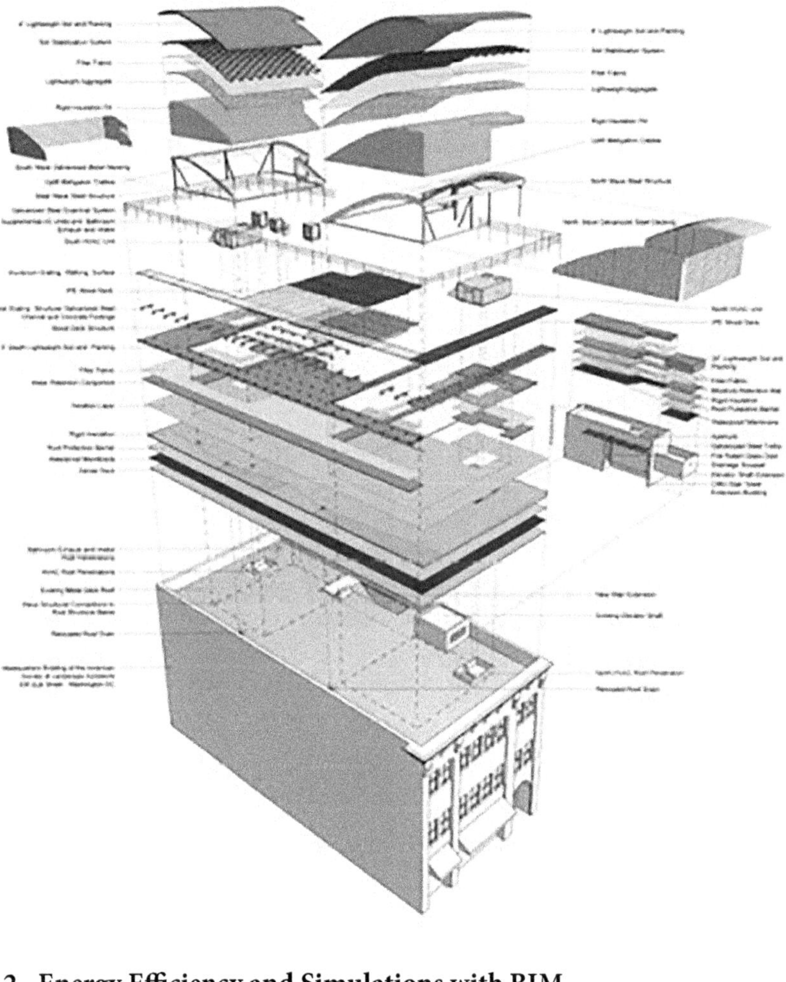

3.2. Energy Efficiency and Simulations with BIM

Using conceptual BIM models and energy analysis tools, such as day light, solar radiation and wind analysis, etc., landscape designers can measure and identify potential loss or gain of energy for different alternatives [13]. Considering the energy consumption and green certificate credits, designs can be optimized to create sustainable environments.

As part of a global trend for sustainable development, Green Campus projects are aiming to support sustainable and climate-friendly campus life with the participation of university staff and students. In order to increase energy efficiency, the environmental performance of campus buildings in terms of water, energy, carbon footprint and waste are recorded with BIM and improvements are determined via energy analysis. Shifting from the architecture to the landscape design of the campus, BIM is beneficial in simulating low energy use via several outdoor features such as storm water infiltration, recycled materials, low consumption and long-life lighting products, and simple tree shading to reduce heat islands [19].

Figure 7: Energy Efficient Lighting in İTÜ Ayazağa Campus[19]

3.3. Sustainable Location and Transportation with BIM

In order to protect the environment and improve quality of life for communities, the efficiency of location and transportation systems are critical. The global policies and commitments to reduce fuel consumption, air pollution, and carbon emissions from vehicles have become critical measures for landscape design [20]. BIM enables the use of various kinds of data such as CAD drawings, GIS, geo-referenced aerial and satellite images, cloud computing, and analytics in the planning and management of infrastructure systems [7]. BIM projects concerning safe cycling and bike sharing, pedestrianization and evacuation etc., are on the agenda of smart cities and institutions for sustainable urban transport and community development.

Figure 8: BIM Transportation Modelling: Proposal for Şemseddin Günaltay Avenue–Fahrettin Kerim Gökay Avenue in Istanbul [21]

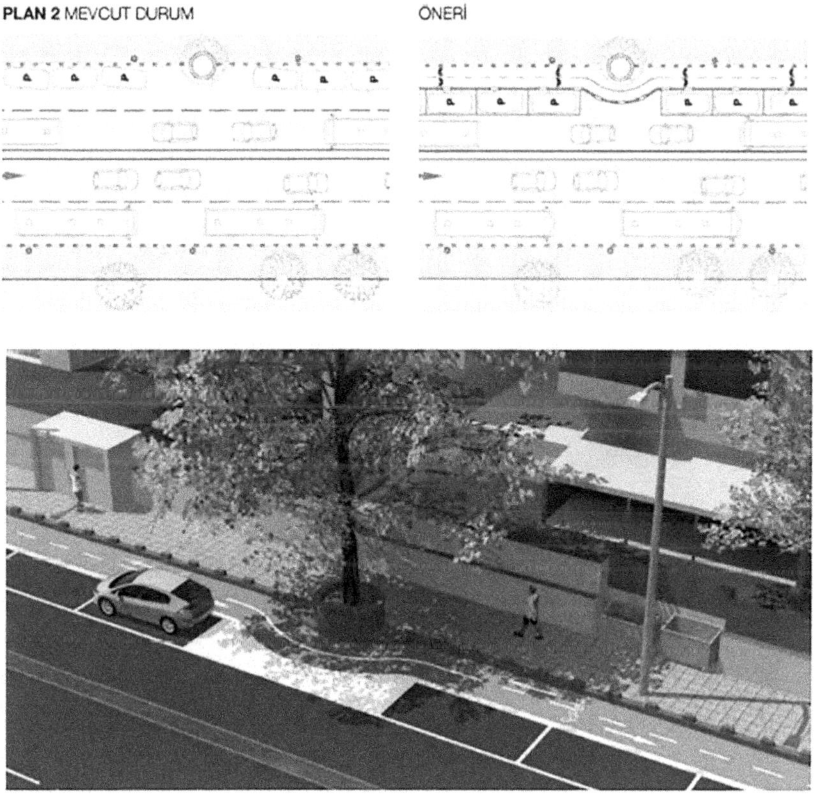

3.4. Water and Rain Water Management with BIM

According to sustainability criteria, rainwater is a resource that provides many environmental and economic benefits. Managing rainwater on site restores natural hydrologic conditions, reduces the possibility of flooding, and creates opportunities for onsite water reuse in applications such as irrigation and landscape features.

BIM programs can automate numerous tasks associated with water management including planting, irrigation and hardscape works.

One of the BIM softwares developed for landscape design can create zones of similar watering needs with a Hydro zone tool and then use built-in worksheets to analyze these for water efficiency, which makes satisfying a site's water budget allowance easier than ever. It also enables automated and manual placement methods for specifying lateral and main piping, sprinkler heads, drip systems, valves, controllers, or other system components that are designed within industry standards. Hence, proposed and as-built systems can be designed, documented, and estimated efficiently [17].

Figure 9: Irrigation Design with BIM [17]

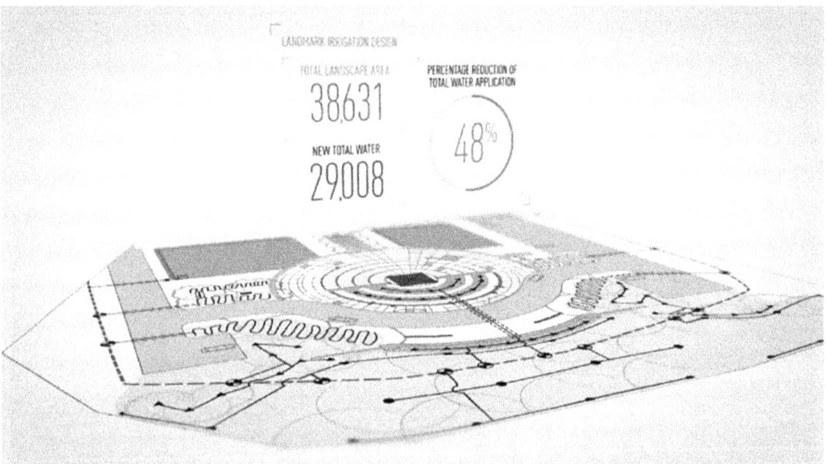

4. Conclusion

This chapter surveyed the utilization of BIM for achieving sustainability and provides an insight into its possibilities in landscape design. Current multidisciplinary platforms and integrated projects require landscape architects to implement BIM in their workflow. Although most landscape architecture-oriented tools are still in progress, BIM promises great possibilities for design, compliance and facility management for landscape design and development [3]. Hence according to some authorities BIM is the future, and those that don't react will be left behind [16].

References

1. The Landscape Institute: *BIM for Landscape*. Routledge, New York (2016).
2. The Limits of BIM, Retrieved on May, 2017 from https://landscapearchitecturemagazine.org/2016/02/16/
3. MHC_GreenBIM_SmartMarket_Report_2010.pdf, Retrieved on May, 2017 from https://www.construction.com/market_research/freereport/greenbim/MHC_GreenBIM_SmartMarket_Report_2010.pdf
4. G. Löhnert, A. Dalkowski, W. Sutter W: *Integrated Design Process, Task 23*, Energy Agency (IEA), Berlin (2003), Retrieved on May, 2017 from http://task23.iea-shc.org/data/sites/1/publications/IDPGuide_internal.pdf
5. What is BREEAM, Retrieved on May, 2017 from http://www.breeam.com/
6. Understanding the Core Concepts of SITES v2 Rating System, Retrieved on May, 2017 from http://www.usgbc.org/education/sessions/understanding-core-concepts-sites-v2-rating-system-9768715
7. Vectorworks Landmark, Retrieved on May, 2017 from http://www.vectorworks.net/landmark
8. Landscape Architecture: A Useful Piece of the BIM Workflow, Retrieved on May, 2017 from http://www.pobonline.com/articles/100788-landscape-architecture-a-useful-piece-of-the-bim-workflow
9. LEED, Retrieved on May, 2017 from http://www.usgbc.org/leed
10. DGNB – German Sustainable Building Council, Retrieved on May, 2017 from http://www.dgnb.de/en/
11. Park Smart, Retrieved on May, 2017 from http://parksmart.gbci.org/
12. BIM and Landscape Architecture, Retrieved on May, 2017 from http://worldlandscapearchitect.com/bim-landscape-architecture/#.WTAm1-uGOUk
13. Autodesk Revit, Retrieved on May, 2017 from https://knowledge.autodesk.com/support/revit-products/getting-started?sort=score
14. Revit and Landscape Architecture: The Issues by Lauren Schmidt http://worldlandscapearchitect.com/revit-landscape-architecture-issues/#.WTAryuuGOUk
15. Topographical Surveys for BIM – What you need to know. Retrieved on May 2017, from https://www.linkedin.com/pulse/topographical-surveys-bim-what-you-need-know-steve-bury
16. BIM and Landscape Architecture-Impact&Issues, Retrieved on May 2017, from https://www.marshalls.co.uk/commercial/blog/article-bim-and-landscape architecture-impact-and-issues-2640

17. Green Roof, Retrieved on May, 2017 from https://www.asla.org/ContentDetail.aspx?id=25362
18. National BIM Library Roof Systems for Green Architecture, Retrieved on May, 2017 from https://www.nationalbimlibrary.com/cefil-uk-ltd/cefil-eco-roof-intensive-roof-system
19. İTÜ Yesil Kampüs, Retrieved on May, 2017 from http://www.yesilkampus.itu.edu.tr/
20. Policies To Reduce Fuel Consumption, Air Pollution, and Carbon Emissions From Vehicles in G20 Nations, Retrieved on May, 2017 from http://www.theicct.org/policies-reduce-fuel-consumption-air-pollution-and-carbon-emissions-vehicles-g20-nations
21. İstanbul'da Güvenli Bisiklet Yolları Uygulama Kılavuzu http://wrisehirler.org/sites/default/files/%C4%B0stanbul%27da%20G%C3%BCvenli%20Bisiklet%20Yollar%C4%B1%20Uygulama%20K%C4%B1lavuzu_0.pdf

Ercan Gökyer[1]

Evaluation of User Perception on the Effects of Landscape Changes: Topography and Ecophysiology

[1]*Bartın University, Faculty of Forestry, Department of Landscape Architecture, 74100, Bartın, Turkey, egokyer@bartin.edu.tr*

1. Introduction

In recent years, human life has been affected by unsustainable urban development as a result of rapid population growth worldwide. Nowadays, in urban areas, people live monotonous lives in restricted environments as result of population growth and urbanization. Besides, changes are continuously occurring in people's surrounding areas. But people do not exactly notice all of these transformations in urban areas. In our surrounding areas, changes emerge related to processes and differences in season. Changes that emerge according to processes are not always perceived. These changes could be understood by performing some analyses and evaluations using landscape ecological approaches in GIS. In cities, seasonal changes are easily understood and observed in green areas. Seasonal changes which are understood with ecophysiological processes and are affected by topographical and climatic conditions are effectively perceived in rural and mountainous areas.

In this chapter we shall see: (1) an explanation of the concept of scale, which has an important influence on the perception of the landscape; (2) a brief description of concepts and technologies related to landscape ecology; (3) a highlighting of the perception of changes and differences that result from topographical and ecophysiological variations.

2. Perception of Scale

Landscape and landscape elements are defined by different scales and contents. At the same time, there are many definitions and terms related to landscape. Watershed, basin, region, ecosystem are some of them [1, 2]. The most obvious difference between these terms and definitions is the scale factor. Depending on the scale differences in the landscape, human perception also varies.

There are different scales to describe landscapes [1]. In landscape ecological studies, the size and scope of landscape elements change depending on the scale. A correctly determined scale is important to landscape ecologists. Spatial data that are required for landscape ecological studies are produced from different sources and mapped on different scales [3]. On broad scales, the number and the variety of landscape elements increase.

Two factors are important for scale. These are grain and extent. Grain means finest level of resolution. The extent means total area or measurement of the study area. In recent years, important progress has been achieved in Geographical Information Systems (GIS) and remote sensing technology. GIS is important tool to produce spatial data for broad scale areas. On the other hand, alternative scale possibilities are provided with GIS and remote sensing technology [4, 5]. In landscape ecological studies, using GIS and remote sensing technology has provided important progress for perceiving and assessing landscape with alternative scale possibilities.

3. Landscape Ecology

Landscape ecology is a subdivision of general ecology that is concerned with numbers. Landscape ecology put forth landscape structure, using both GIS technology and associated tools (landscape metrics). Landscape ecological approaches are expressed through landscape elements and landscape structures with numerical values (number, size, shape, etc.) [2, 6, 7].

In the landscape ecological approach that emphasizes the relationship between spatial patterns and ecological processes [8], landscapes have three essential characteristics. These characteristics are structure, function and change (transformation) [1, 2].

Landscape structure expresses the spatial characteristics of landscape. Landscape metrics which are used in the studies concerned with the landscape structure are essential tools to assess landscape development [9, 10]. Understanding landscape structure not only assesses spatial patterns with numerical values but also provides related ideas for landscape change and landscape function. In that way, the perception and interpretation of landscape function and landscape changes can be objectively carried out.

3.1. Landscape Change

Landscape change is an important research topic in landscape ecology. There are two important factors for landscape change. These are [11,12, 13];

1- Human interventions: road constructions (Figure 1), urbanization, industrial and technological development,
2- Natural disasters: floods, earthquakes, landslides.

Figure 1: Alterations in forest land related to road construction

Figure 1 represents changes that occur due to road construction in the Amasra northwestern part of Turkey. Changes related to the dynamic nature of the landscapes occur continuously. Spatial changes which emerge gradually in the long term are used by landscape ecological approaches and related technologies to perform some analyses. Land use or land cover maps that are reveal spatial characterics of landscapes produced by satellite images (Figure 2) help to understand land mosaic and spatial changes (Figure 3).

In recent years, it is possible to obtain high-resolution satellite images with new satellite imagery technologies. Advances in satellite imagery technology have made an important contribution to the perception and evaluation of landscape and landscape elements.

Figure 2: Different spatial resolution in satellite images

a. Landsat 5 [14] *b. GeoEye–1 (Palm Oil – PNG) [15]*

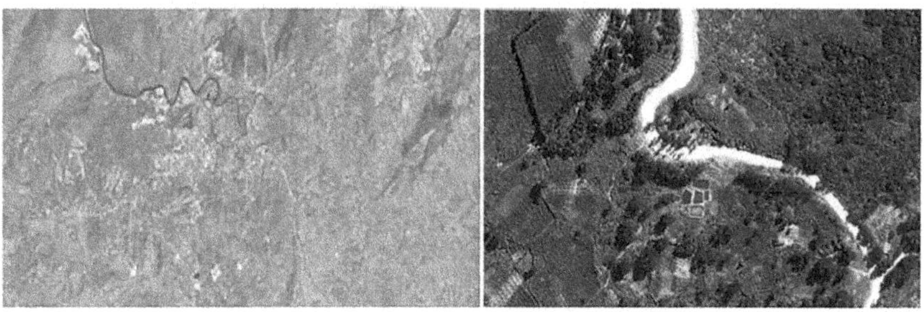

Figure 2 represents two different spatial resolution satellite images. Spatial resolution increases from Figure 2a. to Figure 2b. Figure 2a. represents a 30m spatial resolution Landsat image [14]. Figure 2b. represents a GeoEye–1 (ground resolution is 0.41m in panchromatic mode and at 1.65 m for multispectral resolution) [15]. Landscape elements and details are easily perceived without analysis in Figure 2b.

Seasonal changes are observed for a year depending on the seasonal conditions and the vegetation period of plants. In cities seasonal changes could be observed in green areas. Broad scale seasonal changes should be clearly observed within rural and mountainous areas. Ecophysiological and topographical conditions are effective factors in these changes.

Figure 3: Urban sprawl of the Bartın Province, Turkey (Adapted from, Gökyer 2014b).

a. 1986 *b. 2013*

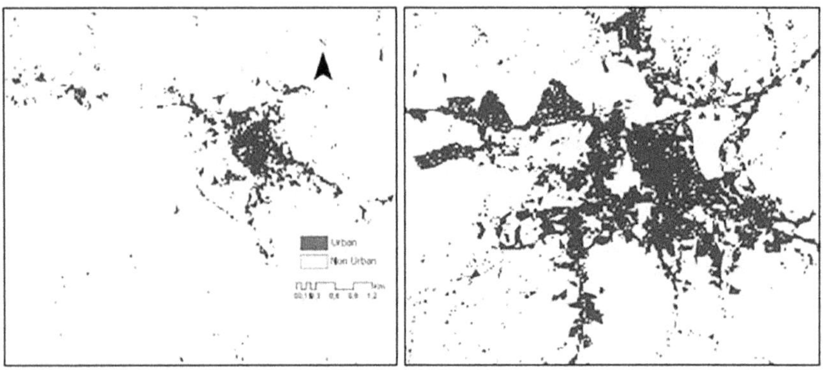

Figure 3 represents the urban development (1986–2013) of Bartın city, Turkey. Measurements with landscape metrics and analyses in GIS indicate that the urban area grew approximately fourfold in the 27 years period [16].

In green areas seasonal changes could be perceived by observing the plants around our environment. Physiological features of plants vary, related to topographical and climatic conditions in the vegetation season [17]. Ecophysiological processes are observed and measured to understand seasonal changes. Thus, the interpretation and expression of seasonal differences and the evaluation and comparison effects of topographical and climatic conditions on plants and habitats can be performed.

4. Topograpy

Data are collected and produced from different sources to use in landscape ecological studies. These sources are satellite images, aerial photos and topographic maps. Maps present the earth's surface graphically on a plane. A topographic map shows the elevations of the objects from the sea level [18]. Using vector data which are produced by drawing the curves on a topographic map (Figure 4) can then produce three important data (altitude, slope and aspect groups) (Figure 5) in GIS platform.

Figure 4:

a. Representation of curves on topographic map b. 3D representation of land.

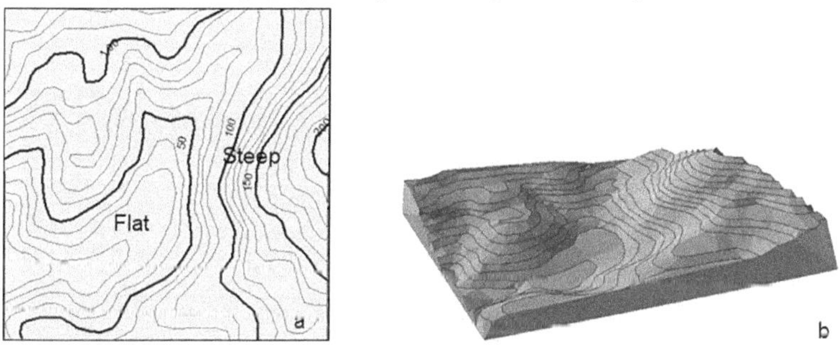

In Figure 4 a. each contour curve represents elevation above sea level. The difference in elevation between intervals is 10m in this case. While closer intervals between curves represent steeply inclined lands, wider intervals represent moderately inclined lands. Gradient changes of land can be seen clearly in Figure 4b.

Figure 5: Maps of altitude, slope and aspect groups

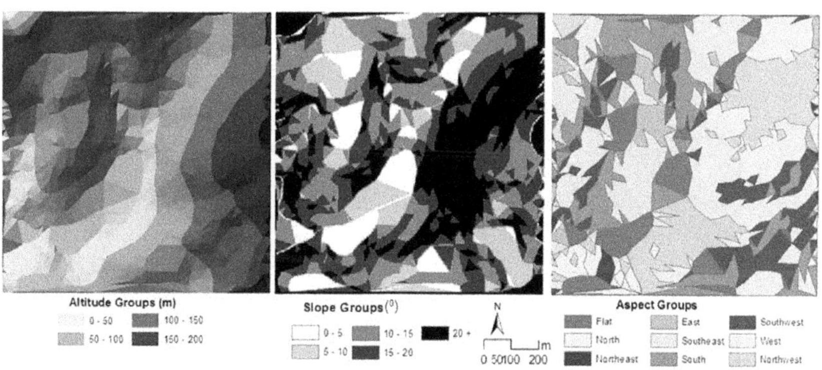

Figure 4 represents three important data which are produced by using contour curves in GIS platform. Altitude and slope groups could be classified using different numerical ranges to put into use the perception and evaluation of spatial characteristics in GIS.

Understanding topographical conditions has made an important contribution to the analysis of the visual quality of land, the distribution of plant species and plant communities. Objects that are viewed within the landscape vary depending on the inclination of the land. Flat and moderately sloping areas that are viewed from surrounding lands with steep slopes show a wide range of perspectives and perceptions (Figure 6). The steeply sloping lands limit the angle of view. Consequently, steeply sloping lands have fewer landscape objects and land use types in a view as distinct from flat and moderately sloping lands (Figure 7).

Figure 6: Flat and moderately sloping lands

Figure 7: Steeply sloping lands

Plant species change in relation to topographical conditions, especially elevation and aspect. Thus, emerging pure and mixed plant communities in forest areas produce particularly different scenes for human perception (Figure 8). The physiological characteristics of plants should be investigated using ecophysiological parameters or processes in order to determine the effects of topography on plants.

Figure 8: Representation of forest zones (a) and mixed plant community (b)

5. Ecophysiology

Ecophysiology uses different instruments and sampling strategies to investigate physiological features of the species on site in a given habitat. Understanding physiological features not only involves measurements in a given habitat but also comparisons which can be made across different sites and habitats [19]. Ecophysiological parameters or processes which are measured and compared on a given site or different sites such as phenology, leaf area development of canopy and understory, annual wood increment, annual litterfall, mineral nutrition of the dominant vegetation, and gas exchange or sap flow, at least of the dominant vegetation, have made an important contribution to determining the physiological features of species and evaluating the effects of global climate changes [20].

Ecophysiological processes which emerge and can be observed from the beginning to the end of the vegetation season such as the occurrence and spillage of leaves, blooming, color changes in leaves, and the occurrence of fruit and seeds have an important effect on human perception. Observing and perceiving these excellent varieties such as flowers, leaves, fruits and seeds has a positive effect and brings serenity to human life. Ecophysiological processes are altered depending on topographical and climatic conditions. For example, at sea level, the vegetation season starts earlier in comparison with high altitudes (mountainous areas). Seasonal changes of plants can be observed at different times by trips from sea level to high altitude areas. However, investigations have been carried out on depending on ecophysiological processes, in order to produce the best growth conditions of plants.

4. General Evaluation

Changes that occur related to topographical conditions and ecophysiological processes are observed both in urban green areas and in rural and mountainous areas. Alterations which are emerging in our surrounding areas are explained through comprehensive investigations using landscape ecological approaches. On the other hand, human life is affected by these alterations. Effects of alterations on human life can be determined and evaluated using these investigations.

However; topographic conditions and ecophysiological processes should be evaluated to perform research related to land use planning [21], landscape and land use changes, and the planning of ecotourism and recreational activities in rural [22] and mountainous areas.

5. References

1. **Ingegnoli, V. (2015)** Landscape Bionomics Biological–Integrated Landscape Ecology. Springer.
2. **Forman, R.T., Godron, M. (1986)** Landscape ecology. 619pp. John Wiley and Sons, New York.
3. **Gergel, S.E., Turner, M.G., (2002)** Learning landscape ecology: a practical guide to concepts and techniques. Springer Science and Business Media.
4. **O'neill, R.V., Smith M.A., (2002)** Scale and Hierarchy Theory (Edited by: Gergel, S. E., and Turner, M. G.,). Springer Science and Business Media.
5. **Greenberg, D.J., Longsdon, M.G., Franklin, J.F., (2002)** Introduction to Geographic Information Systems (Edited by: Gergel, S.E., and Turner, M.G.,). Springer Science and Business Media.
6. **Farina, A. (2000)** Landscape ecology in action. Kluwer Academic Publishers: Netherlands
7. **Dramstad, E., Olson D.J., Forman, T.T.R., (1996)** Landscape ecology principles in landscape ecology and land use planning. Island Press: USA.

8. **Turner, M.G., Gardner, R.H., (2015)** Landscape ecology in theory and practice (Vol. 401). New York: Springer.
9. **Jaeger, J.A. (2000)** Landscape division, splitting index and effective mesh size: new measures of landscape fragmentation. Landscape Ecology. 15: 115–130.
10. **Gökyer, E. (2014a)** Evaluating landscape changes and environmental threats in a coastal landscape case study of Bartın coastal area, Turkey. Fresenius Environmental Bulletin, 23(7 A), 1683–1688.
11. **Antrop, M. (2005)** Handling landscape change. Landscape Change Conference Proceedings. ECLAS 2005. Ankara. 3–15.
12. **Gökyer, E. (2009)** Bartın İli Kent Merkezi ve Arıt Havzası'nda Peyzaj Değerlendirme (Doktora Tezi, Basılmamış), (Landscape Assessment in Bartın City and Arıt Basin) (Doctoral thesis, Unpublished). Ankara Üniversitesi Fen Bilimleri Enstitüsü Peyzaj Mimarlığı Anabilim Dalı, Ankara.
13. **Gökyer, E. (2013)** Understanding landscape structure using landscape metrics. Advances in landscape architecture. InTech.
14. https://www.earthexplorer.usgs.gov
15. http://www.geoimage.com
16. **Gökyer, E. (2014b)** Monıtorıng landscape change and urban growth of Bartın, Turkey, and ıts envıronmental effects. Fresenıus Envıronmental Bulletın, 23(10), 2407–2414.
17. **Öztürk M., Gökyer E., Dönmez Y., Sezgin M., (2016)** Bartın Kenti ve Yakın Çevresindeki Orman Peyzajının Sonbahar-Kış Ekofizyolojik Değişiminin Kent Estetiği Açısından İrdelenmesi (Urban Aesthetics Oriented Evaluation of Autumn-Winter Ecophysiological Dynamics of Forest Landscape within Bartın City and Vicinity). International Winter Cities Symposium, 1268–1280.
18. **Gergel, S.E., and Turner, M.G., Miladenoff, D.J., (2002)** Collecting Data at Broad Scales (Edited by: Gergel, S.E., and Turner, M.G.,). Springer Science and Business Media.
19. **Lüttge, U., Scarano, F.R. (2004)** Ecophysiology. Brazilian Journal of Botany, 27(1), 1–10.
20. **Buchmann, N. (2002).** Plant ecophysiology and forest response to global change. Tree Physiology, 22(15), 1177–1184.
21. **Gökyer, E., Öztürk, M., (2016)** Topographic Evaluation and Suitability of Land Use Characteristics in the Coastal Zone between Amasra and Kurucaşile Cities of Bartın Province, Turkey. International Conference on Natural Science And Engineering (ICNASE'16), 1583–1592.

22. **Gökyer, E., Öztürk, M., Dönmez, Y., Çabuk, S., (2015)** Bartın ili dağlık alanlarında coğrafi bilgi sistemleri kullanılarak ekoturizm faaliyetlerinin değerlendirilmesi (Evaluation of ecotourism activities using geographic information systems in mountain areas of Bartın city). İnönü University Journal of Art and Design, Cilt/Vol. 5 Sayı/No.12, 25–35.

Aslı Korkut[1], Tuğba Kiper[1], Tuğba Üstün Topal[1]

Evaluation of Sustainable Urban Planning Approaches in Terms of the Possibilities for Social Interaction

[1]*Namık Kemal University, Faculty of Fine Arts, Design and Architecture, Department of Landscape Architecture, Tekirdag, Turkey*

1. Introduction

Today, more than half of the worldwide population lives in big cities. There has been a decrease in the quality of the environment and of life due to negative effects such as dense and high-rise housing, an increase in motor vehicle use, noise, air and water pollution, the decrease in green spaces, and stress, particularly depending upon the increase in population density in cities that have entered a rapid growth process along with the industrial revolution [1]. In particular, with the acceleration of urbanisation dating from the second half of the 20th century, metropolitan areas have emerged out of broadened small dwelling units [2,3,4]. Residential structuring has brought common use areas (shopping centres, recreation areas, parks, etc.) to be located away from city centres. This has turned the cities into city blocks and parcels in which people have been alienated through weak social interaction and neighbourhood relationships [5]. Within this process of change, in which urbanisation has become widespread and cities have become decisive, quasi-public areas have changed in quality and residential streets have turned into wide open spaces among high rise buildings. In spite the fact that this building complex type of life is a satisfactory site design as far as the users are concerned, the results it has brought about have not been sufficient. In fact, in the study carried out by Özparlak and Meşhur (2012), it was clearly stated that neighbourhood relationships within home streets were stronger than those in complex buildings [6]. Therefore, approaches that value people and the quality of their lives, and strengthen-social relations, are becoming more and more important. Likewise, the Tenth Development Plan (2014–2018) covers topics such as sustainability, balanced urbanization, the creation of habitable spaces and human development [7]. Most updated approaches are livable cities: the compact city, smart city, eco city, cittaslow city etc. When the livable city is described as an urban system that contributes to physical, social and mental well-being and the personal development of all its inhabitants, it is about delightful and desirable urban spaces that offer and reflect cultural and sacred enrichment. Key principles that give substance to this theme are equity, dignity, accessibility, conviviality, participation and empowerment [8]. The compact city model stands out as the most generic spatial interpretation of the sustainable city and represents an umbrella term for various other related concepts such as the European city model, transit-oriented development, new urbanism, decentralised concentration and smart growth [9]. Smart cities appeared in the late 80s as a means of visualizing the urban context, and they have evolved quickly since then [10]. According to Barrionuevo et al. (2012) [11]; being a smart city means using all available technology and resources in an intelligent and coordinated manner to develop urban centers that are at once integrated, habitable and sustainable. The "Smart Settlement" concept reflects the idea of urban reconstruction that makes cities better for humans and nature [12]. An eco-city is defined as an umbrella metaphor that encompasses a wide range of urban-ecological proposals that aim to achieve urban sustainability. These approaches propose a wide range of environmental, social, and institutional policies that are directed towards managing urban spaces to achieve sustainability. This type promotes the ecological agenda and emphasizes environmental management through a set of institutional and policy-related tools [13]. Cittaslow aims to ensure quality of life in its towns and seeks to create an environmentally sustainable and pleasurable future [14].

Sustainability is the common feature of these planning approaches despite their planning and design principles being issimilar. Starting from this point, as mentioned above, in this research sustainable urban planning approaches and their principles are explained and social interaction and communication facilities are discussed.

2. Sustainability, Urban Planning and Sustainable Urban Planning

International recognition of the fact that environmental protection and natural resources management must be integrated with the socio-economic issues of poverty and underdevelopment culminated in the 1992 Rio Earth Summit. This idea was captured by the definition of "sustainable development", given by the World Commission on Environment and Development, also known as the Brundtland

Commission, in 1987: "development that meets the needs of the present without compromising the ability of future generations to meet their own needs" [15]. Our Common Future (WCED, 1987) [16], or the 'Brundtland Report' as it is commonly known, marked a profound change in our attempts to connect bio-physical environmental, social and economic policy (Figure 1) goals. According to Lundqvist (2007) [17]; the ideas about sustainable development became internationally accepted after the publication of "Our Common Future" in 1987. The conclusion was that sustainable development should be the guiding principle for governments at the local as well as the national level 8. At the UN conference on Environment and Development in Rio de Janeiro in 1992, Agenda 21 was established to manage global issues at a local level 9 [18]. Today the main work towards sustainable urban planning is addressed at the United Nations Human Settlements Programme, UN HABITAT [19]. In association with United Nations Environmental Program (UNEP), the Sustainable Cities Programme (SCP) has been established to improve urban environmental planning and management, a sister programme to Agenda 21. The main principles for sustainable human settlements were established through the second UN HABITAT II Conference in 1996 [20].

Taking the terms 'sustainability' and 'urban' into consideration together is quite a new phenomenon. Sustainable cities form the focus of discussions based upon environmental concerns. Projections of urban planning with few sources and low energy consumption, which include traditional styles, techniques and instruments for people living in rapidly growing cities, have gained importance [1,22,23]. In this regard, urban development has gained recognition though three phenomena, namely constructing wholesome cities, forming livable places, and creating a sustainable urbs [24]. According to Karakurt Tosun, (2013), a sustainable urban plan includes dense and complex land use, design at a human scale, emphasis on public spaces, a coalescence of functions, people-oriented means of transportation, a participatory process, protection of green spaces, provision of technical infrastructure, and coordination of social development. Within this framework, in order to achieve productive use of urban areas, emphasis is placed on high-rise rather than low-rise constructions [1].

Figure 1. Circles of Sustainability [21]

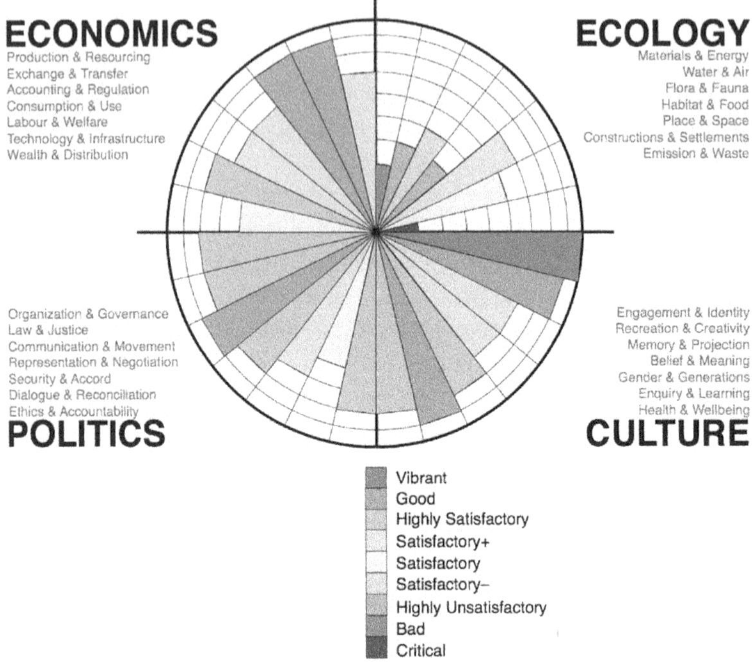

2.1. Sustainable Urban Planning Approaches

According to Becsei (2013), ongoing changes in environmental, economic and cultural conditions demand new approaches to functional and ecological urban development and design [25]. According to Bayram (2001) [26], sustainable cities include approaches which cover sustainable urban development, sustainable human settlements, eco-cities, livable cities and green cities, and which have little differentiation in meaning in each case [27]. In addition to their minor differences, sustainable urban approaches have common characteristics. In the Global Plan of Action, the following aspects of sustainable human settlements are presented: sustainable land use; social development; eradication of poverty; creation of productive employment and social integration; population and sustainable human settlements development; environmentally sustainable, healthy and human settlements; sustainable energy use; sustainable transport and communication systems; conservation and rehabilitation of the historical and cultural heritage; improving urban economies; balanced development of settlements in rural regions and disaster prevention, mitigation and preparedness, as well as post-disaster rehabilitation capabilities [17].

2.1.1. Livable City

The concept of livability should encompass those elements of home, neighborhood, and metropolitan areas that contribute to safety, economic opportunities and welfare, health, convenience, mobility, and recreation. The adjective for a city connotes a desirable quality of life for its citizens,including social activities, attractive public places, and the provision of a certain level of privacy, as well as a sense of community [8]. Ewing (2001) stated in his study that factors such as proximity to the workplace and business districts, the quality of neighbouring schools, and proximity to small-scale businesses people would need in their daily lives, such as security units, restaurants, groceries, and electricians, could be determinant in the definition of the term livable city [28]. In her research (Alberti, 1996) [29] identifies seven concepts of urban quality that she finds relevant to a sustainable urban environment. The seven categories and their explanations can be seen in Table 1.

Table 1. Concepts of urban quality [29]

Environmental Quality	Clean air, water, and soil–adequate food supplies, housing, and infrastructure as well as green areas and open space
Human health	Builds on environmental quality
Efficiency	Resource allocation must maximize the economic output per unit of resource input
Equity	Resource allocation must maximize social benefit per unit of economic output, across social groups, regions and generations
Diversity	Diversity of actors (communities, cultures, and individual's behavior) and diversity of built and natural landscapes is crucial to urban resilience and flexibility
Accessibility	Increased access to goods and services leads to decreased impact on local environment
Learning	Increased learning at local and community level leads to increased flexibility and the ability to modify behavior in response to changing (environmental) circumstances

Şenlier et al. (2013) also emphasised matters such as the characteristics of the residential area, streets/ by-streets, parking lots and pedestrianization, design and construction, the environment and the public as the fundamental criteria of liveability for residential spaces. These fundamental criteria are not considered to be solely related to a rise of quality in housing, space and environment, but are also considered as social thematic criteria taken on a human scale, such as relations among people living on the same site, shopping, entertainment and cultural opportunities, and accessibility to educational and social services, as well as the sufficiency and quality of public transportation [30].

2.1.2. *Compact City*

According to Scheurer (2007), the compact city began to be discussed from the beginning of the 1970s through to the beginning of the 1980s [31]. In international research, the concern with sustainable urban development has grown enormously in the past 25 years. Following the publication of Our Common Future (WCED, 1987), which introduced the idea of sustainable development, a great number of publications have been dedicated to the environmental impacts of urban development and the improvement of life in the city [16]. A couple of years later, the Green Paper of the Commission of the European Communities (CEC, 1990) and the United Nations Earth Summit Agenda 21 (United Nations, 1993) emphasized the role of compact forms of urbanization as a basis for more sustainable urban development [18,32,33]. More recently, the Charter of Leipzig (European Urban Knowledge Network, 2007) and the publication Cities of Tomorrow (European Commission, 2011) emphasized the importance of compact urban developments as an important sustainable strategy for the future development of European cities [34]. Living and working places are not remote in compact cities. Compact city-forming concepts consider cities as: intensive, centralized, dense, concentrated, continuous, bounded, packed, compressed, confined, diversified, integrated, coherent [35] (Figure 2).

Figure 2. Compact city model [36]

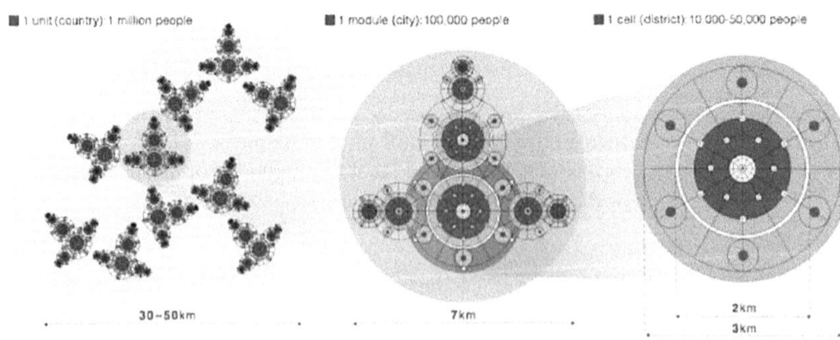

To achieve the goal of sustainable urbanization, the compact model involves an intensification of urban areas, mixed use of land, development of an infrastructure system and green zones, intended to restrict the growth of urban areas [35]. Compact cities have dense and proximate development patterns, are linked by public transport systems, and maintain accessibility to local services and jobs. As such, they play a significant role in responding to the needs of urban areas. Compact cities lessen the impact on the environment, with shorter intra-urban distances and less automobile dependency. They play a part in the economy by increasing the efficiency of infrastructure investment and by giving residents easier access to services, jobs, and social networking [37].

2.1.3. Smart City

'Smart City' becomes prominent as a development term which includes the fundamental qualities of many approaches [12] (Figure 3). It is an approach that emphasises the adjustment processes which will affect the sustainable city approaches which have begun to improve, especially along with the rapid development of technology in the 90's. Basically, this approach is based on a self-management principle without any human interference in the first instance to transportation and energy, or to urban infrastructure and networks [38].

Figure 3. Smart city indicators [38]

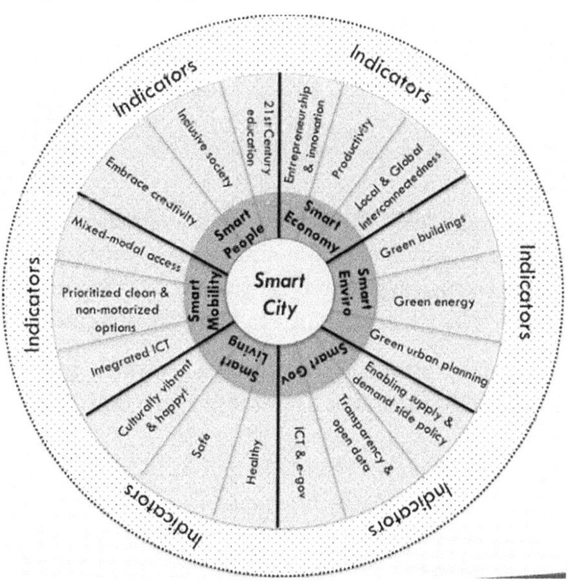

The objective of the smart city approach is to create environmentally friendly cities where information technologies for smart management are used and sustainable development is provided. [12]. Growth in smart cities is provided through the provision of cities with compact neighbourhoods, protected green spaces, and the provision of varieties of transportation through the prevention of uncontrolled urban expansion (Figure, 4).

Figure 4. Smart city definition (EU) [38]

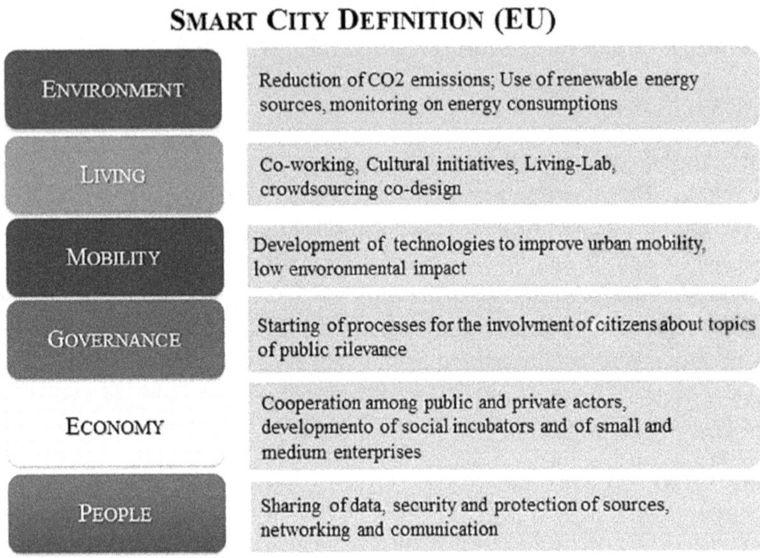

2.1.4. *Eco-city*

According to Jabaroon (2006); the eco-city is an umbrella concept that encompasses a wide range of approaches that aim to make existing cities and urban development more ecologically sound and [13]. Acccording to Roseland (1997) [39], the main principles for developing eco-cities are presented below [17]:

- ✓ Revision of land use priorities to create compact, diverse, green, safe, pleasant, and vital mixed use communities near transit nodes and other transportation facilities
- ✓ Revision of transportation priorities to favor foot, bicycle, cart, and transit over cars, and to emphasize "access by proximity"
- ✓ Restoration of damaged urban environments, especially creeks, shore lines, ridge lines, and wetlands

- ✓ Creation of decent, affordable, safe, convenient, and radically and economically mixed housing
- ✓ Nurturance of social justice and creation of improved opportunities for women, people of color and the disabled
- ✓ Support for local agriculture, urban greening projects and community gardening
- ✓ Promotion of recycling, innovative appropriate technology, and resource conservation while reducing pollution and hazardous wastes
- ✓ Work with business to support ecologically sound economic activity while discouraging pollution, waste, and the use and production of hazardous materials
- ✓ Promotion of voluntary simplicity and discouragement of excessive consumption of material goods
- ✓ Increase in awareness of the local environment and bioregion through activist and educational projects that increase public awareness of ecological sustainability issues

From the planning perspective, Kline (2000), for example, highlights four attributes of the eco-city: ecological integrity, economic security, quality of life and empowerment [40]. The concept of eco-city used here refers to an urban district developed or redeveloped with the aim of promoting the environment in aspects such as the living environment, transport, energy, water treatment, waste management, building and landscape design (Figure 5).

Figure 5. Eco-city indicators [44]

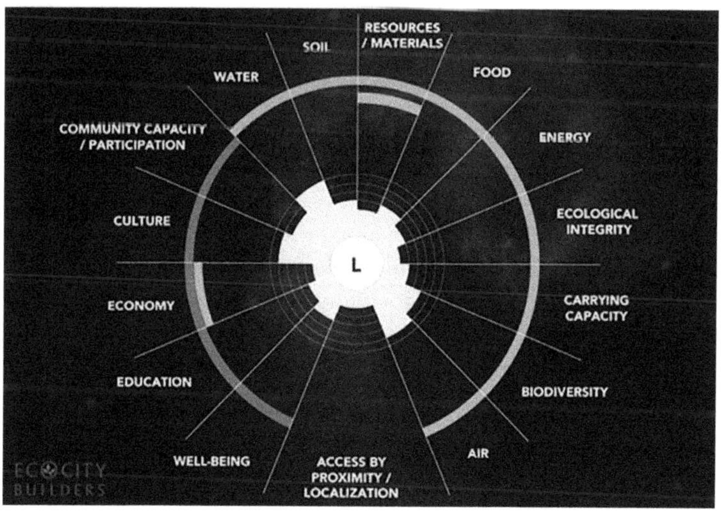

Eco-city development is one way in which governments try to meet these challenges, which among other things are test beds for the development of tools and processes to help find solutions at the city level, with the ultimate goal of reducing the environmental impacts of urban activities and improving regular urban planning and urban development [41].

2.1.5 Slow City (Cittaslow)

Slow city has emerged with the definition of being an international union of cities and towns emphasizing quality of life which protects local identity and characteristics against disidentification due to globalization [42,43].

Cittaslow's agenda focuses on issues concerning quality of life and environment, with an emphasis (through its Slow Food links) on local produce, maintaining local uniqueness, and sustainable urban economies. Member towns must have a population of under 50 000, although larger entities (eg, regional councils and other organisations) can apply the movement's principles as Cittaslow Supporters. The membership accreditation process requires towns to score over 50% in a self-assessment process and inspection of approximately sixty 'requirements for excellence' which fall into the following six areas: (1) environmental policies; (2) infrastructure policies; (3) technologies and facilities for urban quality (eg. plan for controlling noise pollution, promotion of telework, programmes for planting public and private places with important environmentally suitable plants); (4) safeguarding autochthonous production (eg. programmes for educating taste and nutrition in schools in collaboration with Slow Food, census of the typical products of the territory, and support of their commercialisation); (5) hospitality (preparation of slow itineraries of the city); and (6) awareness (education of citizens regarding what a 'slow city' is) [14].

3. Result and Discussion

Due to the increase in the population along with urbanization in big cities over time, city spaces have become unable to cope with their population increase, and have expanded towards the city boundaries. As a result, we see cities in which social relations weaken considerably, and ignore the natural and local values which have emerged. The loss of the effect from a quality and quantity of non-residential communal spaces has also caused weakness in social interaction.

Figure 6. A conceptual model for sustainable cities based on urban planning and urban design considerations [45].

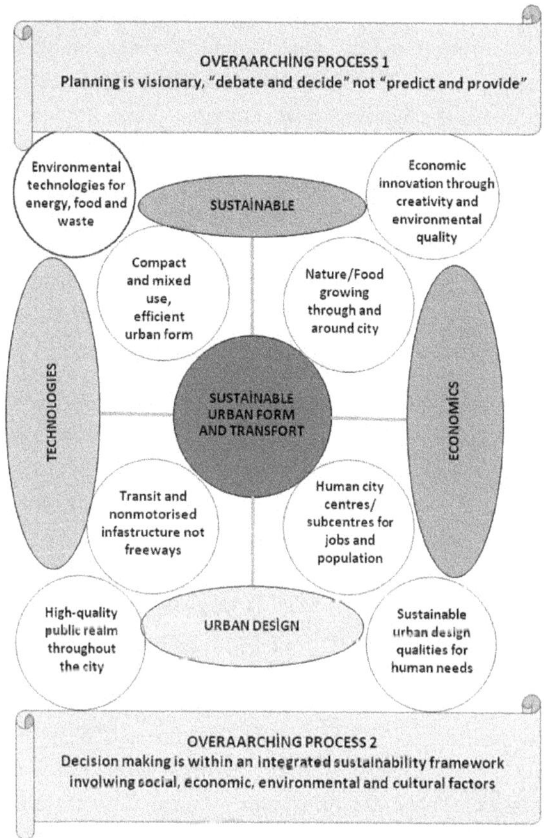

In addition, there have been applications focusing on private life through physical obstacles such as fences, garden walls, security guards, locks and diaphones in order to limit the access of foreigners to living spaces. City block and parcel systems have superceded quarter and street systems in which neighbourhood relations were kept alive. Individualization has gradually come to the forefront along with many factors such as the increase in population, lack of confidence, level of education, gender, professional status, and living conditions. This change experienced in urban spaces over time has led to many new urban approaches. In all these approaches, the common point is a safe settlement understanding based on quality of life, respect for nature, the use of self-sufficient sustainable materials, deprioritization of motor vehicles, an accentuation of the human factor, pedestrianization and cyclability. According to Kenworthy (2004); the findings of this paper can be summarized in Figure 6, which conceptualizes and organizes the ten key dimensions into four critical "Sustainable Urban Form" factors, four essential factors under the heading of "Sustainable Technologies, Economics and Urban Design" and two "Overarching Process" dimensions related to planning and decision-making for sustainable cities [45].

As a result, cities should be designed with an approach in which different functions are included, the human factor is accentuated, socialization is encouraged, environmental values are protected, and livable spaces are created.

References

[1] E. KARAKURT TOSUN: *Sürdürülebilir Kentsel Gelişim Sürecinde Kompakt Kent Modelinin Analizi, Dokuz Eylül Üniversitesi Sosyal Bilimler Enstitüsü Dergisi,* 15(1), s.103–120 (2013).

[2] A. KARADAĞ: *Kentsel Gelişim Süreci, Çevresel Etkileri ve Sorunları ile İzmir, Egekoop,* İzmir (2000).

[3] F. TIN: *Kent Plan Kararları ve İzmir İç Körfezi Üzerine Araştırmalar,* Ege Üniversitesi Fen Bilimleri Enstitüsü Peyzaj Mimarlığı Anabilim Dalı Yüksek Lisans Tezi, İzmir (2006).

[4] E. KARAKURT: *Küreselleşme ve Küresel Yerel Süreçlerde Üst Gelir Grubu Konutu Örneği: Bursa - Bademli,* Uludağ Üniversitesi Sosyal Bilimler Enstitüsü Yayınlanmamış Yüksek Lisans Tezi, Bursa (2002).

[5] N. ŞENSOY, A.A. KARADAĞ: *Sosyal Etkileşimin Komşuluk Düzeyinde Geliştirilmesine Yönelik Bir Araştırma: Ankara Toki Atakent Sitesi Örneği" İnönü Üniversitesi Sanat ve Tasarım Dergisi* Yıllık Özel Sayı 2(6): 279–289 (2012).

[6] F. ÖZPARLAK, M.Ç. MEŞHUR: *Sokaktan Siteye Dönüşen Yarı Kamusal Mekânlar: Komşuluk İlişkileri Üzerine, Mimarlık Dergisi,* Sayı 365 (2012). http://www.mimarlikdergisi.com/index.cf m?sayfa= mimarlik& DergiSayi=379&RecID=2956.

[7] Onuncu Kalkınma Planı Yaşanabilir Kentler ve Kentsel Dönüşüm Özel İhtisas Komisyonu Ön Raporu, Ankara.

[8] S. IDRUS, A. S. HADİ, A. H. H. SHAH, A.F. MOHAMED: *Spatial Urban Metabolism For City, Blueprints for Sustainable Infrastructure Conference,* 9–12 December (2008).

[9] H. GUGGER, G. KERSCHBAUMER: The Compact City: Sustainable, or Just Sustaining the Economy? Harry Gugger and Gwendolyn Kerschbaumer Editors: Ruby, Ilka; Ruby, Andreas; Janson, Nathalie Published in: The economic performance of sustainable construction (ISBN: 978-3-944074-07-8) Berlin: Ruby Press, (2013).

[10] L.G ANTHOPOULOS, A. VAKALI: *Urban planning and smart cities: Interrelations and reciprocities. Future Internet Assembly,* LNCS 7281: 178–189, (2012).

[11] J.M. BARRIONUEVO, P. BERRONE, AND J.E. RICART: "Smart Cities, Sustainable Progress," *IESE Insight* 14: 50–57 (2012).

[12] S. SINMAZ: *Yeni Gelişen Planlama Yaklaşımları Çerçevesinde Akıllı Yerleşme Kavramı ve Temel İlkeleri, MEGARON,* 8(2), s.76–86 (2013).

[13] Y. R. JABAREEN: *Sustainable urban forms: their typologies, models, and concepts. Journal of Planning Education and Research,* 26: 38–52 (2006). http://jpe.sagepub.com/cgi/content/abstract/26/1/38.

[14] S. PINK, L. J. SERVON: *Sensory global towns: An experiential approach to the slow city movement. Environment and Planning A, 45,* 451-466 (2013).

[15] http://www.un.org/jsummit (son erişim tarihi: 09.04.2017)

[16] World Commission on Environment and Development, Our Common Future. Oxford University Press, USA, (1987).

[17] M. LUNDQVİST: *Sustainable Cities in Theory and Practice A Comparative Study of Curitiba and Portland. Fakulteten för samhälls- och livsvetenskaper, Karlstads Universitet* (2007).

[18] Agenda 21: http://www.un.org/esa/sustdev/ documents/ agenda21/ index.htm, 2007–05-06.

[19] United Nations Environment program / GRID- Arenal, Maps and graphics: http://maps.grida.no/go/graphic/curitiba_city_map, 2007–06-08

[20] United Nations Sustainable Human Settlements: UN HABITAT II http://ww2.unhabitat.org/declarations/habitat_agenda.asp, 2007–06-0.

[21] https://en.wikipedia.org/wiki/Circles_of_Sustainability (son erişim tarihi: 15.04.2017).

[22] A. ATIL, B. GÜLGÜN, İ. YÖRÜK: *Sürdürülebilir Kentler ve Peyzaj Mimarlığı, Ege Üniversitesi Ziraat Fakültesi Dergisi,* İzmir, 42(2):215–226 (2005).

[23] E. KARAKURT: *Kentsel Mekanı Düzenleme Önerileri: Modern Kent Planlama Anlayışı ve Postmodern Kent Planlama Anlayışı, Erciyes Üniversitesi İktisadi ve İdari Bilimler Fakültesi Dergisi,* Sayı 26, Kayseri (2006).

[24] A. ÖZCAN: *Ekolojik Temele Dayalı Sürdürülebilir Kentsel Gelişme: Malatya Kent Örneği Üzerinden Bir Değerlendirme, 38. ICANAS (Uluslararası Asya*

ve Kuzey Afrika Çalışmaları Kongresi, 10–15 Eylül 2007) Bildiriler, Atatürk Kültür, Dil ve Tarih Yüksek Kurumu, Ankara (2007).

[25] S. BECSEI: *Green City – Sustainable Urban Planning.* Ahk Conference On Energy Efficient Construction And Refurbishment Riga 26th February B.A.E.R. (2013) becsei+hackenbracht architecture+environment–research. www.ahk-balt.org/…/08_Stephan_Becsei_BAER_26022013.

[26] F. BAYRAM: *Sürdürülebilir Kentsel Gelişme: Araçlar, Yaklaşımlar ve Türkiye, Cevat Geray'a Armağan, Mülkiyeliler Birliği Yayınları:* Ankara, 25: 251–265 (2001).

[27] K. H. YAZAR: *Sürdürülebilir Kentsel Gelişme Çerçevesinde Orta Ölçekli Kentlere Dönük Kent Planlama Önerisi,* Ankara Üniversitesi, Sosyal Bilimler Enstitüsü Yayınlanmamış Doktora Tezi, Ankara (2006).

[28] R. EWING: *Is Los Angeles-Style Sprawl Desirable? American Planning Association Journal,* 63(1), 107–126 (1997).

[29] M. ALBERTİ: *Measuring urban sustainability. Environmental Impact Assessment Review,* (16), 381–424 (1996).

[30] N. ŞENLIER, R. YILDIZ, T. SALIHOĞLU: *Farklı Konut Yerleşim Dokularında Yaşanabilirlik, 25. Uluslararası Yapı ve Yaşam Fuar ve Kongresi,* Bursa, 35–43 (2013).

[31] J. SCHEURER: *Compact City Policy: How Europe Rediscovered Its History and Met Resistance, Journal of The Urban Reinventors,* 2 (2007).

[32] Commission of the European Communities (CEC), *Green Paper on the Urban Environment, Brussels, European Commission* (1990).

[33] United Nations Earth Summit Agenda 21: *The UN Programme of Action from Rio. New York: United Nations* (1993).

[34] K. NABIELEK: *The Compact City: Planning Strategies, Recent Developments And Future Prospects In The Netherlands* METU, Ankara, 11–15 July AESOP 26th Annual Congress, (2012).

[35] M. MIKAEILI, Y. MEMLÜK: *Ekoloji ve Çevre Açısından Kompakt Kent Kavramı ve Uygulama Örnekleri, Anadolu Doğa Bilimleri Dergisi,* 4 (2): 37–50, (2013).

[36] http://www.shimz.co.jp/english/theme/dream/greenfloat.html (son erişim tarihi: 20.01.2017).

[37] http://www.oecd.org/greengrowth/compact-city-policies 9789264 167 865 – en.htm (son erişim tarihi: 20.04.2017)

[38] http://mwf2014.museumsandtheweb.com/paper/smart-technology-for-smart-regeneration-of-cultural-heritage-italian-smart-cities-in-comparison/ (son erişim tarihi: 25.03.2017)

[39] M. ROSELAND: *Eco-city Dimensions: Healthy Communities Healthy Planet*, Gabriola Island, B.C., New Society Publishers, (1997).
[40] E. KLİNE: *Planning and creating eco-cities: indicators as a tool for shaping development and measuring progress. Local Environment*, 5(3): 343–350 (2000).
[41] Y. YIN: *Environmental integration in sustainable urban planning from an institutional perspective.* Doctor thesis. Royal Institute of Technology, Stockholm, Sweden (2014).
[42] E. B. KESKIN: *Sürdürülebilir Kent Kavramına Farklı Bir Bakış: Yavaş Şehirler (Cittaslow)". Paradoks Ekonomi, Sosyoloji ve Politika Dergisi*, 8(1), s. 81–99 (2012).
[43] V. SIRIM: *Çevreyle Bütünleşmiş Bir Yerel Yönetim Örneği Olarak "Sakin Şehir" Hareketi ve Türkiye'nin Potansiyeli, Tarih Kültür ve Sanat Araştırmaları Dergisi* 1(4), s.119–131 (2012).
[44] http://www.ecocitystandards (son erişim tarihi: 10.04.2017)
[45] J. KENWORTHY: *The eco-city: ten key transport and planning dimensions for sustainable city development. Environ. Urban* 18(1):67–85, (2006).

Murat Atasoy[1], Filiz Guneysu Atasoy[2]

Sustainable Ecosystem Services

[1]*Dr, Research Assistant at the School of Forestry and Wildlife Sciences, Auburn University, Auburn, AL 36849–5418, USA, email: mza0055@auburn.edu*

[2]*Dr, Research Assistant at the Agricultural Economics and Rural Sociology, Auburn University, Auburn, AL. 36849–5418, USA, email: fzg0004@auburn.edu*

1. Introduction

The ecosystem services mainly approach concepts integrating the public into forest management and maintenance activities to provide welfare for people. One of the main objectives of ecosystem services is to provide the desired services for humans in qualitative and quantitative ways. Forested landscapes are one of the main aspects of ecosystem services, and currently, tropical and economically developing countries are the landscapes most undergoing deforestation because of a lack of ecosystem services [1]. In fact, the majority of forest degradation is in primary forests, and this significant havoc can have a negative impact on global carbon services, biodiversity, and communities [1]. Therefore, ecological services and their implications for governments and land owners can increase the forested landscape cover around the world.

One of the other objectives of ecosystem services is to sustain natural resources for the future. By implementing the regulations derived from ecosystem services, sustainable ecosystem resources such as clean water and soil can be protected from environmental pollutants. To increase the benefits achieved from using these natural resources, environmental agencies significantly address the protection of the richness of biodiversity and the valuation of ecosystem services. It is also important to note that classes of ecosystem services such as supporting, provisioning, regulating, and cultural services are linked to each other, and therefore, the enhancement throughout the provision of one of these services can be beneficial for the maintenance of every ecosystem service.

Moreover, Fule et al. (2007) stated that provision of ecological services by land owners can be enhanced when land owners are paid; for instance, the value of pine landscapes can increase if carbon sequestration or aquifer recharge service is paid to landowners in addition to selective timber harvesting, hunting rights, or further economic usage of ecosystem services. Therefore, ecosystem services are prominently linked to forested areas and their management practices, considering the ecological and economical values of these unique ecosystems. This research aims to establish the importance of ecosystem services provided by forested landscapes and especially their effects on economic valuation and human well-being. To address the current issues and the historical background associated with forested ecosystems and ecosystem services, various studies and reviews are included in this research.

2. Ecosystem Services, Sustainable Forest Management, and Human Well-being

Ecosystem services can be classified as provisioning services (food, fresh water, wood for fuel, etc.), regulating services (climate regulation, disease regulation, water regulation, etc.), cultural services (spiritual and religious, recreation and ecotourism, aesthetics, etc.) and supporting services (soil formation, nutrient cycling, and primary production) [3]. The classification of ecosystem services are shown in Figure 1. Landell-Mills (2002) stated that ecosystem variability, resilience, and thresholds are the three main and crucial characteristics of ecosystem services.

Figure 1. Classification of Ecosystem Services [4]

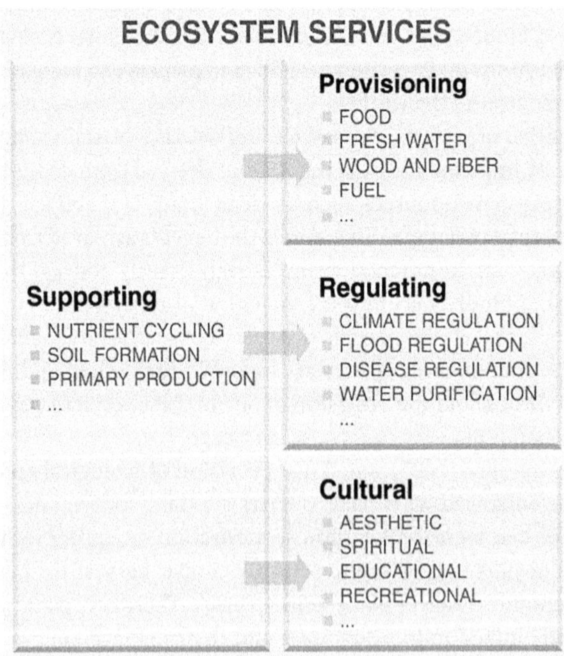

According to the Millenium Ecosystem Assessment (MEA) (2003), theoretical and empirical recent studies [6; 7] have shown that there is a strong relationship between biodiversity and the functionality of ecosystem concept. MEA (2003) additionally reported that ecosystem components are vulnerable to outside effects which directly impact the biodiversity of diverse systems. The concept of vulnerability and resiliency can be identified as closely relevant to each other and these concepts mainly focus on studying human, environmental, and human-environment systems to reveal the comprehensive properties of the systems and relationships with their environment [7]. In order to protect and sustain these systems, payments for environmental services (PES) and sustainable forest management (SFM) programs have been designed, and they primarily focus on the affluence of humans by reducing poverty and managing forest lands to protect diversity and environmental services [9]. Engel et al. (2008) described PES as "*a voluntary transaction where a well-defined environmental service (or the land use likely to secure that service) is being 'bought' by a (minimum one) service buyer from a (minimum one) service provider if and only if the service provider secures service provision*." The restricted benefits obtained from sustainable forest management (SFM) and other indirect tools subsequently emerged as the initiatives for PES and direct tools [9].

Sustainable ecosystem services provide desired services for humans in qualitative and quantitative ways [5]. The generation of a featured ecosystem service is required to be entirely sustainable while yielding that service and perpetuating its functions [5]. Diaz et al. (2007) stated that the correlation between ecosystem properties and ecosystem services may be measured and quantified by ecosystem properties such as plant functional diversity (FD). Plant trait values have a profound influence on ecosystem properties which has a related result on ecosystem services [11]. Moreover, Dobbs et al. (2011) summarized the ecosystem functions as physical, chemical, and biological processes which utilize the ecosystem properties for humans and yield sustainable services. The impacts of urban forest structure, urban morphology, and socioeconomic factors on urban forest environmental service goods can be derived by analyzing urban forest cover and soil indicators [12]. Also, the most effective ecosystem service indicators are tree cover, soil pH and soil organic matter which are prominently influenced by land use change and urbanization.

Furthermore, landscape specific concepts play an important role in landscape patterns of ecosystems and their services which indicate that different landscapes proclaim different types of ecosystem services and landscape patterns shape and maintain the ecosystem service[8]. Moreover, Diaz et al. (2007) indicated that the social response to the incorporation of functional plant diversity with ecosystem services can be based on quantitative assessments of services and the consequences of various land use change impacts on these ecosystem services. For instance, land conversion from agriculture to urban land use can decrease plant functionality because nutrition and soil resources dramatically decline due to environmental degradations such as the building of structures and the contamination of water resources. To derive a sustainable forest management for human well-being, ecosystem services such as water purification and plant protection can help forest managers and land owners. Therefore, public participation and governmental organizations maintain the landscape patterns by conducting regulation services.

Wu (2013) discussed the ecosystems and their services which provide an ultimate goal of sustainability for human well-being and protection of the earth's lifecycle. Human well-being is also prominently correlated with needs for a good life, freedom, health, individual security, and social connections among humans [5]. Costanza et al. (2007) argued with the definition of human well-being and stated that quality of life can be the fundamental concept of which human needs are fulfilled consistently with human well-being. They determined a single and integrating overall metric that uses both the fulfillment and importance of scores of ecosystem properties. More importantly, human well-being can be improved by balancing how well a need is met and how important the need is [13]. On the other hand, Dobbs et al. (2011) stated that human well-being may be negatively influenced by forest management requirements such as perceived risk of crime, pollutants, and direct costs in urban forest landscapes. The direct payments for ecosystem services can also implement various regulations (command and control regulations, integrated conservation and development projects, social markets, etc.) on the wellbeing concept of human wellbeing: an argument for the fairness of payments of environmental services [9].

3. Valuation of Ecosystem Services

Ecosystems and their services play an important role in human, animal, plant and microbial population dynamics where there is a conceptual value of these ecosystems and their existence in society [14]. Assessing the value of ecosystems has a considerable effect on human wellbeing and ecosystem conditions are highly relevant to wellbeing factors [5]. According to Campbell and Tilley (2014), environmental accounting is one of the factors that impact valuing ecosystem services, and this approach was created to provide a valuation external to the economic and intrinsic requirements for thermodynamics. Batabyal et al. (2003) pointed out the scarcity value of ecosystem services which include the destruction of the ecosystem by human activitie;, thus the capacity of environmental and social systems that control the flow of ecosystem services can be scarce. In fact, the necessity of monetary valuation in ecosystem services raises awareness of the importance of communication for the sustainability of ecosystem services [17]. Moreover, an economic valuation of ecosystem services can profoundly produce intelligible and comparable impacts on the costs of intervention [5]. Besides, economic values of ecosystem services are affected by methods and subjective assessments, yet economic instruments do not reflect the problems associated with social fairness and balance within the usage of ecosystems [17].

Furthermore, human well-being consists of various elements such as fundamental needs for quality of life, freedom and choice, health, better social interactions, and safety. Batabyal et al. (2003) asserted that by quantifying the scarcity value of an ecosystem service, changes in environmental systems can be calculated and managed properly. The economic valuation paradigm and human well-being are linked to each other by involving the individuals' reflection on ecosystem changes and preferences for goods or services which eventually create social values [14]. According to Campbell and Tilley (2014), environmental accounting and human well-being allow for the connection between nature's production and benefits that people derive from it, which can be also determined as public value. The total economic value approach uses monetary and market-based units as a measurement tool for ecosystem services and human well–being assessment methods [17]. Different use and non-use values of ecosystem services are the nexus between environmental assets and human well-being [17]. In many cases, according to DeFries et al (2005), the valuation approach is primarily used for examining trade-offs, various ecosystem management techniques, and services that are derived from them.

NRC (2005) reported that quantification of ecosystem values may be useful in decision making through trade-offs, estimating natural resource damage assessment or close cases, and compromising these with environmental assets and national income accounts. Considering the role of valuation in policy making, strict utilitarianism can be a key factor indicating that economic efficiency-related decisions can lead to an increase in the net benefits which society consequently attains [14]. The embedding effect on public goods and its relationship with a contingent valuation method might also decrease the outcomes of an embedding effect [18]. For instance, Loomis et. al., (1993) conducted willingness to pay respondent surveys by including the satisfaction of providing money for a beneficial outcome in a contingent valuation of forest protection, and the study results showed that the presence of embedding effects was not generally correlated with contingent valuation method studies when regional context was introduced to survey respondents.

4. Economic Assessment

Defries et. al., (2005) determined the concept of total economic value in ecosystem services by grouping these into direct use, indirect use, option, and non-use values. If ecosystem services can be consumed or utilized by humanity in a direct way, they are described as direct use. This refers to both consumptive and non-consumptive uses which involve some form of direct physical interaction with environmental goods and services [19]. Recreational activities, resource harvesting, drinking clean water, and breathing unpolluted air are some examples of the direct use of ecosystem services. To provide a sustainable landscape, these resources should be primarily protected by enhancing direct and indirect use sources of ecosystem services. On the other hand, indirect use values are obtained from ecosystem services that utilize outside benefits, option values consisting of values derived from ecosystem services which can be used in the future as optional, and non-use values are existing values though they are not consumed as a resource [4]. In literature, the ecosystem services are evaluated as assets from an economical perspective. Campbell and Tilley (2014) applied ecosystem service calculation methods by investigating the effects of carbon sequestration, nutrient uptake, soil building and erosion avoidance, air pollution mitigation, pollination and biodiversity on ecosystem services between a typical Maryland forest cover and as an alternative land use example, a suburban development. The authors found that the government of Maryland in the United States gains $5 billion in benefits from ecosystem services annually and people invest between $300 and $750 million yearly to maintain these services in Maryland. In this way, there are no differences between any kind of assets and ecosystem services in the economy, and there is a market for either goods or services that are produced by ecosystems.

In fact, the environmental assets of ecosystem services require special attention because the assets or services include 'non-renewable resources' [20]. This is particularly beneficial when obtained from the regularity (policy regulations) and habitat of ecosystems services. For example, ecosystem functions can be gas regulations, climate regulations, water regulations, soil formation, and waste treatment, etc. [21]. Drainage and natural irrigation, flood mitigation and groundwater recharges are additionally considered as benefits of ecosystem services where the water regulation is an ecosystem function (ecosystem regulation). Another example is that pollution control and detoxification are benefits of ecosystem services where waste treatment is evaluated as an ecosystem function (ecosystem regulations) [22].

Another issue related to the valuation of ecosystem services is that their environmental future value is uncertain. Since ecosystems are in fixed supply and the possibilities for restoration are difficult, so their benefits will decline over time [19]. From an economical perspective, the value of goods or services obtained from the ecosystem will increase when they decrease over time. Heal et al (2005) stated that the evaluation of irreversible loss through ecosystem services is complicated due to their valuation being largely unmarketed. In addition, Barbier (2007) stated that their services may not be efficient if they are not valued based on the aggregate willingness to pay for these services. Various valuation methods are used in the literature to estimation of the ecosystems services' values, and these are shown in Table 1. The Hedonic Price model is one of the most efficient methods which has been used by economists for valuation of ecosystem services [22–24]. Production function and replacement cost models are commonly used in the relevant literature.

Furthermore, Boyd (2007) stated that the identification of direct and indirect impacts of ecosystem services is helpful for natural capital accounting systems or economic valuations. Therefore, the ecosystem services are categorized into three main groups which are: intermediate services (nutrition cycling, pollination, and primary product), final services (clean water provision, food production and water regulation), and benefits (drinking water, fruit, flood protection) [26]. In addition, the qualifiers of intermediate services and final services are applicable to conventional economic accounting systems [26]. As an example, clean drinking water can be beneficial where the provision of clean water is a final service. Moreover, according to Pearce (2007)'s study, every ecosystem service must appeal as one which is operational as a decision support system. These are given as follows:

1. Marginality
2. Infrastructure (insurance value)
3. The ability to capture nonmarket ecosystem services through institutional arrangement.

According to Brauman et al, (2007), trade-offs can be revealed during the provision of ecosystem services. Quality, quantity, location, and the timing of flow may also lead to changes in value for other components of ecosystem services. Monetary valuation, on the other hand, provides a better understanding of the policy-making process in terms of maintaining water quality, quantity, and common metric features [28]. Therefore, the marginal value of an ecosystem service in a watershed can also be discrete, considering liter availability which consequently changes the total value of the service.

Table 1. Various valuation methods applied to ecosystem services

Valuation Methods	**Types of Value Estimated**	**Common Types of applications**	**Ecosystem Services Valued**
Travel Cost	Direct Use	Recreation	Maintenance of beneficial species, productive ecosystems and biodiversity.
Averting Behavior	Direct Use	Environmental effects on human health	Pollution control and detoxification
Hedonic Price	Direct and Indirect Use	Environmental impacts on residential property and human morbidity and mortality	Storm protection; flood mitigation; maintenanceof air quality
Production Function	Indirect Use	Commercial and recreational fishing; agricultural systems; control of invasive species; watershed protection; damage costs avoided	Maintenance of beneficial species; agricultural productivity; prevention of damage from erosion and siltation; groundwater recharge; drainage and natural irrigation; storm protection; flood mitigation
Replacement Cost	Indirect Use	Damage Costs avoided; freshwater supply	Drainage and natural irrigation; storm protection; flood mitigation
Stated Preference	Use and Non–use	Recreation; environmental impacts on human health and residential property; damage costs avoided; existence and bequest values of preserving ecosystems	All of the stated above

Source: Adapted from Heal et al. (2005).

In general, the preservation of ecosystem services requires an understanding of the context of markets for ecosystem protection which is primarily associated with land management responsibility, water flows and quality [3]. For instance, in terms of defining watershed commodities, there are five categories as follows:

1. Water flow regulation
2. Water quality maintenance
3. Erosion and sediment control
4. Land salinization reduction/water table regulation
5. Maintenance of aquatic habitats.

In addition to the preservation of ecosystem services, the markets for ecosystem services and watershed services are mainly managed by the private sector which comprises nearly 60% of recorded buyers and over 65% of the recorded sellers. The public sector side of either of the ecosystem services managements is considerably lower compared with that of the private sector in the market. The mechanism for these service payment contexts includes direct negotiation between buyers and sellers, intermediary-based transactions, pooled transactions, internal trading, over-the-counter trades/user fees, clearing-house transactions, auctions, and retail-based trades [3]. Furthermore, the production of an attribute in either a riparian area or wetland can affect the valuation of the service on that watershed in a major way. The fact that people usually prefer not to pay for ecosystem services is associated with a willing-to-pay strategy, so there is no direct measurement method for these types of services [28]. According to Lele (2009), the valuation of watershed ecosystem services by observing one type of land cover or feature can be vague for tropical forests because the forest cover transition or land conversion varies from site to site. For instance, deforestation can reduce agricultural incomes due to a low base flow in Indonesia, yet forest regeneration may decrease agricultural incomes in the Western Ghats, India, because in agriculture, watershed management is directly linked to water augmentation in downstream tanks [29].

5. Watershed Ecosystem Services

Water-related services have been increasingly considered by people over the decades due to water supply for households and alleviation of flood damages. In fact, since various components are involved in hydrologic ecosystem services, there can be five categories emphasizing the relevance of hydrology services: development in water extractive supply systems, improvement of in-stream water services, flooding damage management, production of water–relevant cultural features, and water-related supporting services [28]. According to Calder et al. (2004), forests are fundamental components of water-related services and the environment, and they facilitate an increase in rainfall and runoff, control flows, mitigate erosion, reduce floods, decontaminate water supplies, and improve water quality. Although these benefits provide a regulation and maintenance service for watersheds, during the dry seasons, evapotranspiration from forest foliage and high infiltration rates in soil organic matter, with deep roots of trees, may reduce water retention [30]. Lele (2009) argued that a forest watershed service is a complex structure with ecosystem processes, impacts, outcomes, and benefits. The author also reported that structural change in tropical forests may have various implications for the conceptualisation of forest watershed services. Examples of this are:

- A process change varies because human effects can be determined as the process itself rather than the service.
- The context should be rearranged based on the models that hydrologists and eco-hydrologists apply for.
- The value esteemed for watershed services for a forest type can be identified by changes in human well-being downstream caused by different land use.
- To fulfill the human well-being and watershed ecosystem management relationship there is a need for man-made capital, technological tools and institutions. Economic welfare changes with implications for the various watershed processes and stakeholders that are responsible for these.

In addition to conceptualizing forested watershed services, the public perception and its relationship with forests and water environment is important for land and water policy implications [30]. Calder et al, (2004) reported that for a particular site, it may be complicated to ensure the effects of forests on annual flow range. For instance, studies at Mokobulaan in the Transvaal resulted in afforestation, with the *Eucalyptus grandis* considerably reducing the amount of annual flows by 300–380 mm, with a 200–260 mm decrease during the summer season [30]. More importantly, biodiversity and ecosystem services are linked to one another, and public perception plays a major role in understanding and managing ecosystem services in human-dominated watersheds [31].

Bai et al. (2011) applied an Integrated Valuation of Ecosystem Services and Tradeoffs (INVEST) method to estimate habitat quality across the landscape and the magnitude of degradation of the watershed vegetative cover in Baiyangdian, China. They modeled the watershed to evaluate the relationship between biodiversity persistence, resilience, breadth, and habitat quality over time by testing carbon sequestration, water quality, soil retention, water yield, and pollination in the watershed. As a result of the research, they found that there was a positive relationship between biodiversity hotspots and an increase in ecosystem service outcomes such as water yield, soil retention, carbon sequestration, and their perception by the public around the area. Brauman et al. (2007) discussed the importance of forest cover in the context of watershed services management and indicated that the total volume of surface or ground level water in forested land may be lower than shrub or grass covered watersheds.

6. Carbon Sequestration and Ecosystem Services

Greenhouse gases such as CO_2 have considerably affected climate change caused by anthropogenic activities all over the world. For more than two decades, governments and environmental protection agencies have been intensively working on reducing the level of anthropogenic releases of greenhouse gases. Forests have a positive effect on maintaining the carbon cycle by sequestration and storage of carbon and exchanging the biomass fuels with existing fossil fuels in landscapes [32]. Gutrich and Howarth (2006) stated that the implications conducted by the Kyoto Protocol such as projects and policies improved the net carbon storage in forests. According to Smith and Scherr (2003), the Kyoto Protocol was conducive to the commitment of industrialized countries to reducing emissions by providing various carbon sequestration-induced forest management activities. The authors additionally noted that recently political and technical constraints are the cause of much of the debate on the carbon emission effects of forestry, and social issues such as livelihood play an important role in understanding the Clean Development Mechanism (CDM) of the Kyoto Protocol.

According to Sohngen and Brown (2006), pine plantation has recently become an expanding trend in the Southern United States, where it is estimated that pine plantations have grown by nearly 12 million ha land cover for the past 40 years. The conversion of hardwood forest lands into pine plantations may decrease carbon storage in the Southern United States, and ecosystem services such as carbon storage are mainly affected by the economic forces that these land conversions drive over time [35]. Moreover, in the United States, pine plantation management programs for timber production have become a popular trend due to constraints on public forest land harvesting and increased regulation services enforced by the government. Landowners are also considering various silvicultural techniques such as genetically modified trees, bedding, and herbicides to derive high productivity timber rates, yet the emergence of the carbon market can have prominent effects on the implication of these techniques on pine plantation stands [32]. For instance, Woodbury et al. (2007) claimed that in 2005, 159 million metric tons of carbon were sequestered by the plantations and 36.5% of this consisted of wood products. Additionally, forest land owners in the Southern United States supplied nearly 267 million cubic meters of timber products out of 440 million cubic meters of total annual growing stock of timberlands; therefore, carbon sequestration estimates from wood products in private forest lands play a major role in calculating the carbon stocks [37].

The amount of carbon sequestered can be determined by estimating the carbon in standing biomass, carbon remaining on the belowground after rotation, and the wood products derived from harvested wood. However, carbon sequestered by pine forests may have some methodological obstacles such as determining the carbon pools to be involved, fire and pest outbreaks, and carbon expansion in the systems which would be further compromised by international policy agreements. Incentives for converting agricultural lands to forests, increasing forest management practices, and determining better forest operations throughout the ground carbon pools can help to maintain the balance of carbon sequestered from pine species globally [38].

7. Biodiversity and Ecosystem Services

The two terms, biodiversity and ecosystem services are considered as the same fields in some cases, and biodiversity is commonly evaluated within the ecosystem services concept. However, both biodiversity and ecosystem services can be different subjects if ecosystem services are not properly managed [39]. The connection between biodiversity and the policy field of ecosystem services such as nature conservation is generally misunderstood and nature conservation provides an implication for ecosystem services on global forest cover increase [40]. In fact, land use alterations such as the human conversion of natural landscapes can result in a significant decrease in biodiversity and associated ecosystem services [4]. The role of biodiversity through sustaining the ecosystem services depends on the scale of the ecosystem, yet landscape level biodiversity loss is vague due to a lack of long term and large scale management strategies [41]. To meet the requirements of management strategies, ecosystem services and a regulation of the management plans are fundamental aspects of increasing and sustaining the richness of biodiversity.

The sustainability of biodiversity components such as genes and traits may vary based on the ecosystem service selection, so that the goals of management strategy are prominent in order to sustain the ecosystem process for a sustainable future in terms of biodiversity [39]. Moreover, the value of biodiversity includes cultural, social, and aesthetic benefits which can depend highly on the management strategy of ecosystem services [42]. In order to sustain biodiversity and its components, local land managers and land owners initiate management strategies to protect the richness of biodiversity. In fact, the fauna and flora components of biodiversity can be protected by the direct and non-direct use of ecosystem services by designing ecologically preserved areas around the world. These areas are unique landscapes because endangered species can also be observed and protected from extinction. In this case, governments usually implement management strategies and provide initiatives to increase the richness of the biodiversity, which is another aspect of direct use ecosystem services. Therefore, the cultural, social, and aesthetic benefits of biodiversity can be achieved through using the initiative programs of both governmental and non-governmental organizations.

8. Conclusion

In order to maintain ecosystem services and their products, forest managers and land owners play a major role in the continuum of services. The fact that categories of ecosystem services and their association with forest management practices can enhance the sustainable landscapes of the future if the benefits of involvement in resource managements is comprehensively understood by the public. Therefore, education and outreach efforts can improve the clarity of the link between the human benefits of the forested landscapes and the positive outcomes of sustainable ecosystem services.

Furthermore, the valuation of biodiversity services, economic and cultural for example, can provide more incentive programs for the regeneration of biodiversity and forest management. As an example, the Natural Resource and Conservation Service (NRCS) provides funds of $10.6 million for the regeneration and restoration of the longleaf pine ecosystem, whereas the shortleaf pine restoration and plantation service receives $100,000 funding annually from the NRCS in the United States (NRCS 2016). Thus, incentive programs supporting biodiversity conservation are substantial in order to increase management activities.

In addition to biodiversity services for sustainability programs, carbon sequestration and water quality services have a significant role in restoration and regeneration of forested ecosystems. Also, a reduction in carbon levels will increase forested landscapes; thus, environmental services and agencies can investigate more outcomes of carbon sequestration and greenhouse gas mitigation effects. Moreover, watershed protection, water quality, and a decrease in surface runoff can be attained by increasing the richness of biodiversity. The aboveground biomass of forested stands can also help to increase vegetative cover in urban areas if management practices are frequently and properly applied, which can be also classified in the regulation services of ecosystems.

To achieve a sustainable ecosystem, protection and maintenance of natural resources are key factors which are also mainly affected by ecosystem services and their implications. Public participation meetings and informative publications can additionally help with understanding how ecosystem services can properly improve the quality of landscape features such as water, soil, air, and vegetation. By improving the quality of these features, human well-being can significantly increase and the welfare of the public can reach a better level economically. Therefore, the generation of ecosystem services is required to be wholly sustainable while yielding that service and perpetuating its functions.

References

[1] Elias, Patricia E., and Linda S. Heath. "REDD+ and Ecosystem Services: Relevance to the Forestry Profession." *Journal of Forestry* 113.2 (2015): 268–270.

[2] Fulé, Peter Z. "The Longleaf Pine Ecosystem: Ecology, Silviculture, and Restoration-Edited by Shibu Jose, Eric J. Jokela, and Deborah L. Miller." *Restoration Ecology* 15.4 (2007): 745–746.

[3] Landell-Mills, Natasha, and Ina T. Porras. "Silver bullet or fools' gold?: a global review of markets for forest environmental services and their impact on the poor." (2002): 236–250.

[4] DeFries, Ruth, et al. "Analytical approaches for assessing ecosystem condition and human wellbeing." *Ecosytems and Human Wellbeing: Current state and trends, by Millenium Ecosystem Assessment.* Washington: World Resources Institute (2005).

[5] Assessment, Millennium Ecosystem. "MEA!+ (2003) Ecosystems and Human Wellbeing: A Framework for Assessment+." *World Resources Institute." Island Press: Washington, DC.*

[6] Benayas, José M. Rey, et al. "Enhancement of biodiversity and ecosystem services by ecological restoration: a meta-analysis." *science* 325.5944 (2009): 1121–1124.

[7] Bullock, James M., et al. "Restoration of ecosystem services and biodiversity: conflicts and opportunities." *Trends in Ecology & Evolution* 26.10 (2011): 541–549.

[8] Wu, Jianguo. "Landscape sustainability science: ecosystem services and human wellbeing in changing landscapes." *Landscape Ecology* 28.6 (2013): 999–1023.

[9] Wunder, Sven. "Are direct payments for environmental services spelling doom for sustainable forest management in the tropics?." *Ecology and society* 11.2 (2006).

[10] Engel, Stefanie, Stefano Pagiola, and Sven Wunder. "Designing payments for environmental services in theory and practice: An overview of the issues." *Ecological economics* 65.4 (2008): 663–674.
[11] Díaz, Sandra, et al. "Incorporating plant functional diversity effects in ecosystem service assessments." *Proceedings of the National Academy of Sciences* 104.52 (2007): 20684–20689.
[12] Dobbs, Cynnamon, Francisco J. Escobedo, and Wayne C. Zipperer. "A framework for developing urban forest ecosystem services and goods indicators." *Landscape and urban planning* 99.3 (2011): 196–206.
[13] Costanza, Robert, et al. "Quality of life: An approach integrating opportunities, human needs, and subjective wellbeing." *Ecological economics* 61.2 (2007): 267–276.
[14] NRCS. "Natural Resource and Conservation Service." Retrieved from http://www.nrcs.usda. gov/wps/portal/nrcs/detailfull/national/home/?cid=nrcsdev11_023913 (2016).
[15] Campbell, Elliott T., and David R. Tilley. "Valuing ecosystem services from Maryland forests using environmental accounting." *Ecosystem Services* 7 (2014): 141–151.
[16] Batabyal, Amitrajeet A., James R. Kahn, and Robert V. O'Neill. "On the scarcity value of ecosystem services." *Journal of Environmental Economics and Management* 46.2 (2003): 334–352.
[17] Häyhä, Tiina, et al. "Assessing, valuing, and mapping ecosystem services in Alpine forests." *Ecosystem Services* 14 (2015): 12–23.
[18] Loomis, John, Michael Lockwood, and Terry DeLacy. "Some empirical evidence on embedding effects in contingent valuation of forest protection." *Journal of Environmental Economics and Management* 25.1 (1993): 45–55.
[19] Barbier, Edward B. "Valuing ecosystem services as productive inputs." *Economic Policy* 22.49 (2007): 178–229.
[20] Just, Richard E., Darrell L. Hueth, and Andrew Schmitz. *The welfare economics of public policy: a practical approach to project and policy evaluation.* Edward Elgar Publishing, (2005).
[21] De Groot, Rudolf S., Matthew A. Wilson, and Roelof MJ Boumans. "A typology for the classification, description and valuation of ecosystem functions, goods and services." *Ecological economics* 41.3 (2002): 393–408.
[22] Heal Geoffrey, M., et al. "Shrader-Frechette Kirstin. Valuing ecosystem services: Toward better environmental decision making." (2005).

[23] AMIII, Freeman. "The measurement of environmental and resource values: theory and methods." *Resources for the Future, Washington DC* (2003).
[24] Pagiola, Stefano, Joshua Bishop, and Konrad Von Ritter. "Assessing the economic value of ecosystem conservation." (2004).
[25] Boyd, James. "Nonmarket benefits of nature: What should be counted in green GDP?." *Ecological economics* 61.4 (2007): 716–723.
[26] Fisher, Brendan, et al. "Ecosystem services and economic theory: integration for policy-relevant research." *Ecological applications* 18.8 (2008): 2050–2067.
[27] Pearce, David. "Do we really care about biodiversity?." *Environmental and resource economics* 37.1 (2007): 313–333.
[28] Brauman, Kate A., et al. "The nature and value of ecosystem services: an overview highlighting hydrologic services." *Annual Review of Environmental Resources.* 32 (2007): 67–98.
[29] Lele, Sharachchandra. "Watershed services of tropical forests: from hydrology to economic valuation to integrated analysis." *Current Opinion in Environmental Sustainability* 1.2 (2009): 148–155.
[30] Calder, Ian, et al. "Forest and Water Policies. The need to reconcile public and science perceptions." *GeologicaActa: an International Earth Science Journal* 2.2 (2004): 157–166.
[31] Bai, Yang, et al. "Spatial characteristics between biodiversity and ecosystem services in a human-dominated watershed." *Ecological Complexity* 8.2 (2011): 177–183.
[32] Alavalapati, Janaki RR, and G. Andrew Stainback. "Effects of carbon markets on the optimal management of slash pine (Pinus elliottii) plantations." *Southern Journal of Applied Forestry* 29.1 (2005): 27–32.
[33] Gutrich, John, and Richard B. Howarth. "Carbon sequestration and the optimal management of New Hampshire timber stands." *Ecological Economics* 62.3 (2007): 441–450.
[34] Smith, Joyotee, and Sara J. Scherr. "Capturing the value of forest carbon for local livelihoods." *World development* 31.12 (2003): 2143–2160.
[35] Sohngen, Brent, and Sandra Brown. "The influence of conversion of forest types on carbon sequestration and other ecosystem services in the South Central United States." *Ecological Economics* 57.4 (2006): 698–708.
[36] Woodbury, Peter B., James E. Smith, and Linda S. Heath. "Carbon sequestration in the US forest sector from 1990 to 2010." *Forest Ecology and Management* 241.1 (2007): 14–27.

[37] Smith, W. Brad, et al. "Forest resources of the United States, 2007: a technical document supporting the forest service 2010 RPA Assessment." *General Technical Report-USDA Forest Service* WO-78 (2009).
[38] Johnsen, Kurt H., et al. "Meeting global policy commitments: carbon sequestration and southern pine forests." *Journal of Forestry* 99.4 (2001): 14–21.
[39] Mace, Georgina M., Ken Norris, and Alastair H. Fitter. "Biodiversity and ecosystem services: a multilayered relationship." *Trends in ecology & evolution* 27.1 (2012): 19–26.
[40] Turner, Will R., et al. "Global Conservation of Biodiversity and Ecosystem Services Bio Science. Online at http://www. bioone. org/doi/abs/10.1641/B571009." *Accessed April* 28 (2007): 2009.
[41Worm, Boris, et al. "Impacts of biodiversity loss on ocean ecosystem services." *science* 314.5800 (2006): 787–790.
[42] Swift, Mike J., A-MN Izac, and Meine van Noordwijk. "Biodiversity and ecosystem services in agricultural landscapes—are we asking the right questions?." *Agriculture, Ecosystems & Environment* 104.1 (2004): 113–134.

Aysel Yavuz[1]

Recent Urban Spaces Produced by the Consumer Society: Shopping Malls

[1]*Assistant Professor, Karadeniz Technical University, Faculty of Forestry, Department of Landscape Architecture, Turkey*

1. Introduction

The recent radical expansion in the urban development process has introduced previously unimaginable transformations in lifestyles, and quality urban life has become a commodity for those with a high income [1]. The new concept of urbanism, which has its own ideology, has become a value in itself, and this formation has introduced a new industry. Instead of houses, homes with gardens and condominiums for individuals' housing needs, life centers, etc. have started to become more common. Compact life centers that include shopping malls were produced with advertising pitches such as "the new art of life", "a new lifestyle is born", "you leave your coat at the cloakroom at the entrance, the kids at the day care center at the mall, and then you can relax and shop, meet friends, have a drink together … " [2, p. 43]. In the creation, development and expansion of the needs that could be called luxurious and 'in fashion', the capitalist system and the consumption culture were effective [3]. We have arrived at a point where consumption surrounds all human life, all activities are linked in accordance with the same unitary style, and ways of rewarding people are pre–arranged, climatized and cultivated [4]. This meant that consumption would become an increasingly important part of the lives of people everywhere in the world [5]. Economic movements such as the desire of the consumer to purchase goods and services provided by modern capitalism, along with the sustainability of income through the production and sale of these goods for a market, occur in a cultural context [6]. We consume several goods and services that we need for survival or just for our pleasure. To obtain these goods and services we often go to places such as streets, restaurants, parks, shopping malls, etc. [5].

Contemporary shopping has become more than just a necessity to meet our needs. Consumer spaces have become public spaces which individuals frequent for recreation, socialization and communication, and shopping has become a type of weekend entertainment [7]. It was demonstrated at the Athens convention that spending leisure time in a productive way supported the physical and psychological health of urbanites. Furthermore, leisure activities were categorized under three groups: as the activities that could be conducted in the vicinity of individual's residence, hence "daily activities"; activities that could be conducted by traveling out of the city or the region, hence "weekly activities"; and the activities that could be conducted by visiting new cities, hence "annual activities" [8].

Figure 1. Shopping mall examples: İstanbul, İstiklal Caddesi, [Photo by A.Yavuz)

Lefebre observed these spaces as leisure centers, leisure societies, and luxury and pleasure cities, and stated that this was especially indicated by the rhetoric in the advertisements [2]. Shopping centers are more complex trade communities and leisure spots [9] than ordinary commercial outlets, and they combine functional diversity, vitality, comfort and quality. Big stores that previously were located in urban centers have begun to leave town centers due to a lack of space for their stores and parking facilities. With the number of vehicle owners steadily increasing, it became impossible for the people to leave their vehicles while meeting their basic needs. Besides, when shopping, people wanted a place to sit down, to have a cup of tea or eat their food, continue shopping after resting a while, conduct other activities before or after shopping, and share a social environment. In short, they wanted to conduct all their activities and fulfill all their needs in one place. While working people could fulfill these needs after work or on weekends, large shopping malls were able to extend this time slot and serve customers until late at night, attracting more users. At the beginning, while the main purpose was "shopping", today objectives such as leisure, entertainment and recreation, etc.

have become prominent, and have also come to be excuses for shopping. Shopping centers have become strong points of attraction as people want quality and comfort while conducting these activities.

Figure 2. Shopping mall examples: İstanbul, Ataköy A⁺ AVM, (Photo by A.Yavuz)

On the other hand, the characteristics that transfer and turn shopping malls into centers of attraction every day, changing and reshaping human needs, has reduced the use of pedestrian shopping streets and the time spent in these. The need to restore the significance of the pedestrian shopping streets that have lost their activity to a preference for shopping malls in urban life is considered important for the social dimension of urban life as well as our cultural values.

2. Changing Consumption Habits and Spatial Mobility in the Historical Process

Throughout history, social exchanges have been spaces that manage urban life, nourished by commercial activities [10]. The fact that radical changes in the form of social relations were reflected in significant changes to the urban space could be considered as a necessity of the social process.

As a result of the necessity of continuing social life, humankind has considered it necessary to gather regularly in order to conduct certain social interactions such as trading goods and transferring knowledge and experiences throughout history [10]. Without humans and their conditional relationships, all buildings and roads would have been abandoned heaps of inorganic matter that could not have been perceived any differently. The characteristic that renders all physical construction is the human and the social experiences that produce a human's physical environment as well [11]. Agoras, the first examples of collective activities, saw people gathering to give long and enthusiastic speeches. The only furniture needed was a rostrum for the speakers and a seating area for the community leaders. Furthermore, the stage was removed and temporary sales counters were set up to convert the scene into a market when necessary. Festivals and religious ceremonies were held in the same space. For collective shows, a certain part of the agora was set up to roughly resemble a theater [12]. Agoras, due to their multifunctional characteristics, were the focal points of the cities where trade flourished [10].

The change in the shopping areas that started with open and closed markets, and continued with agoras and stoas, was extended to the arcades, passages, large stores and urban or peripheral shopping centers, along with the rise of capitalism in the 19th century [7]. The first examples of consumer spaces that were transferred into indoor social spaces spread to the whole world due to globalization, although the first examples appeared only in America [5,7]. In particular, Asia's economic boom is reflected by the development of modern temples of consumerism. Nine out of ten of the biggest shopping centres in the world are now located in Asia. Much more than simply being opportunities for shopping, these centres are increasingly turning into big amusement parks and lifestyle centres with shops attached. For example, the Golden Resources Mall offers 557 419 m^2, which is about the size of 75 football pitches [13].

In the early 1960s, shopping malls began to be built only in the suburbs. In the 1970s, at an international conference on urbanism in Versailles, a German delegate gave a significant speech stating the need to preserve the attraction of urban centers against suburban shopping malls. The significance of urban centers and pedestrian streets was emphasized vis- à-vis the potential problems (consumption madness, etc.) [14].

The first mall in Turkey was Galleria in Ataköy, Istanbul, based on "The Galleria" shopping mall in Houston. The Galleria shopping mall, located in the Ataköy Tourism Center complex, was built in 1988. Promoted as a demonstration of modern life, Galleria was unrivaled for years in Turkey. Today, there are 385 shopping malls in Turkey and the forecast for 2023 is around 475. The total shopping mall turnover in Turkey started increasing in 2012; however, the rate of increase has decreased during the past year. In the last 10 years, USD 14 billion of the USD 150 billion foreign investments in Turkey were investments in shopping malls [15]. An example of these large shopping malls provides a one-stop shopping opportunity for anything imaginable, as shown in Figure 3, [16]. There is a global increase in specialized shopping malls [5].

Figure 3. Shopping mall examples:
a: The West Edmonton Mall, Canada *b: The Dubai Mall, Dubai [16]*

In 21st-century Turkey, material and spiritual values are consumed rapidly, and the shopping habits of individuals have changed, with the expansion of the populist culture which has become prominent in the last 20 years. Large shopping malls have become consumer spaces where commerce is offered to make up for losses in the cities.

Figure 4. Shopping mall examples: İstanbul, Zorlu Center, (Photo by Yavuz, A.)

Open space designs were constructed in indoor spaces where several functions such as eating, shopping, leisure and entertainment were offered at the same time. The shopping venues constructed with new and contemporary building material such as galleries, escalators, ramps, panoramic elevators, special water games, different lighting fixtures, natural light and technological transparent top covers, combined with food and entertainment spaces, have been transformed into spaces of rich visual and functional resources [10]. In addition to their attractiveness, these spaces, which structurally and functionally promise wholeness and completeness in a single space [17], could keep people shopping for hours with the help of the above-mentioned attraction centers [10]. The Dubai mall includes one of the largest aquariums in the world. The SM Mall of Asia in Metro Manila, the Philippines, contains an IMAX theater with one of the world's biggest 3D screens. The mall also features an Olympic-sized ice skating rink. It can accommodate both recreational and competitive figure skating, as well as ice hockey. The West Edmonton mall includes the largest indoor amusement park in the world, the largest indoor waterpark in the world and the Ice Palace [16].

In these spaces, individuals' movements, desires and pleasures are being shaped and even directed. Often, users are not even aware of this fact [17].

Figure 5. Shopping mall examples: İstanbul, Forum İstanbul (Photo by A. Yavuz)

These spaces, which gradually become an urban meeting space and have taken over several functions of streets, are becoming increasingly significant in contemporary Turkey [10] and are increasingly becoming descriptive and symbolic urban spaces.

Figure 6. Shopping mall examples: Brussels, Galeries Royales Saint Hubert (Photo by E.Tanır)

On the other hand, the city changes, frays and functionally assumes a somewhat disorganized structure based on consumerism. The urban development model based on this approach, which could have dangerous consequences in the long run, should not be ignored. Natural areas that are significant for the urban future such as suburban reserves, agricultural-forest land and water basins would decrease and perhaps disappear.

Another potential negative consequence of this process is that the conventional urban centers become deserted, while the centers become suburban, creating their own attraction centers, contrary to the old approach [18]. As the buildings in the urban center become vacant, these spaces become candidates for future depression zones, causing chaos, complications, and multiple social imbalances. In particular, the streets would be lost and the pedestrian shopping streets would turn into empty or dead spaces, losing their vitality and functionality.

Consumerism, which symbolizes a related identity and lifestyle, is centralized, and in this process excludes low-income groups in particular [18], whilst developing towards building only large shopping malls which appeal to upper income groups [19]. Despite all the possibilities and diversity of this approach, it is impossible to argue that shopping malls are more democratic than streets due to the fact that they exclude certain segments [18]. The creation of social groups that can and cannot access large shopping malls has further solidified the distinctions that exist between the lowest and highest income groups [19]. Shopping centers are in a sense contradictory spaces. This contradiction is based on the dimming of the boundaries that can be drawn between public and private spaces, both structurally and functionally [17].

If urban spaces create more complex social relationships, draw boundaries between social groups and create problems of social justice, it is necessary to reassess the processes that trigger change [19].

Figure 7. Shopping mall examples: Trabzon, Forum AVM, (Photo by A.Yavuz)

Conclusion

Contemporary individuals have created a new cultural habit by directing their leisure activities from open green spaces to the indoor spaces of shopping malls, and preferring participative and mobile activities. This could be easily recognized when we examine the individuals' living space, hobby, leisure, entertainment and even holiday preferences. Hence, since it is not possible for industry and trade to remain indifferent to this rising tendency amongst individuals, it has become necessary to create incentives to turn this development into an advantage for the environment and individuals, rather than investing heavily in this phenomenon.

The failure of adequate interventions in ecological transformation in spite of environmental damage that could not be limited on a global scale demonstrates an approaching danger for humanity. This is not only a material problem; it is a spiritual and moral issue that involves the perception of and changes to the relationship of an individual with nature. Making drastic changes in consumerist culture, productivity culture and institutional organizations, as well as lifestyles, has become imperative [1]. It is necessary to transform the social structure that the capitalist system created and shaped as a result of the Industrial Revolution into a production society rather than a consumer society [20]. It is clear that shopping malls are supported by a rhetoric that consciously encourages consumption in exchange for status or psychological satisfaction, by influencing people excessively and distancing them from fulfilling only their basic needs. Since it is not a coincidence that global powers made shopping malls a center of attraction and consumption, necessary regulations should be enacted and planning strategies developed as soon as possible. For psychological satisfaction and recreational needs, it is very important to keep urban open spaces, streets and squares alive for social interaction, and to preserve urban values for the sake of cultural sustainability.

References

1. D. HARVEY: *Asi Şehirler, Şehir hakkından kentsel devrime doğru, çeviren:Ayşe Deniz Temiz, Meltis Yayınları, İstanbul,* (2013)
2. H. LEFEBRE: *Şehir hukkı, Sel Yayıncılık, İstanbul* (2015).
3. Ü. ARKLAN: *Tüketim ve cazibenin mekansal izdüşümü olarak alışveriş merkezleri. Erciyes iletişim dergisi,* Akademia 2 (2012).
4. J. BAUDRILLARD: *Tüketim toplumu, Çeviren, Ferda Keskin, Ayrıntı yayınları, İstanbul,*(2008).
5. G. RITZER: *Büyüsü bozulmuş dünyayı büyülemek, Çeviren, Şen Süer KAYA, Ayrıntı Yayınları, İstanbul,*(2000).
6. R. BOCOCK: *Tüketim, Çeviren İrem Kutluk, Dost Kitabevi, Ankara,* (2005).
7. Y. SAYAR, D. SÜER: *Alışveriş Merkezleri Türkiye Örnekleri, Egemimarlık,* 40(41), 14. (2002).
8. L. CORBUSIER : *Atina Anlaşması, Çeviren, Ayda Yörükan: Yapı Kredi Yayınları, İstanbul,(2015).*
9. D.L. UZZEL: *The Myth of The Indoor Cıty, Journal of Environmental Psychology,*15, 299–310 (1995).
10. R. ZENGEL: *Tarih İçinde Değişen Tüketim Mekanları, Egemimarlık,* 40 (41), 10 (2002).
11. M. ÖZBEK: *Kamusal Mekanlar ve Kent Haritaları, XXI: Mimarlık Kültürü Dergisi,* 10, 12–13. 2001.

12. R.E. WYCHERLEY: *Antik Çağda Kentler Nasıl Kuruldu?, Nur Nirven, Nezih Başgelen, Arkeoloji ve Sanat Yayınları, İstanbul.* (1986).
13. http://www.detail-online.com/article/the-10-biggest-shopping-centres-in-the-world-16291/
14. J. G. HAJDU: *Pedestrian Malls in West Germany, Peceptions of Their Role and Stages in Their Development, JAPA*, 54 (3), 325–355 (1988).
15. http://www.ayd.org.tr/MicroSites/Conference_IX/Files/Hulusi_Belgu_TR.pdf
16. http://www.touropia.com/largest-malls-in-the-world/
17. P.B GÖKARIKSEL: *Ne Kamusal Ne Özel Alan: Alışveriş Merkezleri, XXI:Mimarlık Kültürü Dergisi*, 10, 16 s. (2001).
18. H.T. ŞENGÜL: *Tüketim Toplumu, Tüketim Kültürü ve Tüketim Merkezleri, Egemimarlık*, 40 (41), 8 (2002).
19. İ. Ö. SÖNMEZ: *Yapısal Dönüşümler Sürecinde Yerel ve Yerel Üstü İlişkilerin Mekansal Yansımaları-Tüketim Mekanları, Egemimarlık*, 40 (41), 32 (2002)
20. H. ŞAFAK: *Tüketim Toplumu, Çevre Performans Đndeksi Ve Türkiye'nin Çevre Performansının İndekse Göre Değerlendirilmesi, Tarih Kültür ve Sanat Arastırmaları Dergisi*, 1(4), 132–148 (2012).

Assist. Prof. Dr. Elif Tokdemir Demir[1]

Writing in the Disciplines: Aiding Discipline Specific Writing with a Data Driven Learning Approach

[1]*Ufuk University, Vocational School, Department of Applied English and Translation Studies, Ankara/Turkey, e-mail: elif.demirel@ufuk.edu.tr*

1. Introduction

1.1. Corpora and Language Learning

A corpus (plural corpora) can be defined in linguistic terms as "a collection of spoken and written texts, organized by register and coded for other discourse considerations" [1]. Corpora have been compiled for several reasons since the 1960s and 70s. Initially corpora were used in preparing dictionaries, general usage manuals, reference works, and word-lists; later they were for preparing teaching materials and internationally recognized language tests. Examples of early corpus compilation projects from the 1960s and 70s are the Brown corpus which is a collection of written American English consisting of 15 text categories, 500 text samples, and a total of 1 million words; the LOB (Lancaster Oslo Bergen) corpus of British English and the London-Lund corpus of spoken British English with 100 text samples. Corpora compiled in the 1980s and early 1990s include the COBUILD corpus, the Bank of English and the BNC. More recent corpora of the 1990s include the Longman CANCODE corpus, the COCA corpus and the MICASE corpus. Table 1 shows a list of major corpus compilation projects since the early days of computerized corpora in the 1960s. Today, the potential of corpora for facilitating language learning has been widely recognized. One way in which corpora contribute to the teaching and learning of languages is through advancing our knowledge of language itself and how it works [2]. The purposes for learning language, especially for learning the English language, have also changed over the decades. Today people learn English for a variety of reasons other than solely to communicate orally with the speakers of English. An important branch of English teaching has become the EAP: English for Academic Purposes.

Table 1: Examples of corpora since the 1960s

Early Corpora (1960s–70s)	**Corpora of the 1980s and early 1990s**	**Recent corpora since 1990s**
Brown	COBUILD	Longman CANCODE
LOB	Bank of English	COCA
London-Lund	BNC	Time
	ICE	MICASE
	Wellington Corpus	T2K-SWAL
	Helsinki Corpus	Santa-Barbara
	ARCHER	Longman Conversation
		ICLE
		LINDSEI
		BAWE
		MICUSP

To this end, corpus linguistics could aid learners in learning discipline-specific vocabulary in particular. But how this might be achieved is the question. Often corpora are designed for researchers and may not be so straightforward for learners to use and benefit from. Therefore several approaches have been developed to put corpora into use by language learners.

1.2. Approaches to Corpus-Based Teaching

The two main corpus-based teaching approaches which have been developed are the indirect approach and the DDL (Data driven learning) approach. The indirect approach [3] describes the kind of input provided by the teaching/learning materials which follows an "indirect approach" in that it presents the learner with "what" to learn. DDL, on the other hand, advocates a more direct approach which shows the learner "how" to learn it [4].

Currently, many nonnative speaking (NNS) researchers write their research in English for a variety of reasons including reaching a wider audience and recognition in the academic world. This requires them to learn the conventions of writing and the appropriate use of discipline-specific vocabulary with their collocations. One advantage of the DDL approach is that it can be applied to all disciplines provided that a representative corpus is compiled. The current chapter demonstrates the use of a DDL approach in the teaching of academic writing in the disciplines, specifically in the field of "landscape architecture" since in this type of approach possibilities are sought for encouraging learners to interact directly with corpora. This is due to the fact that the spirit of DDL itself is experimentation with data.

1.3. DDL: Data Driven Learning

DDL is a flexible approach to teaching language through the use of experimenting with language with the help of a corpus. It is flexible in the sense that it is not limited to any subject matter or language. In order to apply DDL in any given context, one of the most important requirements, however, is to have a representative corpus of the register at hand which displays the linguistic and lexical characteristics of that specific register. Thus, the purposes of this chapter are twofold: the first one being to display the potential of a DDL approach in aiding the teaching of discipline-specific vocabulary in its context as well as how to use it in writing, and the second one being to show an application of the approach using Landscape Architecture discipline as an example. With this purpose in mind, a specific corpus of research articles on landscape architecture was compiled. The DDL approach was followed in order to guide novice researchers in their academic writing practices in the field of landscape architecture. Guidance was provided specifically on the use of key phrases and their collocations in the field. The key phrases to be included in the DDL activities were detected by using the AtnConc concordancer [5].

Usually, researchers in Turkey who do not study English as their research field are not provided with sufficient systematic guidance on how to write and publish in their fields. These researchers are under pressure to get published in English and they usually learn by doing. This causes them to waste valuable energy and time on the way. The DDL approach could teach them how to analyze specific language data and develop their writing skills through exploration.

1.4. Online Corpora and DDL

One way of providing help to this group of students is to use available online corpora such as the BNC [6]. A quick training session on basic features could prove to be very useful for students for reference purposes. This website allows you to quickly and easily search the 100 million word British National Corpus (1970s-1993). The BNC was originally created by Oxford University Press in the 1980s-early 1990s, and now exists in various versions on the web. Since concordancing programs have become available to teachers and students, their possibilities have been offering new directions for developing teaching materials which enable students to make direct discoveries about language [4, 7] and as an aid to course design [8]. In the BNC, in order to learn more about the contextual use of a vocabulary item, learners can search for words and phrases using different options. The options are between searches by exact word or phrase, wildcard or part of speech, or combinations of these. As with other interfaces to the BNC, one can search by words (mysterious), phrases (fairly certain or white + noun), lemmas (all forms of words, like sing or tall), wildcards (un*ly or r?n*), and more complex searches such as un-X-ed adjectives. For example, if one wants to search for all adverbs starting the prefix 'un-' and ending with the suffix '-ly', the following wildcard search can be used: 'un*ly'. Figure 1 shows the most frequent 10 results obtained with the wildcard search 'un*ly'. As can be seen in Figure 1, this wildcard search yields adverbs such as unlikely, unfortunately, undoubtedly, usually and unexpectedly.

*Figure 1. Results of the wildcard search 'un*ly' on BNC*

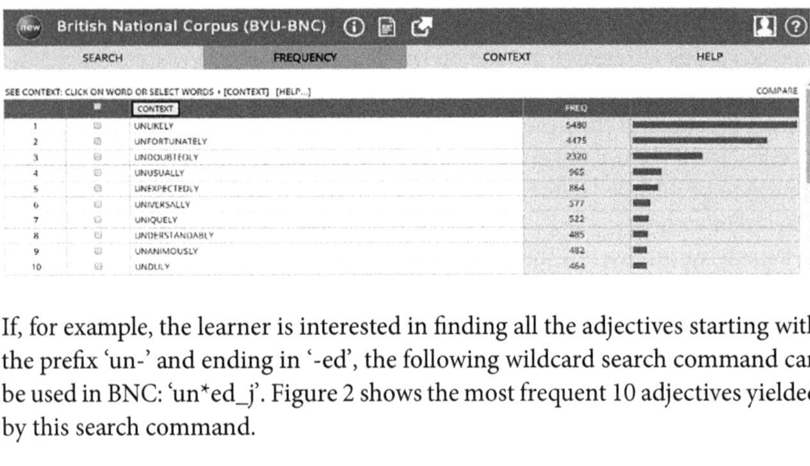

	CONTEXT	FREQ
1	UNLIKELY	5480
2	UNFORTUNATELY	4475
3	UNDOUBTEDLY	2320
4	UNUSUALLY	965
5	UNEXPECTEDLY	864
6	UNIVERSALLY	577
7	UNIQUELY	522
8	UNDERSTANDABLY	485
9	UNANIMOUSLY	482
10	UNDULY	464

If, for example, the learner is interested in finding all the adjectives starting with the prefix 'un-' and ending in '-ed', the following wildcard search command can be used in BNC: 'un*ed_j'. Figure 2 shows the most frequent 10 adjectives yielded by this search command.

*Figure 2. Results of the wildcard search 'un*ed_j' on BNC*

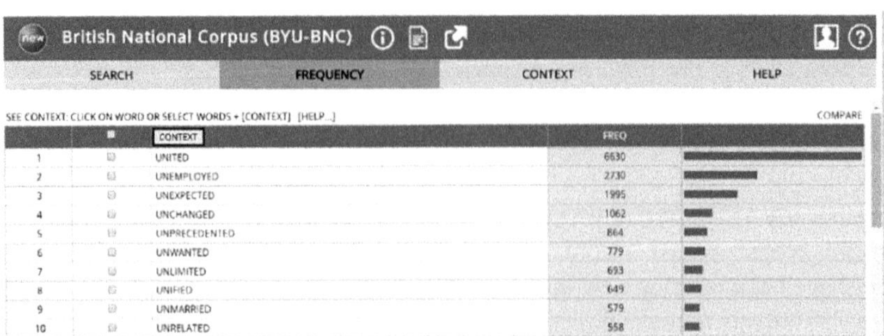

	CONTEXT	FREQ
1	UNITED	6630
2	UNEMPLOYED	2730
3	UNEXPECTED	1995
4	UNCHANGED	1062
5	UNPRECEDENTED	864
6	UNWANTED	779
7	UNLIMITED	693
8	UNIFIED	649
9	UNMARRIED	579
10	UNRELATED	558

When learning vocabulary, it is very important to learn the collocations of vocabulary items, in order to be able to use the vocabulary items correctly in context. In this respect, corpus searches can help learners to find common collocations of a word. To exemplify this, on BNC it is possible to search for surrounding words (collocates) within a ten-word window (e.g. all nouns somewhere near paper, all adjectives near woman, or all nouns near spin).

Figure 3. Results of the adjective collocation search for 'smile' on BNC

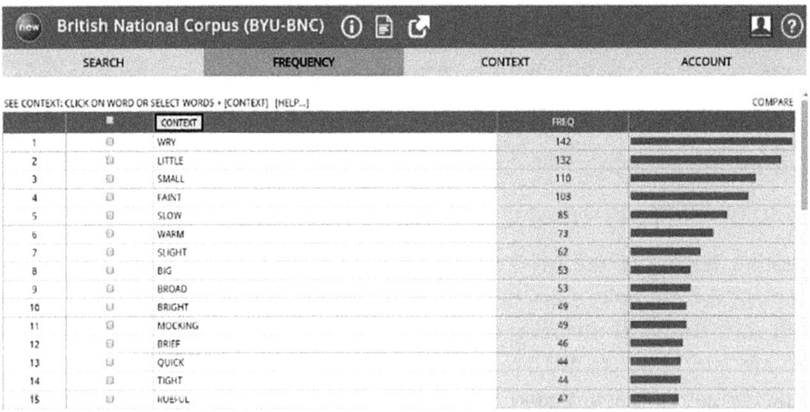

For example, you can search for the most common nouns near thick, adjectives near smile, nouns after look into, or words starting with clos* near eyes. Figure 3 shows the results of a search on BNC for adjectives which collocate with smile.

1.4.1. Register Comparison with BNC Corpus

With this architecture and interface, learners can also easily find the frequency of words and phrases in any combination of registers that they define (spoken, academic, poetry, medical, etc). This allows the user to see how words and phrases vary across speech and many different types of written texts. We can easily find which words and phrases occur much more frequently in one register than another, such as -ness words in poetry, adjectives in tabloid newspapers, nouns in advertisements, good + [noun] in fiction, or verbs in the slot [we * that] in academic writing. This can also be applied easily to collocates, such as nouns with the verb break in news, adjectives with woman in fiction, nouns with passionate (fiction vs. newspaper, or academic vs. fiction), or nouns near chair in academic register vs. fiction. Figure 4 shows the comparison of 'we * that' across six registers:

spoken, fiction, magazine, newspaper and non-academic. The last category, misc., refers to entries in the corpus which cannot be classified into a certain register.

*Figure 4. Register comparison search for [we * that] in BNC*

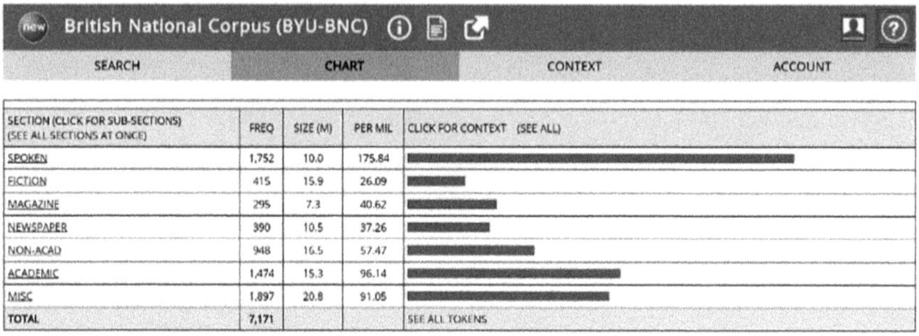

SECTION (CLICK FOR SUB-SECTIONS) (SEE ALL SECTIONS AT ONCE)	FREQ	SIZE (M)	PER MIL	CLICK FOR CONTEXT (SEE ALL)
SPOKEN	1,752	10.0	175.84	
FICTION	415	15.9	26.09	
MAGAZINE	295	7.3	40.62	
NEWSPAPER	390	10.5	37.26	
NON-ACAD	948	16.5	57.47	
ACADEMIC	1,474	15.3	96.14	
MISC	1,897	20.8	91.05	
TOTAL	7,171			SEE ALL TOKENS

1.4.2. Semantically-based Queries with BNC Corpus

Another feature of the BNC is to put out semantically-based queries to the corpus. For example, users can compare the most frequent nouns that appear with small and little, the most frequent adjectives with men and women, or the most frequent nouns with utter and sheer. It is also possible to find the frequency and distribution of synonyms of a given word, such as beautiful for example or the verb clean, see which synonyms are more frequent in competing registers (such as synonyms of strong in academic and news), and use synonyms as part of a more complex query (such as synonyms of clean with nouns). Finally, users can create "customized lists" for any category that interests them, and then re-use these in subsequent queries (such as colors + clothes, or words related to beautiful + forms of woman). All of these queries provide an insight into the English language and provide learners with hands on experience in analyzing the language. A benefit of this kind of hands-on approach is that the information learnt in this manner is stored more permanently in the memory.

2. The Corpora used for the Present Study: The Focus Corpus and the Reference Corpus

In order to detect keywords effectively in a corpus, a reference corpus is needed. In this study two corpora compiled by the researcher were used, one as the focus corpus and the other as a reference corpus. The focus corpus used for the study consists of current published research articles in the field of landscape architecture. The reference corpus consists of articles in various sub-disciplines of the social sciences. The focus corpus for the study, namely LandCorp (Landscaping Corpus) consists of academic research articles published in major journals in the discipline of Landscape Architecture discipline.

Table 2. The distribution of Research Articles in the Focus Corpus: LandCorp

Subject	# of texts	Apprx. # of words
Construction, Documentation and Administration	10	107 017
Design and Planning	10	70 591
Management	10	74 671
Implementation	10	87 172
Natural and Cultural Systems	10	92 531
Public Policy and Administration	10	82 531
Site Design and Implementation	10	71 079
Environmental Monitoring	10	67 563
Total	80	653 155

A total of 80 research articles from eight major journals on different research areas of Landscape Architecture were included in the corpus. The different research areas included in the corpus are listed in Table 2. The focus corpus consists of approximately 650 000 words.

The reference corpus used in the study, namely the AAC (American Academic Corpus) consists of articles published in the social sciences disciplines in the US. The distribution of research articles according to the journals included in the study is provided in Table 3 below. The reference corpus consists of 2 600 000 words.

Table 3. The distribution of Research Articles in the Reference Corpus: AAC

Journal	#of texts	Apprx.# of words
American Educational Research Journal	56	663 226
American Journal of Community Psychology	27	227 816
American Journal of Distance Education	37	149 723
American Journal of Economics and Sociology	81	576 513
American Journal of Philology	44	517 669

Journal	#of texts	Apprx.# of words
TESOL Quarterly	24	208 021
Arts and Social Sciences Journal	9	51 049
Educational Planning	4	25 963
Educational Research Quarterly	10	32 219
ELT Journal	49	187 705
Total	341	2 639 904

3. Steps of Concordancing in the Teaching ofAcademic English

In the present study two approaches towards the use of concordancing in the teaching of discipline-specific academic vocabulary have been followed. The first one of these is the one proposed by [9] accordingly, the following steps are proposed in the use of concordancing for the teaching of academic vocabulary [9]:

1. Students are presented with multiple examples of the vocabulary items in context.
2. Students examine these one-line concordances, highlighting word groups surrounding the key word to discover how they are used, and then answering questions on their use in context.
3. The presentation of concordances are followed by exercises enabling students to ensure that they are using the item appropriately.
4. These tasks are then followed by creative and improvisational activities.
5. The learning pathway for students provides a chain of tasks or activities in which the successful completion of each task facilitates the completion of the succeeding activity.

The second approach followed in the present study is the Research-Practice-Improvisation approach [4]. The approach has four major steps around which exercises are organized. The steps are shown in Table 4 below:

Table 4. The Steps of the Research-Practice-Improvisation approach by Johns (1991)

Step 1	LOOK	Students look at concordances for the key term and words surrounding it, thinking of meaning.
Step 2	FAMILIARIZE	Students familiarize themselves with the patterns of language surrounding the key term by referring to the concordances as they complete the tasks.

Step 3	PRACTICE	Students practice key terms without referring to the concordances.
Step 4	CREATE	Students create their own pieces of writing using the terms studied to fulfill a particular function of academic writing

4. Detecting Keywords and Keyness in Corpus Linguistics

Before the activities were designed for landscape architecture specifically, the target vocabulary had to be determined so that the vocabulary studies would be useful for this discipline. For this purpose a keywords analysis was performed using the LandCorp as a focus corpus and the AAC as a reference corpus. Keyword analysis is a common method in corpus linguistics and usually the top 100 keywords are examined. Keywords can be defined as words whose frequency is unusually high in comparison with some norm: a reference corpus [10]. Keyness of a keyword, on the other hand, represents the value of log-likelihood or Chi-square statistics, in other words, it provides an indicator of a keyword's importance as a content descriptor for the appeal. The significance (p value) represents the probability that this keyness is accidental [1]. Thus, a word is said to be key if its frequency in the text when compared with its frequency in a reference corpus is such that the statistical probability as computed by an appropriate procedure is smaller than or equal to a p value specified by the user [11].

4.1 Results of the keyword analysis

The keyword analysis yielded the first 100 words in the corpus which could function as keywords in the context. Table 5 shows the first 20 keywords obtained as a result of the keyword analysis. As shown on Table 5, the strongest 5 keywords in the context of landscape research articles are landscape, environmental, design, planning and water. The first step in designing the vocabulary exercises was to detect the most common keywords in the corpus: that is, to decide what to teach that is most beneficial for the students; or as in this case, Turkish landscape researchers who wish to write their research in English.

After the most important keywords were determined, the activities were designed around these keywords following the DDL approach. In order to be able to use vocabulary in context correctly and efficiently, it is important to know the collocations of vocabulary items. Therefore, the collocations of the vocabulary items to be included in the activities were found using the AntConc Programme with the LandCorp corpus.

Accordingly, Table 6 shows the collocations of the keyword 'landscape' used before and after the keyword in context. Thus, the most common collocations of the keyword 'landscape' used before it are 'cultural' and 'urban' (e.g. cultural landscape, urban landscape) and the most common collocations used after it are 'architecture' and 'metrics'.

Table 5. Results of the Keyword Analysis: List of first 20 Keywords

	Keyness	**Keyword**
1	10 742 405	Landscape
2	7 639 737	Environmental
3	3 451 265	Design
4	3 116 914	Planning
5	2 671 947	Water
6	2 645 130	Urban
7	2 046 530	Organic
8	2 022 143	Soil
9	1 811 537	Management
10	1 728 160	Architecture
11	1 616 813	Landscapes
12	1 478 334	Land
13	1 386 889	Area
14	1 353 361	Areas
15	1 282 538	Tourism
16	1 281 831	Species
17	1 274 472	Plants
18	1 170 136	Acids
19	1 067 163	Vegetation
20	976 640	Concentrations

Table 6. Collocations of the Keyword "Landscape"

Collocations used before the keyword				**Keyword**	**Collocations used after the keyword**		
cultural	urban	European forest	different within natural	LANDSCAPE	architecture metrics performance structure ecology	Character policy assessment planning diversity	Design

5. Application of Research–Practice–Improvisation approach by Johns (1991) to the Teaching of Landscape Architecture Vocabulary

As was mentioned in the previous section, for the first step of the DDL approach; that is the 'Look' stage, students look at concordances for the key term and words surrounding it, thinking of meaning. Therefore, the concordance lines obtained from the AntConc concordancing program for the selected keywords were prepared as the first stage of the activity. Thus, the students were provided with sentences chosen from the corpus in which the keyword landscape was used with landscape highlighted and in the middle. The rationale behind providing these lines is showing the learners the vocabulary item in its context and giving them a chance to explore its usage. Figure 4 shows the handout prepared for the 'Look' stage of the activity with the instructions given to the learners.

Figure 5. Concordance Lines for 'landscape' used in the First Step of the DDL Activity: 'Look' Step

From a design perspective, plant materials have three major functions in the	landscape	: aesthetic, structural and utilitarian.
For example, landscape metrics are used for modelling the impacts of landscape change on soil erosion groundwater recharge, water quality or	landscape	aesthetics.
This may help to develop a deeper understanding of the link between 2-D and 3-D	landscape	analysis
The results of this study will help planners,	landscape	architects, and administrators for planning and designing more ecologically and socially sustainable neighborhoods.
The project also included the development of a	landscape	assessment model identifying areas of landscape sensitivity and value.
As the students had invested much time in generating the	landscape	biography, there was little time left to explore the possibilities of future use and management.
These trends in	landscape	change are in the same direction in each country, but changes in natural and cultural heritage also differ between countries, due to the different starting points.
Furthermore, there were also favored	landscape	character area types where the size of the designated public green areas and possible destinations did not explain their popularity: built-up and non-built-up grey areas (ID23, 25), parks connected to water (ID37) and non-built-up areas with half-open vegetation (ID36).
This expert-based	landscape	characterization serves as a framework within which subsequent participatory planning processes are conducted, similar to the situation Caspersen (2009) describes for Denmark.
As such,	landscape	designs have been used to impose a time regimen onto the land and its inhabitants.

In order to help learners more with the use of the vocabulary, the vocabulary items were also color coded; the keyword was highlighted in green, nouns were highlighted in purple, adjectives were highlighted in blue and verbs were highlighted in yellow. The same was done for the other two keywords: 'environmental' (see Figure 6) and 'organic' (see Figure 7). At this stage, the learners were asked to examine the concordances and try to see some patterns in the use of the vocabulary items. At this first stage, the aim was to help learners get some exposure to the target vocabulary items by reading the sentences which displayed it in their context. Thus, they could obtain important information about the vocabulary item such as: which different meanings can be attributed to it, which words come before it, which words come after it, which words are used to modify it. Here, a limited number of concordance lines are provided to exemplify the activity, but the number of concordance lines can be increased or decreased according to the specific vocabulary item.

Figure 6. Concordance Lines for 'environment' used in the First Step of the DDL Activity: 'Look' Step

Since the London Smog (1952) there has been a long series of	environmental	accidents and disasters which the media have reported.
This study examines how corporate	environmental	performance affects not only the level of detail of a company's environmental disclosures, but also what information is disclosed.
The ability to link CNT concepts to geographically defined spaces could also potentially be useful for purposes of	environmental	management, and for conservation planning in particular.
The basic manifestation of anthropocentrism in	environmental	ethics is that very few constraints are imposed upon our treatment of the environment.
They showed that good performers disclosed more	environmental	information, based on a sample of 191 US firms with high-polluting industries (pulp and paper, chemicals, oil and gas, metals and mining, and utilities).
Action and concern about	environmental	problems should be targeted towards the solution of these problems.
The key	environmental	issues are the significant nuisance caused by construction works; the cultural heritage impact on Signal Hill and Signal Tower;...
Environmental decision making, environmental management and	environmental	policy were for many years dominated by the unproven conviction that environmental problems (if they exist at all) were the unavoidable side effects of scientific and technological progress.
Schultz (2000) argues that the value people give to an object depends on the extent to which they include that object within their sense of self, and that pro-	environmental	behaviour is more likely with increased connectedness to nature or place.
The last part of this paper looks at	environmental	policy issues where the ethical component is changing the direction of the discussion.

Figure 7. Concordance Lines for Environment used in the First Step of the DDL Activity: 'Look' Step

Clay particles tend to absorb an	organic	substance on their surface.
It is often assumed that the Amish practice	organic	farming and that they represent an appropriate model for ecological agriculture based on sustainability and stewardship.
Low molecular weight	organic	acid absorption in forest soils: effects on soil solution concentration and biodegradation rates.
In the whole lacustrine ecosystem, all kinds of	organic	acid behavior are intermediated by microbial activity to a great degree.
These	organic	acids also have been hypothesized to play a key role in long-term pedogenic processes such as podsolization.
Aqueous-phase oxidation is a potentially important source of	organic	aerosol and could explain the atmospheric presence of oxalic acid.
While Hasanağa Park is a more natural park, embracing traces of	organic	agriculture traces, Başayak Park is a more formal and organized park, designed with an understanding of modernist planning during the Post-Republican Period.
Bacterial production on humic and nonhumic components of dissolved	organic	carbon.
An informal garden has more naturalistic,	organic	forms that are normally found in gardens that mimic nature
What is more,	organic	garbage and sewage might be locally processed and returned to the adjacent land thus cutting out the expenditure of energy presently devoted to pumping or trucking these materials away from the city centers.

At the second stage of the Research-Practice-Improvisation approach [4] activities, or the 'Familiarize' step, students familiarize themselves with the patterns of language surrounding the key term by referring to the concordances as they complete the tasks given to them. In Figure 8, you can see one of the tasks in which the learners are asked true--false questions about the use of the keyword 'environment'. This stage helps students pair up words with their collocations and to sort out non-collocations.

Figure 8. Task designed for the Second Step of the DDL Activity: 'Familiarize' Step

Step 2: FAMILIARIZE for the keywords: 'landscape', 'environmental' and 'organic'

1) Which of the following statements do you think are true? Tick your answer in the box.

A landscape could be modelled	□True	□False
A landscape could be analyzed	□True	□False
A landscape could be developed	□True	□False
A landscape could be designed	□True	□False
A landscape could be favored	□True	□False

2) Which of the following statements do you think are true? Tick your answer in the box.

One could discuss/ talk about environmental media	□True	□False
One could discuss/ talk about environmental issues	□True	□False
One could discuss/ talk about environmental accidents	□True	□False
One could discuss/ talk about environmental keys	□True	□False
One could discuss/ talk about environmental nature	□True	□False

The third step of the Research-Practice-Improvisation approach is the 'Practice' step. At this step, learners practice key terms without referring to the concordances with the help of specially designed tasks. Figure 9 shows a sample activity designed for this stage of the DDL activities for practicing Landscape Architecture keywords. In these activities, keywords have been omitted from sentences taken from the corpus and learners are asked to find the correct keyword which completes the sentence. Here, in order to complete this task, learners have to rely on their memory and the information they have gathered in the first two steps. Therefore, it is very important that the activities are done successively.

Figure 9. Task designed for the Third Step of the DDL Activity: 'Practice' Step

ngs drawn from a case study of regional	....	policy in Germany (see Gailing, 2012). The Spree
iosphere reserve, but also a socially constructed	....	where institutional problems of interplay arise (
structures (such as species or ecosystems) and on	....	as a traditional aesthetic ideal influenced by l
e as a traditional aesthetic ideal influenced by	....	painting and the precepts of Romanticism. Accordi
ties. Due to the fact that a socially constructed	....	is a complex common good, it is impossible to des
mative for the social construction of a specific	....	informal institutions in the field of the const

The last step of the Research-Practice-Improvisation approach is the 'Create' step where learners create their own piece of writing using the terms studied to fulfill a particular function of academic writing. This step is the most challenging one since it requires production by the learners. At this stage, the activity designed for the field of Landscape Architecture has been adopted to provide some contextual guidance for the learners. Thus, learners are first given short texts about a subject in Landscape Architecture. The texts are chosen from the LandCorp corpus. Each text is chosen so that it has one of the keywords studied in the previous activities as its focus. In other words, the text develops around those keywords. After learners read the texts given to them, they are asked to summarize the texts making sure that they include the keywords in their summary. The purpose of asking students to use the keyword in the summary is to challenge them to use the vocabulary items they have learnt in their own context in a suitable way. Figure 10 shows a sample activity of the 'Create' step.

Figure 10. Task designed for the Fourth Step of the DDL Activity: 'Create' Step

Step 4: CREATE for the keywords 'landscape', 'environmental' and 'organic'

Read the following paragraphs and summarize them. In your summary, specifically include the highlighted words.

Blue Stick Garden, Photo, Jean-François Vézina, June 2006, Montréal

Beginning in the 1980s, this generation of landscape architects designed landscapes that embraced pop culture, patently ignored organic forms, and unapologetically employed contemporary technology. Known for substituting sticks for flowers, lipstick for trees, and ping pong balls for leaves the Quebec landscape architect, Claude Cormier, for example, demonstrates that designed landscapes need not be distant scenes of lost nature. Rather he insists that landscapes are social and sensorial, and important products of mass culture. As part of this culture, they are market driven, swayed by consumption, and very much a part of our contemporary times.

A text is given to the students about landscape architecture; specifically about landscape design in this activity. The learners are asked to read the text, and summarize it, making sure that they include the three keywords 'landscape', 'environmental' and 'organic' in their summaries. Having completed the first three steps of the activities, the learners are supposed to feel confident enough to start creating their own sentences or short texts using the keywords of the study. At the 'Create' step, gradually, the length of the productions can be increased from sentence level to paragraph level and eventually to text level.

6. Conclusion

Using a DDL approach to discipline-specific writing provides several benefits for learners. First of all, it provides opportunities for the learners to explore language specific to their field in a familiar context. With a progression of activities from controlled to creative, it develops learners' confidence in using discipline-specific vocabulary. On the part of the teacher, the approach also provides benefits. For example, using a specifically compiled corpus and concordancing programs, it is possible to detect which language items to focus on by using keyword analysis. First of all, using a specifically compiled corpus and concordancing programs, it is possible to detect which language items to focus on by using keyword analysis. The teacher has the opportunity to custom design materials which could cater for needs of very specific learner groups. With this chapter, the aim was to display the potential of the DDL approach for teaching discipline-specific vocabulary with the special case of Landscape Architecture. A relatively small-size corpus, LandCorp was compiled for the sample activities designed in this chapter. However, corpus design allows for development and expansion of the corpora through constantly adding new content. In this way, it would be possible to focus on other areas such as the specific syntax used in academic research articles within a specific field. By making corpora online and available to other researchers, it is also possible to create a synergy between researchers.

One possible limitation which needs to be addressed is that corpus linguistics is not very well known by researchers outside the field of linguistics. As has also been mentioned, software in the field is not necessarily designed with students in mind and concordancers are usually written for language researchers, using jargon known only to linguists and language specialists [12]. Therefore, hands-on training is needed for prospective material developers who wish to develop DDL activities for their field and learners who wish to use concordancing to advance their knowledge of discipline-specific language. However, interest in this field is growing and discipline- specific corpora have been compiled for many fields of research which researchers can benefit from. In addition, the concordancing programs are developed to be fairly easy to use for people outside the field of linguistics such as AntConc, Concordance, Corpus Presenter, WordSmith Tools, MonoConc Pro and ConcApp, to name just a few.

References

[1] BIBER, D., CONNOR, U., & UPTON, T. A. : *Discourse on the move: Using corpus analysis to describe discourse structure (Vol. 28). John Benjamins Publishing* (2007).

[2] BOULTON, *A.: Bringing corpora to the masses: Free and easy tools for interdisciplinary language studies. Corpora, Language, Teaching, and Resources: From Theory to Practice*, 69–96 (2010).

[3] RÖMER, U. : *Pedagogical applications of corpora: Some reflections on the current scope and a wish list for future developments. Zeitschrift für Anglistik und Amerikanistik, 54(2), 121–134* (2006).

[4] JOHNS, T.: *Should you be persuaded: Two samples of data-driven learning materials. Retrieved on May 15, 2017, from* http://moodle.ils.uw.edu.pl

[5] ANTHONY, L. : *AntConc (Version 3.2. 2)[Computer Software]. Tokyo, Japan: Waseda University, (2011).*

[6] LEECH, G. : *100 million words of English: the British National Corpus (BNC). Language Research, 28.1: 1–13.* (1992)

[7] TRIBBLE, C., & JONES, G. : *Concordancing in the Classroom. London: Blackwell Publishing (1990).*

[8] FLOWERDEW, J. : *Concordancing as a tool in course design. System 21.2 231–244 (1993).*

[9] THURSTON, J., & CANDLIN, C. N. : *Concordancing and the teaching of the vocabulary of academic English. English for Specific Purposes, 17(3), 267–280 (1998).*

[10] SCOTT, M.: *PC analysis of key words—and key key words. System 25.2 233–245* (1997).

[11] SCOTT, M.: *Comparing corpora and identifying key words, collocations, and frequency distributions through the WordSmith Tools suite of computer programs. Small corpus studies and ELT, 47–67* (2001).

[12] CARGILL, M. & ADAMS, R. *Data-driven learning of discipline-specific research English: developing a concordancing approach for individual use, paper presented at Applying Applied Linguistics: 29th Annual Congress of the Applied Linguistics Association of Australia, Adelaide, 15–17 July.* (2004).

Murat Özyavuz[1], Beste Karakaya Aytin[2], Deniz Gözde Ertin[3]

Social Dimension of Squares' Contributions to Urban Life

[1]*Namık Kemal University, Faculty of Fine Arts, Design and Architecture, Department of Landscape Architecture, Değirmenaltı Campus, Tekirdağ mozyavuz@nku.edu.tr*

[2]*Trakya University, Faculty of Architecture, Department of Landscape Architecture, Macedonia Campus, Edirne bestekarakaya@trakya.edu.tr*

[3]*Trakya University, Faculty of Architecture, Department of Architecture, Macedonia Campus, Edirne dgozdeertin@trakya.edu.tr*

People have created open and closed spaces in cities designed to encourage gathering and coexistence. Every city has urban spaces where coexistence occurs. These spaces have important functions in the city. The first public outdoor spaces, which take the form of squares, reflect the city's cultural heritage and are the cultural environments that link the past and the future. In the historical process, open spaces/squares, which can be called agora, forums, plazas, campos, piazas, grand places, are assembly spaces which are shaped according to the conditions of the day and where many functions are realized. Squares as meeting places are the original urban items that form the focal points in the cities. The squares that have important functions for society and have witnessed important events have the feature of being the center of the city at the same time as being the elements that constitute the urban pattern. Squares which are the "activity center" in the center of dense urban areas [1] provide the basis for communication and activities carried out by people for social integrity and the revitalization of urban life. At the same time, the squares make important social contributions to urban life as public outdoor spaces where individuals can meet and interact with other people.

1. Squares as a Public Outdoor Spaces

Public space can be characterized as "a common ground linking society where people perform functional and ceremonial activities in their normal daily routines or festivals (festivals, holidays)" [2]. Public space extends from the streets to the squares and parks and the surrounding buildings and forms the most important parts of the cities [3].

Squares, defined as the first urban spaces that people discovered and used and constitute the most important public focal points of the city as well as serving as meeting places for the citizens throughout history, are the primary example of public outdoor spaces. A public outdoor space is the common living space where urban people come together. Squares are the most significant elements of a city, providing cultural and social activities such as gathering, meeting, entertainment and rest, allowing people to socialize.

Squares are polyphonic spaces that allow various social and cultural classes to communicate and where people with different characteristics recognize each other and share their experiences.

1.1. The Concept of Square

Square, "Platea", which means open space or an extended street in Latin, was born in English and French with the word "Place". Spanish "Plaza" and Italian "Piazza" come from the same origin [4]. The square is an urban space where the limiters are the architectural structures or green elements in the city, the doors are access roads and the ceiling is the sky [5]. Fauole describes the square as an empty space defined by the designed environment and expressly states that a place must be pedestrianized before it can be defined as a square [6]. Alexander (1977) refers to the square as "the largest room in the public space of the city" and states that "although it is a physical element in the concrete structure of the city, it is an essential social component of the city" [7]. Squares have fulfilled very important functions socially and have the feature of being "city centers"; they were platforms for important events of the period. According to the functions they have, they are places that give the city its identity.

1.2. Types of Squares

Squares are classified in different ways by researchers in urban patterns according to their formal, symbolic and functional characteristics. Zucker categorized urban squares into five types according to forms; closed, dominated, nuclear, grouped and amorphous squares (Figure 1) [8].

Camillo Sitte (1965) classified squares according to the positioning and configuration of the space formed in front of the dominant element of the square, finding two types: deep and wide squares [10]. The other classification of squares is Rob Krier's. Krier (1979) focused on elementary geometry to categorize urban squares. He classified urban squares as formal and geometric, as triangular, square and circular spaces and as different versions of these geometric forms [11].

Figure 1. Types of Squares

Closed Square: Piazza Annunziata, Italy [9]

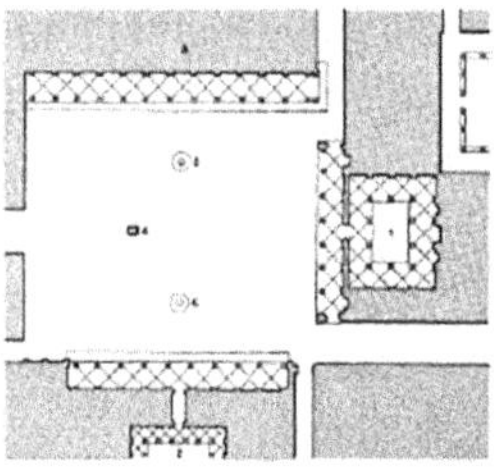

Dominated Square: Piazza San Pietro, Italy [9]

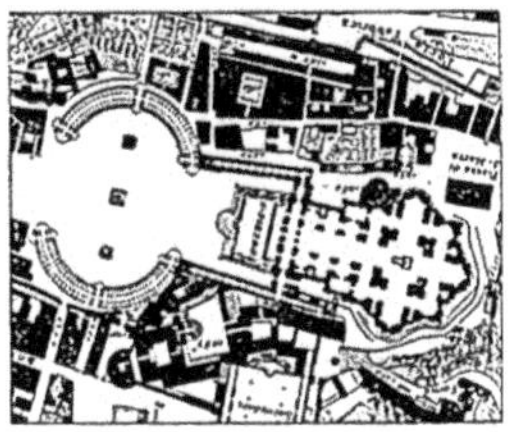

Nuclear Square: Piazza Del Popolo, Italy [9]

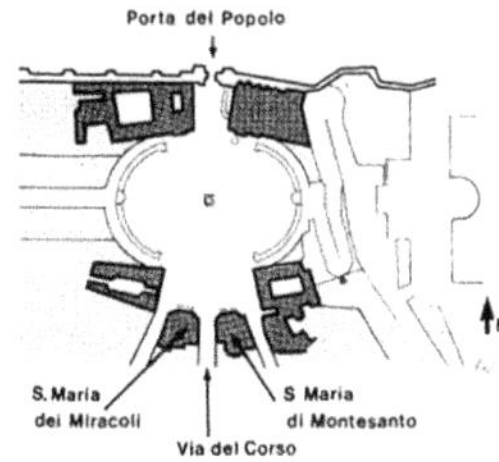

Grouped Square: Piazza Navona, Italy [9]

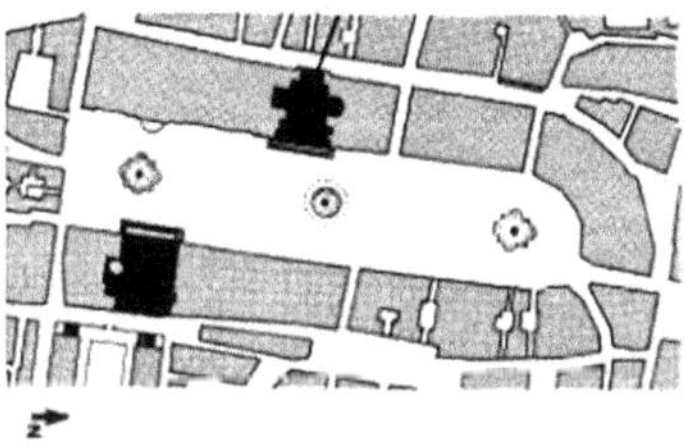

Amorphous Square: Old Town Square, Prague [URL 1]

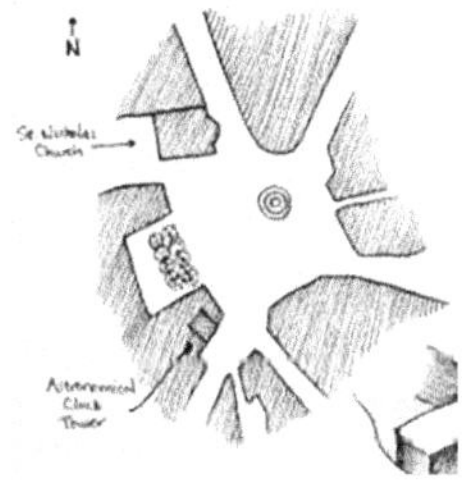

2. In The Historical Process; Squares

From past to present thc most extensive usage areas of urban life are public outdoor spaces. One of the most effectively used areas of urban public outdoor spaccs is the square. Urban squares are the significant spaces that reflect the culture of the city, providing physical, socio-cultural and economic contributions to the city. These places have been the basic element of cities from ancient periods to today and have been shaped by society's culture, beliefs and social situation. Today, however, it has been observed that squares have changed due to factors such as industrialization, urbanization and globalization. In this context, the changes of the squares in the historical process have been examined in two categories: the pre-industrial revolution and the industrial period.

2.1. Pre-Industrial Revolution Period

A square, of which the first example was found in the Greek cities and called the "Agora", was initially a wide road and a place where individuals gathered. Squares also gained and expanded the quality of the marketplace due to the trade and manufacturing freedom of the 5th century AD [12]. Agoras are open areas within the city where people are gathered to choose leaders, make decisions, and watch festivals or sporting shows held on certain days. The "Forum" was replaced in Rome by the agora of the ancient Greek city. The Forum is a place where people gather, courts are actualized and markets are established. While the agora was an urban space used for democratic life activities, the forum featured a place where the power of the administration, which wanted to dominate the world, exhibited its power. In the forum's surroundings, there were buildings such as theaters, circuses and hammams open to general and all public use [13]. In medieval cities, squares did not have a specific geometric form in the urban pattern and were small spaces on a human scale.

During the Renaissance period, innovations and changes occurred in all fields of art. The effects of this change was also shown in architecture and urban planning. Spaces such as squares, palaces, churches and administrative buildings were arranged with this effect. These spaces which are surrounded by magnificent buildings that show the power of the leadership have a geometrically shaped, symmetrical and axial arrangement [12].The wide and neat roads and vistas which were arranged in the Renaissance period became more evident in the urban pattern of the Baroque period. Arrangements that are composed of radial-shaped roads and emphasize the center are the most important feature of Baroque designs [14].

2.2. Industrial Revolution Period

With the advent of the Industrial Revolution, a rapid change began to take place in the cities. Social and symbolic functions gathered in a single center before the industrialization periodstarted to occur in different places as a result of the changes in the social structure, and are reflected in the city center and the development of transportation facilities by the Industrial Revolution [15]. In the 19th century, industrial cities with immigration and growing populations hosted many foreigners whocame to the city, as well as new urban transportation systems such as horse carts and tramways. These two factors created confusion in the city square and streets; squares began to become part of the urban transport system [16].

At the beginning of the 20th century a functional approach was adopted for cities and buildings, but neither the psychological nor social aspects of building and urban design were mentioned; squares which were the focal points and meeting areas of cities up until the Industrial Revolution became undesirable places [17]. The introduction of motor vehicles to urban space led to a reduction in the socialization areas for citizens, a movement of people away from urban spaces, and an impediment to the movement of people within urban spaces, alienating urban people from these areas. With the disappearance of life between the buildings, shopping centers and communication instruments became opportunities for people to establish relationships [17].

Nowadays, there are still squares that reflect the various periods, bearing the traces of history. However, we observe that these squares often lose their function. In the past, spaces used for gathering and socializing lost many of their original functions with the development of technology and came to serve many different purposes. On top of these squares, new squares are designed as intersection points in which traffic is often organized and social activities take place, in many cities they are transit points where cafes and restaurants are sometimes located. In addition, with the differentiation of purpose of creation of squares in the past, squares began to be given space in the shopping centers or plazas where a lot of people would gather. With time and a changing lifestyle, squares were designed to serve different purposes.

3. Contributions of Squares to Urban Life

Squares are an important part of everyday life and provide high quality to urban life. Squares have a lot of benefits to urban life such as the physical, ecological, economic, political, aesthetic and social. Squares can establish a connection, as well as proportions between the different spaces of the city thanks to the central qualitics of the urban pattern [9]. At the same time, squares contribute to urban life by providing traffic safety, and meeting the requirements of parking lots around the urban squares is one of the substantial issues of urban planning and design [18]. Squares are public spaces accessible to everyone and they offer the opportunity to act freely, and for this reason squares are also defined as places of democracy [19]. Squares are urban centers where citizens engage in commercial activities. Markets have been one of the essential elements of squares throughout history [20]. Squares increase the visual landscape value and contribute to aesthetics and the image of the city. And also squares also have ecological benefits for urban life. Especially, green areas within squares help to improve air quality, provide habitats for living organisms, reduce noise pollution, and offset negative effects of urban heat islands and surface water runoff. Notwithstanding all of these benefits and functions, the most important benefits of squares are the social ones.

The most important function of squares is to create a social life/social pattern that contains celebrations, dialogue, social actions and passive communication as well as children playing between buildings in the city (Figure 2). This social life allows people to communicate and socialize with each other and creates a common identity by having many people together in a public outdoor space. [17].

Figure 2. Old Market Square and Trafalgar Square-London, festivals, concerts and various activities [URL 2, 3]

Squares with sociability characteristics are important for the socialization of individuals. In addition, the sociability of these spaces promotes a spirit of solidarity and personal growth for all citizens, regardless of gender, race, ethnicity, age or social and economic status [21, 22]. The most important issue for squares in the past and the present was the existence of the human because the human was at the heart of the squares. (Figure 3). A public space devoid of human beings loses its social dimension over time.

Figure 3. San Marco Square, Venezia- Carnival Masks [URL 4]

Squares are spaces that catalyze social life in cities [23]. Ray Oldenburg (1997), an American urban sociologist, identifies squares as "third places" or "great good places" where people can gather and interact. Oldenburg defines the first and second places, which are people's home and work [24]. He emphasizes that the squares are essential for the social vitality of the community as the third places. The third places welcome everyone in anytime to meet, and also make citizens feel at home. These places have a social identity which makes some people join them as members of third places and creates a sense of place for their users [25]. Squares form a platform that allows social activities to take place in urban life. The relationship of different groups in the city is necessary for social life to come into existence. In the squares, people have the opportunity to discover themselves, others and their surroundings. The squares also have a political role in urban life, creating a basis for discussion of different ideas. From this aspect, the squares contribute to the development of democracy. The squares involving all segments of society are of vital importance in the development of urban identity and culture, especially in the establishment and maintenance of social relations in the context of various activities and actions. The squares fulfill an important social task by preparing an environment for mass education, thinking, discussion and information sharing in urban life, as well as strengthening the social dimension of urban life.

4. Conclusion

In the historical process, new residential areas and public outdoor spaces were needed based on population growth. Squares, which were the main public out door spaces that existed with both physical and social dimensions, are the most important tools for identifying cities and supporting public life in the economic, social and cultural life of cities.

Squares must be organized as qualified spaces that meet the requirements of the users in order to provide the contributions intended for urban life. Thus, squares may be used and social contributions may be reached at the highest level. Today the rapid development of technology in particular makes people socialize through the media instead of socializing in public spaces.

In fact, the "square" is a place which allows people to socialize face to face with others, at the border of where architecture intersects with sociology during the historical process. Today, these areas, which are used only for activities such as concerts and ceremonies, and which have pedestrian access, should be regarded as a space allowing for the analysis of social alienation.

References

1. LYNCH, K. A Theory Of Good City Form. Cambridge, Ma: Mıt Press, (1981)
2. KOSTOFF, S., The City Assembled, Thames And Hudson, London, (1999)

3. MADANIPOUR, A., Design Of Urban Space, Wiley, New York, (1996)
4. MARCUS, C. C., FRANCIS, C., People Places: Design Guidelines for Urban Open Space, Van Nostrand Reinhold, New York, (1990)
5. WEBB, M., The City Square, Thames And Hudson, London, (1990)
6. FAUOLE, P., Squares in Contemporary Architecture, İng. çev. Leslie A. Ray, Waanders Publishers Architectura & Natura Press, Amsterdam, (1995)
7. ALEXANDER C, ISHIKAWA, S and SILVERSTEIN, M., A Pattern Language - Towns, Buildings, Constructions. Oxford University Press, (1977)
8. ZUCKER, P. Town and Square: From the Agora to the Village Green, Cambridge, London, (1970)
9. MOUGHTIN, C., Urban Design: Street and Square, Third Edition, Architectural Press publications, p. 87–90, (2003)
10. SITTE, C City Planning According To Artistic Principles, Columbia Universities studies in art history and archaeology, New York: Random House, (1965)
11. KRIER, R., Urban Space, Foreword by Colin Rowe, London, Academy Editions, (1979)
12. GIRITLIOGLU, C., Şehirsel Mekân Öğeleri Ve Tasarımı, İ.T.Ü. Mimarlık Fakültesi Baskı Atölyesi, İstanbul, (1991)
13. SUHER, H. Kent Planlama Sürecinde Kent Meydanlarının Yeri Ve İşlevi, Peyzaj Mimarlığı Tmmob Peyzaj Mimarları Odası Yayını, Mayıs-Haziran Sayısı, (1998)
14. GUNAY, B., SELMAN, B., "Kentsel Görüntü Ve Kentsel Estetik Örnek Kent: Ankara", Kent, Planlama, Politika, Sanat (Der. İ. Tekeli), S. 277–319, Odtü Mimarlık Fakültesi Yayını, Ankara, (1994)
15. TER OZKAN, U., Konya Kenti Açık Sit Alan Varlığı İçerisinde Tarihi Kent Merkezinin Kentsel Tasarımı Üzerine Bir Araştırma, Doktora Tezi, S: 56–61 Ankara Üniversitesi, (2002)
16. SENNETT, R., Kamusal İnsanın Çöküşü, Çev. Serpil Durak, Abdullah Yılmaz, Ayrıntı Yayınları, İstanbul, (2002)
17. GEHL, J., Life Between Buildings (5th Edition), Danish Architectural Press, Copenhagen, (1996)
18. DONMEZ, Y., CABUK, S., OZTURK, M., & GOKYER, E. Safranbolu Kentsel Sit Alanında Otopark Sorunu ve Çözüm Alternatifleri. Journal of Bartın Faculty of Forestry, 18(2), 137–145., (2016)
19. CARR, S., FRANCIS, M., RIVLIN, L., & STONE, A. Public Space. Cambridge, England: Cambridge University Press, (1992)

20. MEMLUK, M.Z., Designing Urban Squares, INTECH, http://www.intechopen.com/books/advances-in-landscapearchitecture, p. 519–522., (2013)
21. EFROYMSON, D. THI KIEU THANH HA, T. THU HA, Ph. Public Spaces: How They Humanize Cities. Dhaka: HealthBridge – WBB Trust, (2009)
22. KURNIAWATI, W. Public Space for Marginal People. Procedia – Social and Behavioral Sciences: (36): 476–484, (2012)
23. CROWHURST, L., SUZANNE AND LENNARD, HENRY L. Genius of the European Squares, International Making Cities Council: Gondolier Press, (2008).
24. OLDENBURG, R, The Great Good Place – Cafés, Coffee shops, Bookstores, Bars, Hair salons, and Other Hangouts at the Heart of a Community. Cambridge: Da Capo Press, (1997)
25. KARIMNIA, E., In Search of Sociable Squares From a Human and Placemaking Perspective, KTH, Royal Institute of Technology, Department of Urban Planning and Environment Division of Urban and Regional Studies, Master Thesis, Stockholm, p. 15–16, (2012)

Internet References

URL 1. http://kimdaul.blogspot.com.tr/2010/11/fall-sketch-trip-part-2-prague-czech.html

URL 2. https://stuartfrew.wordpress.com/2016/09/05/the-left-lion-and nottinghams-old-market-square/

URL 3. https://tr.pinterest.com/pin/455074737317971976/

URL 4. http://www.thousandwonders.net/Piazza+San+Marco

Nazan Kuter[1], Mihriye Çakmak[2]

Universal Design in Public Outdoor Space

[1]*Cankiri Karatekin University, Faculty of Forestry, Department of Landscape Architecture, 18200, Cankiri, Turkey.*

[2]*Cankiri Karatekin University, Graduate School of Natural and Applied Sciences, Department of Landscape Architecture, 18200, Cankiri, Turkey.*

Introduction

Public spaces have played an important role both socially and politically, and they have fundamental functions for cities throughout history. Major public spaces are thought to have originated in Ancient Greece and Roman cities. The "*Greek Agora*" of the ancient times is regarded as the first example of public outdoor spaces that are usually located in the center of the city and they function as both the political and economic centers. The history of public spaces continues with the "*Roman forum*" which was the main space for public life. Roman forums enabled commercial, religious and political activities, sports facilities and informal meetings [1]. Public spaces can be defined as places of interaction and social contact among different groups and individuals. Public spaces such as streets, squares and parks can shape social interaction and human exchange, and they are the basic functional parts of urban space. They are the indispensable components of urban settlements and the main tools which enhance urban quality. Additionally, they are the arenas, ideally owned by the public and managed under public interest. The "*ideal public space*" should be freely accessible and open to free actions of society deriving from its ontological attributes. The quality of the space is often measured as the quality of urban life, and it is not only physical entities that provide activities, but they also possess meanings that transcend time and they create identities for cities and a sense of place [1,2].

The concept of "*quality of life*" stems from the idea of providing people with better lives by increasing their material and immaterial wealth [3]. The term "*quality of life*" refers to the physical, psychological, and social domains of health, seen as distinct areas that are influenced by a person's experiences, beliefs, expectations, and perceptions [4]. Quality of life has gained significant importance as a research topic in sociology as well as in several other disciplines such as economics, psychology and political science in the last decades [5].

Livability refers to sustaining "*long time wellbeing or quality of life*", and it refers to the environmental and social quality of an area as perceived by residents, employees, customers and visitors. Livability that emerged in the 1960s led to the development of the discourse of "*sustainability*" in the 1980s. The concept of livability has also started to be used by two urban design movements, namely "*New Urbanism*" and "*Smart Growth*" which appeared in the 1990s. Livability comprises "*safety and health*" (such as traffic safety, personal security, public health), "*local environmental conditions*" (such as cleanliness, noise, dust, air quality, water quality), "*the quality of social interactions*" (such as neighborliness, fairness, respect, community identity and pride), "*opportunities for recreation and entertainment*", "*aesthetics*", and "*existence of unique cultural and environmental resources*" (such as historic structures, mature trees, traditional architectural styles). The concept of livability directly benefits people who live in, work in or visit an area, increases property values and business activity, and can improve public health and safety. To create livable cities and public urban spaces, the quality of public urban space design becomes important [6]. Urban green spaces contribute to the livability and quality of life with their environmental, physical, social, recreational and psychological functions.

Urban green spaces, as one of the layers of the urban pattern, are one of the most important elements of the urban landscape, and they have various benefits to urban life varying from ecologic, economic, biological, technical, and structural to social and psychological [3]. The main objective to be taken into consideration in the design of public outdoor spaces is that these areas should be made physically easy to access for children, the elderly and the disabled. Accessible facilities are defined as being "facilities that are designed to be accessible and understandable to all users of a building or external environment" [7]. Accessible buildings, environments and means of communication do not only serve those with different and various degrees of disability, they also facilitate everyday life [8]. Disability is defined as "any restriction or lack (resulting from an impairment) of ability to perform an activity in the manner or range considered normal for a human being [9]. Disabled people include those who have long-term physical, mental, intellectual or sensory impairments which in interaction with various barriers may hinder their full and effective participation in society on an equal basis with others [10]. The number of people with disabilities is growing. According to 2010 Global Population estimates, more than a billion people are estimated to live with some form of disability, or about 15% of the world's population. This estimate is higher than previous World Health Organization estimates, which date from the 1970s and were around 10%. Two main reasons for this increase are *i*) the higher risk of disability for aging older people, and *ii*) the global increase in chronic health conditions associated with disability, such as diabetes, cardiovascular diseases, and mental illness [11].

The biggest obstacle to participation in society for people with disabilities is transportation and physical environmental conditions. Universal design means planning and manufacturing goods, buildings, public outdoor spaces and facilities to be usable by all people to the fullest possible extent. It has developed from a focus on disabling barriers in the environments.

Universal design needs to be situated in different individuals' experience of disabling barriers and enabling environments, because as with all human beings, those with disabilities are different and have different plans and goals [10].

As the number of people with disabilities increases, the importance of accessibility of public outdoor spaces increases, and universal design has been a topic of much discussion in recent times. A growing number of architects, landscape architects, town planners, designers etc. are creating warm, welcoming and accessible public outdoor spaces while embracing universal design as an essential element from day one. Their goal is simply equal access; everyone who visits public outdoor spaces should be able to do so comfortably and efficiently. It is important to take into account cultural, economic, engineering, and environmental, gender and social contexts in developing countries.

Universal Design / Universal Design Principles

Universal design has been described as "the design of products and environments to be usable by all people, to the maximum extent possible, without the need for adaptation or specialized design". Universal design is not just about accessibility for a small proportion of society, but more about providing an inclusive, usable and accessible environment for everyone regardless of their age, size, or abilities and not just for those with disabilities [8,10]. The definition of universal design includes the notions of usability and accessibility. These two concepts are related but different attributes of an entity (product or service). Usability implies accessibility in that if the user cannot physically access the system, then the system is, by default, not usable. But accessibility does not imply usability. For example, one user may be able to physically access a computer program, but it may be too difficult for him/her to learn and therefore be unusable [9].

Universal design is a term that was first used in the United States by architect Ronald L. Mace in 1985, defining it as follows: "Universal design is an approach to design that incorporates products as well as building features which, to the greatest extent possible, can be used by everyone". Another definition is: "Universal design may be defined as the best approximation of an environmental facet to the needs of the maximum possible number of users". A term often used synonymously is "**barrier-free design**", but to many people it is perceived more negatively since it is closely related to the needs of people with disabilities, and so still has a close connection to accessibility issues. Universal design is about democracy in design for everybody; children and adults, elderly people, men and women, people of different nationalities, and so on. A better synonym for universal design is "**design for all**", being the more popular expression in European countries and "**design for all**", is about ensuring that environments, products, services and interfaces work for people of all ages and abilities in different situations and under various circumstances [12,13,14,15]. Universal design has seen a growing importance in recent years as a new paradigm that aims at a holistic approach ranging in scale from product design to architecture, and urban design on one hand and systems of media and information technology on the other [12]. While the concept of universal design emerged primarily for people with disabilities in mind, it helps everyone with needs for support and assistance including the elderly, pregnant women, children and people with a temporary illness or injury. So, the benefits of implementing universal design are wide. Applying the seven universal design principles in Table 1 provides the best possible framework for design professionals to envision design possibilities that all users with disabilities or otherwise can benefit from [14,16].

Table 1 Universal design principles [Adapted from 8,13,14,16,17,18,19]

PRINCIPLES	DEFINITION
Principle 1: Equitable use	Design that is useful and marketable to persons with diverse abilities. **Guidelines:** a) Provide the same means of use for all users: identical whenever possible; equivalent when not. b) Avoid segregating or stigmatizing any users. c) Provisions for privacy, security and safety should be equally available to all users. d) Make the design appealing to all users. e) Design that accommodates a wide range of individual preferences and abilities.
Principle 2: Flexibility in use	**Guidelines:** a) Provide choice in methods of use. b) Accommodate right or left handed access and use. c) Facilitate the user's accuracy and precision. d) Provide adaptability to the user's pace.

PRINCIPLES	DEFINITION
Principle 3: **Simple and intuitive use**	Design that is easy to understand, regardless of the user's experience, knowledge, language skills, or current concentration level. **Guidelines:** a) Eliminate unnecessary complexity. b) Be consistent with user expectations and intuition. c) Accommodate a wide range of literacy and language skills. d) Arrange information consistent with its importance. e) Provide effective prompting for sequential actions. f) Provide timely feedback during and after task completion.
Principle 4: **Perceptible information**	Design that communicates necessary information effectively to the user, regardless of ambient conditions or the user's sensory abilities. **Guidelines:** a) Use different modes of communication (pictorial, verbal, tactile) to present essential information. b) Provide adequate contrast between essential information and its surroundings. c) Maximize the legibility of essential information. d) Differentiate elements in ways that can be described (i.e. make it easy to give instructions or directions). e) Provide compatibility with a variety of techniques or devices to be used by people with sensory limitations.
Principle 5: **Tolerance for error**	Design that minimizes hazards and the adverse consequences of accidental or unintended actions. **Guidelines:** a) Arrange elements to minimize hazards and errors: most used elements, most accessible, hazardous elements eliminated, isolated or shielded. b) Provide warnings of hazards and errors. c) Provide fail-safe features. d) Discourage unconscious action in tasks that require vigilance.
Principle 6: **Low physical effort**	Design that can be used efficiently and comfortably and with a minimum of fatigue. **Guidelines:** a) Allow user to maintain a neutral body position. b) Use reasonable operating forces. c) Minimize repetitive actions. d) Minimize sustained physical effort.

PRINCIPLES	DEFINITION
Principle 7: Size and space for approach and use	Design that provides appropriate size and space–for approach, reach, manipulation, and use, regardless of the user's body size, posture or mobility. **Guidelines:** a) Provide a clear line of sight to important elements for any seated or standing user. b) Make all components comfortable to reach for any seated or standing user. c) Accommodate variations in hand and grip size. d) Provide adequate space for the use of assistive devices or personal assistance.

The aim of universal design is to develop theory, principles and solutions to enable everybody to use the same physical solutions to the greatest possible extent, whether this be buildings, public outdoor areas, means of communication or household goods. The key concepts are equal status, equal treatment and equal merit. Inherent in these notions is the ideal that everybody should have the same possibilities to participate in different areas of life such as education, work and leisure etc. [8].

Today, universal design principles are applied in the design of industrial products and in many designs of urban areas such as public spaces, parks, roads, pedestrian zones. The concept of universal design has been better understood in recent times and the design of barrier-free environments has been implemented in many countries of the world.

According to [7], universal design is important for planning in terms of the following aspects:

- It helps avoid bad development and delivers genuinely sustainable solutions for communities.
- It helps create better places for all abilities and all age groups that are equitable, inclusive, participative and accessible.
- It avoids the need for wasteful and inefficient retro-fitting of solutions, as these matters should be considered at the outset of the design process.
- It informs genuinely integrated strategies for land use, transportation and urban design.
- It creates greater efficiency for public infrastructure investment and produces better economic development models.

- It widens the audience and market for well-considered development projects enhancing commercial viability.
- It helps provide an environment in which people can age and retain their independence.

Public Outdoor Space Design

Disabled people who make up an important part of society encounter problems arising from urban planning and design. At the basis for resolving these problems lies the ergonomic and elaborate planning and design of public outdoor spaces. Table 2 shows the standards of pedestrian sidewalks, ramps, stairs, grade pedestrian crossings, pedestrian underpasses and overpasses, parking lots, open and green spaces, and city furniture.

Table 2. Some standards and design criteria that should be applied in the design of public outdoor spaces for disabled users [Adapted from 20]

Pedestrian Sidewalks	▪ The width of the sidewalk should be at least 150 cm (ideally 200 cm). ▪ On the sidewalk, there should be at least 25 cm beside the property and 50 cm safety strip including the curbstone on the curbstone side. ▪ The inclination of the sidewalk should be less than 2% and surface water should be removed by drainage. ▪ The floor covering materials should not be slippery, should not pose a risk of stumbling, and should not reflect light. ▪ Sidewalks must form a continuous or solid surface at the same level. ▪ The height of the sidewalk should range from 3 cm and 15 cm. ▪ A ramp with a width of 90 cm should be made towards the pedestrian crossing at an 8% slope towards the road elsewhere on the sidewalk. ▪ The surroundings of the obstacles such as trees, shrubs, etc. must be surrounded by sensible surface textures. ▪ In order to be noticed, the city furniture on the sidewalks such as billboards and lighting, and trees, must be placed on a platform with a 10 cm height from the pavement. ▪ Resting benches should be placed on the pedestrian walkway, depending on the width of the road or in a way that does not interfere with pedestrian traffic.
Ramps	▪ The ramps should be at least 90 cm wide. ▪ If there is a ramp exceeding 10 m and a difference of more than 50 cm at a height, or a ramp to a second ramp, at least 250 cm of straight rest areas should be made. ▪ If the ramp changes direction at the landing, there should be at least 150 cm x 150 cm space available. ▪ The ramp should not be steeper than 8%. For ramps longer than 10 m, the maximum slope should be 6%.

	▪ If there is a ramp that is longer than 10 m or a transition from a ramp to a second ramp, at least 250 cm straight resting areas should be constructed. ▪ The surfaces of the ramps should be coated with hard, stable, non-slip covering, and material with minimum roughness. The roughness heights on the surface should not be greater than 2 cm. ▪ The knobs should continue for 45 cm at the beginning and at the end of the ramp. The handrail must be made on one or both sides of the ramp when passing over a 20 cm height level of elevation.
Stairs	▪ The formula of 2 x riser height + 1 x step width = 63 cm should be used with a maximum riser height of 15 cm. ▪ Rugged, anti-slip coating should be used on the walking surfaces of the stairs. ▪ The steps and risers should be shown in a separate color and a 2.5 cm wide non-slip strip at the step tip. ▪ If the height difference is over 180 cm, there should be a 200 cm landing between the stairs. ▪ If the stairs change direction at the landing, there should be at least 180 cm x 180 cm space available. ▪ The width of the stairs should be at least 180 cm from the handrails. ▪ There must be a handrail on both sides of the stairs. ▪ At the beginning and at the end of the stairs there must be sensible (different color and texture etc.) surfaces. ▪ The sensible surface should be at least 60 cm wide. ▪ The stairs should be illuminated in a way that allows night use.
Grade Pedestrian Crossings	▪ Sensible (different color and texture, etc.) surfaces should be used at pedestrian crossings. ▪ Do not put obstacles (electricity, lighting, traffic etc.) that reduce the ability to move on pedestrian crossings, roads and junctions. ▪ The pedestrian crossings should be well illuminated from the top. ▪ Pedestrian crossings should be well marked with place marks (zebra lines). ▪ The ramp and the road surface material should be in different textures in the pedestrian crossings. ▪ Pelican type pedestrian crossings should be up to 120 cm high so that the traffic sign with button (lighted) and voice can be used by the disabled.
Pedestrian Underpasses and Overpasses	▪ Ramps, elevators and escalators not exceeding 8% inclination should be preferred instead of stairs in pedestrian underpasses and overpasses. ▪ Pedestrian underpass and overpasses should be well illuminated. ▪ The coating materials used in the pedestrian crossings should be nonslip and durable.

Parking Lots	▪ 5% of the total number of parking lots should be reserved. ▪ A car parking space must be at least 360 cm wide (ideally 390 cm). ▪ Parking places should be marked with directional signs of disability and there should be a disabled parking mark on the surface. ▪ Disabled parking signs should be visible, readable and illuminated. ▪ The entrance and exit areas of the parking lot should be at the same level with the road or the inclination of the ramp should not exceed 8% and the surface should be covered with separate materials to prevent slipping and to determine the entrance and exit. ▪ A pavement ramp should be made and the height of the curbstone should be 3 cm. ▪ The parking lot ticket machine and parking meters should be between 90 cm and 120 cm high so that they can easily accessible for the disabled.
Open and Green Spaces	▪ The entry of open space or parking space should be sensible and accessible. ▪ The roads in open and green areas should be non-slip. ▪ Floor coverings should be directive and robust.
City Furnitures	▪ The city furniture should be placed so as not to obstruct movement, it should be free from sharp and protruding edges. ▪ The head recovery distance should be higher than 220 cm. ▪ If touchable and contrasting colored signs should be made around the obstacles, their height should not be less than 70 cm. ▪ In the case of surface coatings, tissue differentiation must be created to define the locations of the fixtures. ▪ Colors that will create contrast with the environment should be preferred while choosing city furniture to be perceived as colorful. ▪ The benches for seating should be placed at regular intervals such as 100 m to 200 m, with a 120 cm area for the wheelchairs next to the sitting bench, the bench's height should be 45 cm above the surface and the height of the backrest should be 70 cm. ▪ The height of the tables in the recreation areas should be between 75 cm and 90 cm, and the minimum depth under the table should be 60 cm so that the wheelchair can approach from all directions. ▪ At least one of the open or closed public telephone booths must be arranged in relation to disability issues. ▪ Garbage cans should be mounted at least 40 cm away from the curb stone at the edge of the sidewalk and the height should be at least 90 cm and at most 120 cm. ▪ Mail boxes should be at an accessible height of between 90 cm and 120 cm. ▪ The fountains should be at a height of 85 cm for wheelchair users.

Concluding Remarks

Public outdoor spaces where people come together have played an important role in the formation process of cities and in the development of quality of life. These spaces have been included in urban space as streets, squares, parks, paths, public transport, recreation areas and playgrounds in the historical process and have become one of the most important parts of urban space. For this reason, these spaces must be designed to welcome all people without restrictions and be accessible by all groups of users. People should have every opportunity to connect with nature when visiting the park with their families. Studies have provided convincing evidence that the way people feel in pleasing public outdoor spaces improves their recall of information and creativity.

Design for all refers to the design of products and environments such as buildings, public pathways, streets, squares, public green spaces and technical installations etc. that allow all people to use these products and environments as far as possible without being side-lined and all people can participate actively, whether they have a disability or not. This is because there are no differences due to age or ability in universal design. So, regardless of age, size, obstacles or abilities, it is necessary to create public outdoor spaces that can be used by all. Disabled people encounter various obstacles in urban areas and are therefore forced to live in difficult conditions. Designers need to consider situations where the needs of these people, who have lost a variety of abilities, also change.

Ultimately, successful design which considers universal design principles can create a more usable, accessible, safe and intuitive environment. But, besides universal design principles, other factors such as aesthetics, cost, security, gender and culture are also important and must be taken into consideration. The role of designers, as recreation providers, is not only to create accessible public outdoor space, but also to add specific details for each type of obstacle in public outdoor spaces to their projects.

References

1. N. KULÖZÜ: Transformation of Public Space: The Case of Hacibayram Square. Middle East Technical University, The Graduate School of Natural and Applied Sciences. (Master Thesis), 153p. Ankara (2008).
2. N.O. MEMLÜK: Inclusivity of Public Space: Changing Inclusivity of an Urban Park, Gençlik Parki, Ankara. Middle East Technical University, The Graduate School of Natural and Applied Sciences. (Master Thesis), 170p. Ankara (2012).

3. E. BİNGOL: A Quality of Life Perspective to Urban Green Spaces of Ankara. Middle East Technical University, The Graduate School of Natural and Applied Sciences. (Master Thesis), 125p. Ankara (2006).
4. M. TESTA, D.C. SIMONSON: Assessment of Quality-of-Life Outcomes. The New England Journal of Medicine, 334(13), 835–840 (1996).
5. Ş. KAYA: Quality of Life in Turkey: A Comparative Analysis with the European Union Member States. Middle East Technical University, The Graduate School of Social Sciences. (Master Thesis), 102 p. Ankara (2006).
6. P. GHADIMKHANI: Increasing Walkability in Public Spaces of City Centres: The Case of Tunali Hilmi Street, Ankara. Middle East Technical University, The Graduate School of Natural and Applied Sciences. (Master Thesis), 204p. Ankara (2011).
7. NDA: Building for Everyone: A Universal Design Approach. Series 9: Planning and Policy. Centre for Excellence in Universal Design. National Disability Authority (NDA). 92p. http://universaldesign.ie/Built-Environment/Building-for-Everyone/9-Planning.pdf. (accessed 29.04.2017) (2014).
8. F. ASLAKSEN, S. BERGH, O.R. BRINGA, E.K. HEGGEM: Universal Design Planning and Design for All. 34p., Cornell University ILR School, Oslo, Norway (1997).
9. R.F. ERLANDSON: Universal and Accessible Design for Products, Services, and Processes. 244p, CRC Press, Taylor&Francis Group, Boca Raton, USA (2008).
10. I.M. LİD: Developing the Theoretical Content in Universal Design. Scandinavian Journal of Disability Research, 15(3), 203–215 (2013).
11. WHO: World Report on Disability. The World Bank, World Health Organization (WHO): 24p. http://www.who.int/disabilities/world_report/2011/report/en/ (accessed 18.05.2017) (2011).
12. N. D'SOUZA: Is Universal Design a Critical Theory?. Designing a More Inclusive World. Keates, S., Clarkson, J., Langdon, P., Robinson, P. (Eds.). Chapter 1. pp. 3–9. Publisher Springer-Verlag, London (2004).
13. S. IWARSSON, A. STÅHL: Accessibility, Usability and Universal Design-Positioning and Definition of Concepts Describing Person-Environment Relationships. Disability and Rehabilitation, 25(2), 57–66 (2003).
14. ANONYMOUS: Universal Design Handbook. Building Accessible and Inclusive Environments. Advisory Committee on Accessibility (ACA) Access Design Subcommittee. The City of Calgary Community & Neighbourhood Services (CNS) Social Policy & Planning Division. Edited and design by Creative Services. The City of Calgary, Canada. 100p. https://www.calgary.

ca/CSPS/CNS/Documents/universal_design_handbook.pdf?noredirect=1, (accessed 19.03.2017) (2010).
15. M.E.K. AHMED, A. SUNGUR ERGENOĞLU: An Assessment of Street Design With Universal Design Principles: Case in Aswan/As-Souq. MEGARON. 11(4), 616–628 (2016).
16. ANONYMOUS: Accessibility Design Guide: Universal Design Principles for Australia's Aid Program. A Companion Volume to Development for All: Towards A Disability-Inclusive Australian Aid Program 2009–2014. Australian Government (AusAID). Edited by Clarity Communications, Designed by GRi.D Communications, Printed by Blue Star Print, Canberra, Australia. 139p. http://dfat.gov.au/about-us/publications/Documents/accessibility-design-guide.pdf, (accessed 19.03.2017) (2013).
17. M.F. STORY, J.L. MUELLER, R.L. MACE: The Universal Design File: Designing for People of All Ages and Abilities. Institution: North Carolina State Univ., Raleigh. Center for Universal Design. ERIC Processing and Reference Facility,170p. (1998).
18. M.F. STORY: Maximizing Usability: The Principles of Universal Design. Assistive Technology, 10(1), 4–12 (1998).
19. D. RODMAN: Universal Design Guidelines for Outdoor Spaces: Plan and Design for Choice. 78p., Publishing: City of Pitt Meadows and the District of Maple Ridge, British Columbia (2009).
20. ÖZİ: Yerel Yönetimler İçin Ulaşılabilirlik Temel Bilgiler Teknik El Kitabı, H.Kaplan, Ü.Yüksel, A.B. Gültekin, C. Güngör, N. Karasu, M. Çavuş (Eds.), T.C. Başbakanlık Özürlüler İdaresi Başkanlığı (ÖZİ) Yayınları, General Publication No:49, ISBN: 978–975-19-4679-9. 2010, 122p., Ankara (2010).

Rukiye Duygu Çay[1]

Assessment of the Safety of Playgrounds in Terms of Surface Material and Equipment

[1]*Trakya University, Faculty of Architecture, Department of Landscape Architecture, Macedonia Campus, Edirne*

1. Introduction

Play is one of the most basic requirements that prepares children for adult life, and directly affects physical, social and mental development. Also it is defined as the foundation activity for almost all future learning. Play is important in promoting health. Besides, it is important in establishing social patterns of behavior that allow children to react to their peers. Outdoor play offers more opportunities and varieties than indoor play; biking, running and climbing, use of space, interaction with nature, and experimenting with one's physical abilities can be given as examples.

Physical activity may be influenced by access to safe playgrounds and other recreational opportunities. However, one of the most important factors in selecting playgrounds for parents is that they be safe for their children [1].

Playgrounds provide recreational opportunities for children, away from traffic and other outdoor hazards. In addition, playground activities can develop children's cognitive, physical and social skills. [2].

2. Playground Injuries

Many recreational areas may pose dangers if measures are not taken. Playground injuries occur in several environments including public parks, school playgrounds and private residences [3]. Studies have shown that improper use of surface materials and equipment causes playground injuries [2–7].

Rommesmo, Garden-Robinson [4] revealed that the majority of playground injuries are caused by falls. In addition to this they discovered that;

- More than half of children aged 5 to 9 have injuries associated with playgrounds.
- Children aged 4 and under are more likely to suffer injuries to the head and face.

- On the other hand, children aged 5 to 14 are more likely to suffer injuries to hand and arm.
- Younger children have high risks for injury associated with age-inappropriate playground equipment. Children under the age of 5 especially should play in areas that are designed for their age, with appropriate equipment.
- The type of surface material has effect on injuries in playgrounds. For instance, children who fall on concrete or asphalt floors were found to suffer twice as many injuries as those who fall on shock-absorbing materials.

3. Safety in Playgrounds

Playgrounds make a great contribution to physical, cognitive and social development in childhood, but with a high potentially for injuries [6].

U.S. Consumer Product Safety Commission (RPSC) reported that falls are the most common hazard pattern (44% of injuries), and equipment-related hazards, such as breakage, tipping over, design and assembly followed this (23%). Other hazard patterns include entrapment, colliding with other children or stationary equipment. Playground-related deaths involved the entanglement of ropes, leashes, or clothing, falls, and impact from equipment tipping over or structural failure [8].

However, the following factors are important in creating a safe playground and must be borne in mind;

3.1. Playground Site

Playgrounds may involve many dangerous and negative conditions that can harm children. In order to ensure children's health and safety, if there are roads with traffic, ponds, lakes, slopes, etc. in the vicinity of a playground, precautions should be taken and children should be prevented from having direct access. Furthermore, shaded areas must be created against the effects of sunburn, especially in hot regions.

3.2. Age Groups

For safety, it is necessary to separate pathways and playgrounds with shrubs or benches for different age groups. Older children are more active when playing in playgrounds, so this separation will reduce the chance of injury for younger children. Besides, where access to playgrounds is easy (e.g. unlimited or enforced only by signage), the designer should recognize that since child development is fluid, parents could select a playground slightly above or below their children's capabilities (e.g. two or five years old). This could be easy for supervising then, and provide prevention from hazards to which children of different ages are exposed. For this reason, there is an overlap at age five [8]. A playground should allow children to develop their skills by providing gradual challenges, and these challenges should be proper to age-related abilities. In addition to their physical characteristics, the cognitive and social skills of pre-school and school-age children are also different. Due to these differences relating to type, scale and the layout of equipment, age-appropriate playground designs should be placed in the area (Table 1) [9].

Table 1. Examples of age appropriate equipment [9]

Toddler – Ages 6–23 months	**Preschool – Ages 2–5 years**	**Grade School – Ages 5–12 years**
• Climbing equipment under 80 cm high • Ramps • Single file step ladders • Slides • Spiral slides less than 360° • Spring rockers • Stairways • Swings with full bucket seats	• Certain climbers • Horizontal ladders less than or equal to 150 cm for ages 4 and 5 • Merry-go-rounds • Ramps • Rung ladders • Single file step ladders • Slides • Spiral slides up to 360° • Spring rockers • Stairways • Swings – belt, full bucket seats (24 years) & rotating tire	• Arch climbers • Chain or cable walks • Free standing climbing events with flexible parts • Fulcrum seesaws • Ladders – Horizontal, Rung & Step • Overhead rings • Merry-go-rounds • Ramps • Ring treks • Slides • Spiral slides more than one 360° turn • Stairways • Swings – belt, rotating tire • Track rides • Vertical sliding poles

3.3. Conflicting Activities

The presence of conflicting activities, like active and passive activities, and children running between them, cause many injuries in playgrounds. Separating this area into different sections can prevent these injuries. Active and physical activities should be divided from passive or quiet activities. Active and physical activities should be dispersed to avoid crowding in one area.

3.4. Field of Vision

A playground should be designed to allow parents to look after their children while playing with various equipment. Playground equipment should be visible from the younger children's area so that parents who have multiple children in different age groups can keep track of their older children [9].

3.5. Design

Playgrounds and their structural elements are designed in accordance with the desires, needs and abilities of their users. Variables such as the size and form of the site, materials, the relationship to other structures and the skills of the builders are at the forefront of the design process [9].

4. SURFACE MATERIAL

The surfacing under and around playground equipment is one of the most significant factors in decreasing the possibility of injuries. Falling onto a shock-absorbing surface is less likely to cause serious injuries than falling onto a hard surface. There are two types of ground material that can be used in playgrounds: loose-fill and synthetic materials [10]. Loose-fill materials such as sand, pea gravel, wood chips, and shredded rubber are relatively easy and cheap to install but require regular maintenance. Synthetic materials, such as rubber mats and tiles, poured rubber and artificial grass with rubber infills are more expensive to install. They do not require maintenance but they can be damaged in intense weather conditions and wear down over time [11]. Examples can be given of the shock-absorbing effect of the materials:

<u>Asphalt and concrete</u> have no shock-absorbing abilities, so they are not appropriate surface materials for playgrounds.

<u>Grass</u> is used at the base of some play equipment but weather conditions and wear can greatly reduce the effect of shock-absorption.

<u>Loose-fill materials</u>, such as sand, pea gravel, bark and wood chips and shredded tires have sufficient shock–absorbing qualities with proper depths and maintenance.

<u>Rubber material</u>s require minimum overhaul after installation to maintain the shock–absorbing quality [9].

Different materials have different shock-absorbing effects in the playground. Seaver [11] stated that the U.S. Department of Justice published 2010 ADA

Standards for the utilization of surface materials. According to this, sand is only permitted in play areas where the potential fall is from a height of under 150 centimeters. Wood chips and shredded rubber are the only loose-fill materials that can be used where the fall height reaches up to 3 meters. Also, synthetic rubber and grass surfaces can be used potential fall heights of up to 3.5 meters. These measures may decrease depending on the intensity and quality of the material [10]. Like Altmann, Ashby [5] indicated that the Consumer Product Safety Commission (CPSC) recommends a minimum of 30 centimeters of loose-fill material under and around equipment in playgrounds, to reduce the potential risk of injuries from falls. However, due to wear and the formation of some compression, these levels should not fall below 22.5 centimeters (Table 2).

Table 2. Minimum compressed loose-fill surfacing depths (Adapted from Consumer Product Safety Commission [9])

Length (cm)	of	Loose-fill Material	Protects to	Fall Height (m)
22,5		Shredded rubber		3
22,5		Sand		1,2
22,5		Pea gravel		1,5
22,5		Wood mulch		2,1
22,5		Wood chips		3

Bark and wood chip is one of the shock-absorbing materials that require daily inspection contamination, regular raking and topping up to maintain the recommended depth (300 mm is the minimum depth). It should be sifted clean of foreign and sharp objects. Reichelderfer, Overbach [12] defined sand as a proper material, both shock-absorbing and playable. However, there is the round granular sand required for a shock-absorbing surface, and then angular sand for playing with. Round granular sand should not contain particles more than 3 mm in diameter. It should be laid and maintained to a depth of 300 mm. Also raking and topping-up processes should be implemented regularly.

In playgrounds rubber can be applied in the form of wet-pour or rubber tiles. They are not appropriate for use in areas where flexibility and change are required. Materials should be regularly inspected for signs of wear.

Besides all this shock-absorbing material, hard surfaces can be used to access routes for both pedestrians and vehicles, form drainage networks and restrain wear [12].

It should be noted that, as all materials have some advantages and disadvantages, none of them can completely prevent injuries.

4. Equipment

Studies have shown that injuries in playgrounds are most often associated with falling from play equipment [7, 13, 14]. The most common locations for playground equipment injuries are school playgrounds, public parks or playgrounds and residential locations. Falls are mostly from climbing equipment, slides and swings [5].

Children's injuries in playgrounds are most often caused by falls, being pushed by other children, jumping from equipment, being struck by moving equipment, caught in equipment, hit against equipment while playing, colliding with stationary equipment, and using broken, fallen or collapsed equipment [5].

Equipment should be inspected regularly if it poses a danger, such as broken parts, protruding bolts, and sharp edges. Also, equipment should not have spaces that could endanger children. For example, guardrails must have openings of more than 9 cm and less than 23 cm, as between ladder rungs.

Children of different ages have different needs and abilities. When selecting playground equipment, the age range of children must be considered. Playgrounds should promote children's development, but should be in scale with children's sizes, abilities, and developmental levels. In addition, playground equipment should be designed in such a way that it can be used by children with disabilities.

The US Consumer Product Safety Commission [9] indicated some playground equipment that is not recommended for use on public playgrounds:

- Trampolines
- Swinging gates
- Giant strides
- Climbing ropes that are not secured at both ends
- Heavy metal swings (e.g. animal figures) – these are not recommended because of their heavy rigid framework.
- Multiple occupancy swings – swings that are intended for more than one user are not recommended (except tire swings) because their greater mass presents a risk of impact injury.
- Rope swings – Swinging ropes are not recommended because they may fray or form a loop, so they can cause strangulation hazards.

4.1. Playground Equipment

4.1.1. Stationary Equipment

For safety, the use zone should extend a minimum of 180 cm in all directions from the perimeter of the equipment. Two pieces of stationary equipment that are adjacent to one another can overlap if the play surfaces in each surface are less than 75 cm above the protective surface. Also, if the play surfaces on either structure exceed a height of 75 cm, the minimum distance between the equipment should be 275 cm [9].

4.1.2. Slides

Children can use slides with many positions, rather than sitting or facing forward while sliding. These prejudicial positions should be discouraged to reduce injuries. Besides, bare metals should not be used on platforms, slides and steps. This is because if the metal parts are exposed to direct sunlight they may cause serious burn injuries.

Slides should have a platform of adequate length to provide the transition from top to base surface. The platform should be at least 48 cm deep for toddlers, 35 cm for preschool-age and school-age children; its width should be at least equal to the slide width. It should not be curved or sloping, and should be surrounded by guardrails or barriers [9].

The use zone of slides should extend a minimum of 180 cm from the perimeter of the equipment.

Embankment slides should have chutes with a maximum height of 30 cm above the underlying ground surface. This can reduce the risk of falling injuries. In addition, short spiral slides (one 360 degree turn or less) should be used in playgrounds for toddlers and preschool-age children because they have less ability to maintain balance and postural control [8].

Slides should have an exit region to provide children with a means to maintain their balance and make the transit from sitting to standing easier. This region should have rounded or curved edges to prevent lacerations or further injuries.

4.1.3 Swings

Every child likes to swing in playgrounds. It is one of the safest pieces of equipment if properly installed with the right precautions. However, some children do not use this equipment as they ought. School-age and preschool-age children tend to jump off the swings, some of them tend to swing on foot and also sometimes more than one child tends to use a swing. These patterns of behaviors should be precluded in order to reduce injuries.

The use zone in front and behind the single-axis swings should be larger than the sides, in order to prevent children from hitting a swing in motion. It is recommended that these front and behind use zones have a minimum distance of twice the height of the pivot points above the surface material. Besides, the slides' use zones should have a minimum of 180 cm from the perimeter of the swing structure. For the multi-axis swings, the use zone should have a minimum of 180 cm + length of suspending members from the pivot point (Figure 1) [9]. Measures and parts (seat, belts, etc) of swings change to ensure the safety of children from different age groups (Figure 2, Table 3) [8].

Table 3. Minimum clearances for single-axis swings (Adapted from US Consumer Product Safety Commission [8])

Dimension	Toddler	Preschool-age	School-Age
D1	50 cm	75 cm	75 cm
D2	50 cm	60 cm	60 cm
D3	60 cm	30 cm	30 cm
D4	50 cm	50 cm	50 cm

Figure 1. Use zones for single-axis and multi-axis swings (Adapted from US Consumer Product Safety Commission [9])

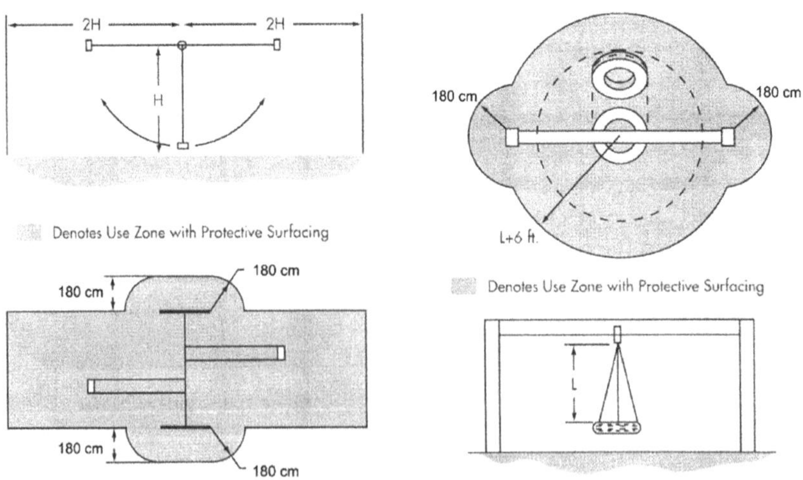

Figure 2. Minimum clearances for single-axis swings (Adapted from US Consumer Product Safety Commission [8]

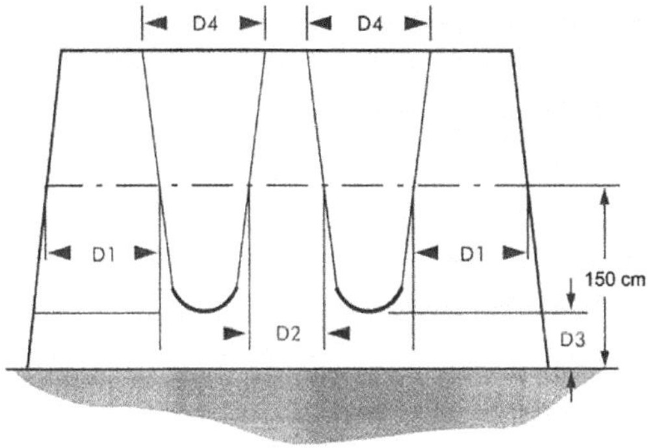

4.1.4. Merry-Go-Rounds

Merry-go-rounds are one of the most enjoyable pieces of equipment for children in playgrounds. Children may sit or stand on the platform while it rotates. However, some children tend to jump from the merry-go-round while it is in motion. It is not safe to use this equipment unsupervised, especially for young children. Also, merry-go-rounds should not be used in playgrounds designed for toddlers [8].

4.1.5. Composite Play Structures

All suggestions mentioned above are also applicable to composite play structures. However, attaching swings to composite structures is not recommended [9].

The use zone should have a minimum of 180 cm from the external perimeter of structures that involve two or more complementary play components such as slides, horizontal ladders and climbers, attached together. This zone can be extended further in front of the slide, where slides are attached to a platform higher than 180 cm from the protective surface [8]. Wardle [15] summarized a few points to be considered regarding play equipment and playgrounds, as follows:

Protrusions

Children can be scratched by protrusions (e.g. bolts, bars, pieces of wood, nails) and this may cause injuries. Equipment should be inspected carefully for protrusions and potentially hazardous areas should be fixed.

Ropes, Strings and Cables

Children like to run especially in playgrounds and they do not look around when running. If a cable or a piece of string is in the way, the child can fall or become seriously injured. A playground should not have any ropes, cables or strings at surface or below head level.

Height

Many playgrounds are too high for younger children (up to age six). This height in the equipment may cause children to be hurt. Generally, platforms, bridges or climbers should not exceed 150 cm; also slides should not exceed a height of 180 cm. [16].

References

[1] Cradock, A.L., et al., *Playground safety and access in Boston neighborhoods.* Am J Prev Med, 2005. 28(4): p. 357–63.

[2] Howard, A.W., et al., *The effect of safer play equipment on playground injury rates among school children.* CMAJ, 2005. 172(11): p. 1443–6.

[3] Potter, D., *Risk and Safety in Play.* 2003, Taylor and Francis, London, UNITED STATES.

[4] Rommesmo, S., J. Garden-Robinson, T. Barnhart, *Is your playground safe for kids.* 2008.

[5] Altmann, A., K. Ashby, V. Stathakis, *Childhood injuries from playground equipment.* Hazard, 1996. 29: p. 1–12.

[6] Lillis, K.A., D.M. Jaffe, *Playground injuries in children.* Pediatric emergency care, 1997. 13(2): p. 149–153.

[7] Mott, A., et al., *Patterns of injuries to children on public playgrounds.* Archives of disease in childhood, 1994. 71(4): p. 328–330.

[8] US Consumer Product Safety Commission, *Public playground safety handbook.* 2010, Government Printing Office.

[9] US Consumer Product Safety Commission, *Handbook for public playground safety.* 1997, US Consumer Product Safety Commission.

[10] Marshall, M. *Playground Surfacing Materials ADA-Approved and Non-Appproved.* 2011 April 2017]; Available from: https://www.ncemch.org/child-care-health-consultants/Part2/2-4_playground_surfacing.pdf.

[11] Seaver, E. *PlayGROUND_ How ground surface materials influence children's activities in public playgrounds,* Undergraduate Honors Theses, 2015. University of Colorado at Boulder,

[12] Reichelderfer, T.E., A. Overbach, J. Greensher, *Unsafe Playgrounds*. Pediatrics, 1979. 64(6): p. 962–963.
[13] Chalmers, D., J.D. Langley, *Epidemiology of playground equipment injuries resulting in hospitalization*. Journal of paediatrics and child health, 1990. 26(6): p. 329–334.
[14] Ball, D.J., K.L. King, *Playground injuries: a scientific appraisal of popular concerns*. Journal of the royal society of health, 1991. 111(4): p. 134–137.
[15] Wardle, F., *Outdoor play: Designing, building, and remodeling playgrounds for young children*. Early Childhood News, 1997. 9(2): p. 36–42.
[16] Morrison, M.L., M.E. Fise, *Report and Model Law on Public Play Equipment and Areas*. 1998.

E. Seda Arslan Muhacir[1]

Cultural Ecosystem Services and Human Well-Being

[1]*Suleyman Demirel University, Faculty of Architecture, Department of Landscape Architecture E-mail: esedaarslan@gmail.com*

'No matter who we are, or where we live, our well-being depends on the way ecosystems work.'
Roy Haines–Young and Marion Potschin

1. Introduction

The provision and realization of products required for human well-being such as food, fresh water, raw materials and fuel, as well as the processes necessary for the sustainability of this well-being by ecosystems, depends largely on the structure and functioning of ecosystems. Ecosystem services are the final outputs of ecosystems that directly affect human well-being, in relation to their structure, processes and characteristics [1]. Ecosystem services are also defined as a composite of natural ecosystems combined to maintain human life [2] or as functions of ecosystems that directly or indirectly affect human society [3]. In the Millennium Ecosystem Assessment, which is considered a milestone in the literature on ecosystem services, and which inspired a large number of studies on ecosystem services, ecosystem services are defined as 'the benefits and utilities that people derive from ecosystems' [5]. Benefits from healthily-functioning ecosystems emerge in different forms, such as food, water, clean air, medical raw materials, and recreational and cultural values, and these benefits are of vital importance to humans just as for other living beings on earth [6,7,8]. In this context, there are many reports and studies covering various aspects of ecosystem services, such as ecological, economic, cultural and social services. For example, in the definitions of reports on different typologies pertaining to ecosystem services [5, 9, 10], the references to the relationship between human well-being and ecosystem services is noteworthy [11] (Table 1).

Table 1. Definitions of ecosystem services [11]

Framework	Definition of ecosystem services
Millennium Ecosystem Assessment	'the benefits people obtain from ecosystems'
The Economics of Ecosystems & Biodiversity	'direct and indirect contributions of ecosystems to human well-being'
Common International Classification of Ecosystem Services	'contributions that ecosystems make to human well-being'

In terms of the classification of ecosystem services, there are three international classification systems [5, 9, 10] [12]. Each classification includes suppliers, as well as regulatory and cultural services. The classification made by TEEB divides 22 ecosystem services into four main categories (supplier, regulator, habitat, cultural and social activity services) [9]. The classification by CICES, made through existing classifications, focuses exclusively on the dimensions of ecosystem services [10]. These are the services provided by living organisms or obtained as a result of the combination of living organisms and abiotic processes. The classification in the Millennium Ecosystem Assessment Report, which is the most widely used one in the relevant literature and which covers ecosystem services in the broadest sense, defines four service categories. According to this, ecosystem services are provisioning services that provide for fresh water, food, fuel; regulating services that provide for air quality and climate regulation; supporting services that support other services such as soil formation; and cultural services that provide for services such as recreational tourism, educational value, and sense of space [5]. Because of their indirect effects on people, supportive services are different from other service classes in which people can directly benefit from such provisioning, regulating and cultural services [13,14].

Unfortunately, the benefits provided by ecosystems are noticed by very few people. The reason for this is related to the fact that people do not have enough knowledge about ecosystem services. When the role played by any ecosystem service for human well-being is unknown, it is difficult to determine the value of that service for human well-being [15].

This situation, which applies to each ecosystem service class, is even more difficult for cultural ecosystem services which provide non-tangible benefits.

In the scope of this study, which will address the difficulties in determining the relationship between cultural ecosystem services and human well-being, examples will be provided of prior evaluations in the context of cultural ecosystem services. In this context, the study aims to provide a theoretical contribution to cultural ecosystem services and human well-being relations within the framework of the ecosystem service approach.

2. Benefits of Cultural Ecosystem Services

Cultural ecosystem services acquire aesthetic, spiritual, psychological and other non-material benefits by relating to ecosystems. [16] The Millennium Ecosystem Assessment Report [5] defines cultural ecosystem services as being aesthetic values, recreation and ecotourism, cultural diversity, spiritual and religious values, information systems, social relations, inspiration, and a sense of space (Table 2).

Table 2 Definitions of cultural ecosystem services

[5] Name of service	Definition of service
Cultural diversity	'The diversity of ecosystems is one factor influencing the diversity of culture.'
Spiritual and religious values	'Many religions attach spiritual and religious values to ecosystems or their components.'
Knowledge systems	'Ecosystems influence the types of knowledge systems developed by different cultures.'
Educational values	'Ecosystems and their components and processes provide the basis for both formal and informal education in many societies.'
Inspiration	'Ecosystems provide a rich source of inspiration for art, folklore, national symbols, architecture, and advertising.'
Aesthetic values	'Many people find beauty or aesthetic value in various aspects of ecosystems, as reflected in the support for parks, scenic drives, and the selection of housing locations.'
Social relations	'Ecosystems influence the types of social relations that are established in particular cultures.'
Sense of place	'Many people value the 'sense of place' that is associated with recognized features of their environment, including aspects of ecosystems.'
Cultural heritage values	'Many societies place high value on the maintenance of either historically important landscapes or culturally significant species.'
Recreation and ecoturism	'People often choose where to spend their leisure time based in part on the characteristics of the natural or cultivated landscapes in a particular area.'

Ecosystem services are vital for human beings [6]. Such benefits provided by provisional services include tangible benefits such as food, fresh water, raw materials and fuel, while unintended benefits provided by cultural ecosystems include a sensation of space, educational value, tourism and recreation [17]. Identifying tangible benefits in this context can be addressed at a more basic level than assessing non-tangible benefits. Services such as food, fresh water and biological raw materials can be valued more objectively for societies, but measuring non-tangible benefits can be shaped by subjective evaluations [18, 19, 20]. For example, while the recreational value of a forest ecosystem can be measured, the benefit of the sense of space acquired by each individual can vary from person to person. Non-tangible benefits included by cultural services are [20]:

- Aesthetic and moral values
- A sense of space and connection with space
- Spiritual and religious values

The tangible relations of cultural services to provisional services include

- the interaction norms with the natural environment and values pertaining to the traditional (for example, fruit and nuts gathering)
- factors such as lifestyle and income levels, which are related to business, personal income and health condition, and which affect the local economy
- other factors that affect social networks, land administrators, and the rate and quality of service production.

In this context, the framework established in work which sets out to relate the non-tangible benefits of cultural ecosystem services to physical space is extremely striking [21] (Figure 1). According to such work, the benefits of cultural ecosystems are made up of identities, experiences and capabilities, and these benefits are factors shaping physical spaces. Another factor that shapes physical spaces in the cultural context is cultural experiences.

In this context, it can be said that the determination of the tangible and non-tangible benefits of cultural ecosystem services involves a series of evaluation processes related to a place. These processes depend on data that are shaped according to each place.

Figure 1 A conceptual framework for cultural ecosystem services [21]

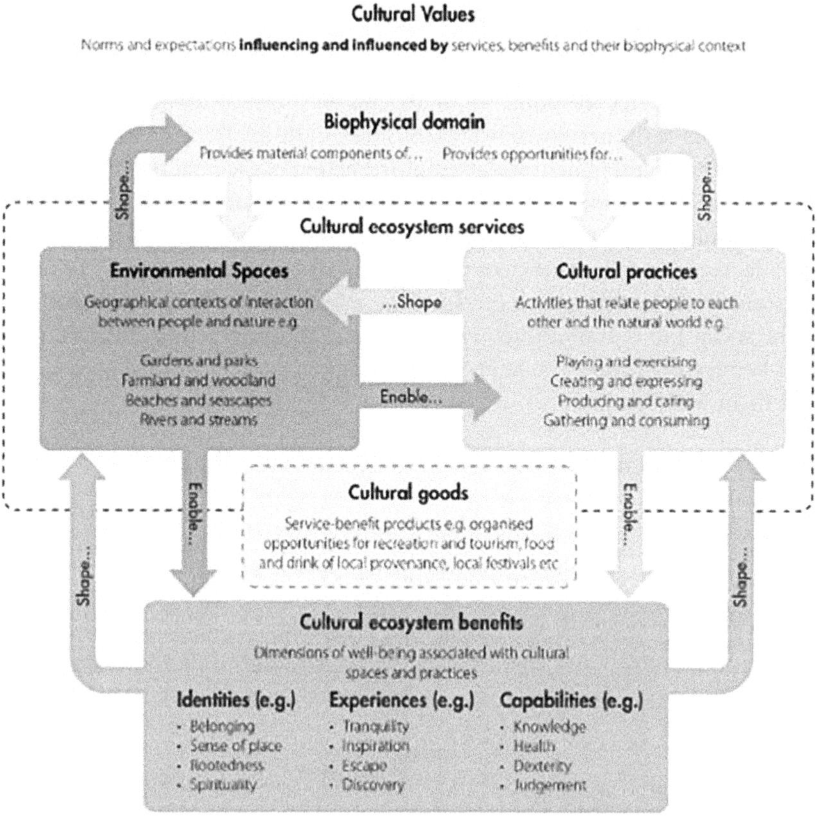

3. Ecosystem Services and Human Well-Being

In developed societies, human beings form an open system that can affect their own well-being and can transform vital energy from nature, or acquire it directly from all sources around them [22].

The existence and sustainability of human well-being is also directly related to the ecosystems and the ecosystem services provided by them. In this context, healthy ecosystems have vital components that support human well-being. The Millennium Ecosystem Assessment Report [5,17] states that human well-being has many components, and defines these components as follows: good life, freedom and choice, health, good social relations, and personal security.

This report also defines and classifies the relationship of human well-being to socioeconomic factors, as well as the overall ecosystem services. In this context, the effect of socio-economic factors on the relationship between ecosystem services and human well-being is examined in Figure 2. The color of the arrows indicates the potential to depend on socioeconomic factors, while the thickness of the arrows indicates the level of the relationship between ecosystem services and human well-being. The extent of the relationship between ecosystem services and human well-being and the potential to depend on socioeconomic factors varies at every level in each ecosystem. It is known that the effects of ecosystem services on human well-being are related to 18 other environmental factors such as technological and cultural factors, in addition to the social and economic factors shown here. This interaction affects human well-being and in this context ecosystem services are also affected by changes in human well-being.

Figure 2 Linkages between ecosystem services and human well-being [5]

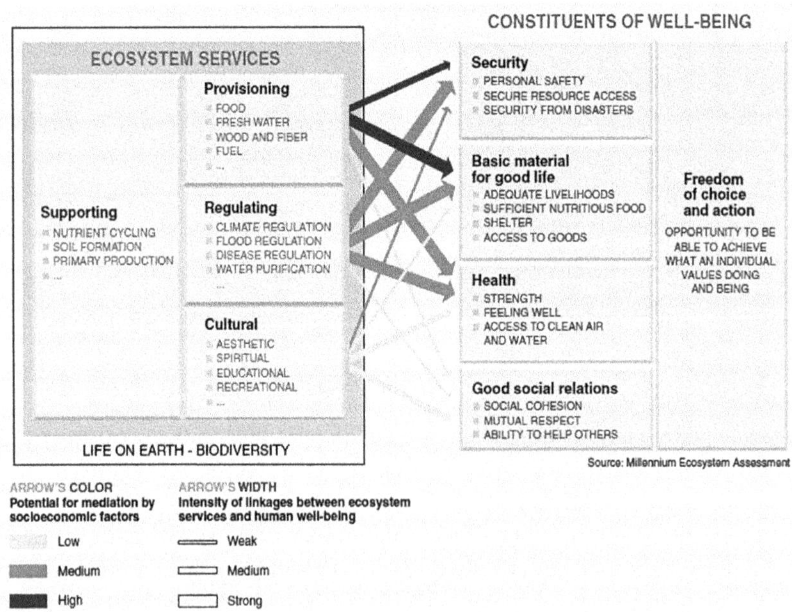

Well-being is provided by most provisional services and is interrelated with four service categories, along with basic requirements [23].

In studies aimed at determining the relationship between human well-being, its components and ecosystem services in different countries, it is seen that low-income people associate well-being with basic needs for survival and a safe living environment for themselves and their children [24,17].

In this context, ecosystems, their services, and human well-being relations are complex structures that are composed of processes which can be shaped by changes in ecosystems over time [17].

4. Cultural Ecosystem Services and Human Well-Being

The sustainability of human well-being undoubtedly depends on the healthy functioning of ecosystems. Human beings need tangible benefits that are basic human needs such as food and fresh water, which are provided by provisional services. They also need non-tangible benefits such as inspiration, aesthetic values, recreation and ecotourism, which are provided by cultural ecosystems [5,17,21,25].

For example, a forest ecosystem provides tangible benefits such as food, water, fuel, forest products, as well as non-tangible benefits such as inspiration, aesthetic values, recreation and ecotourism.

To quantify and evaluate these benefits, there is of course a need for a quantitative and objective assessment tool for ecosystem services with regard to the area-based structures and the determination of concrete benefits. This is because measuring non-tangible benefits is very difficult as so much is percieved subjectively.

Research on cultural ecosystem services has been on the rise since 2008, and these studies range from those that explore the cultural ecosystem services in depth to those that only examine current controversies [26]. Additionally, around 80% of studies in this compilation indicate that around half of recreational and ecotourism services concentrate on aesthetic values.

It is stated that among the studies on cultural ecosystems, only studies on tourism and aesthetics are not devoid of numerical evaluations and models [27,11].

As a result, it is necessary to establish a relationship with each ecosystem under study in order to measure the direct or indirect relations between cultural ecosystem services and human well-being. In this context, it may be possible for digital evaluation methods to obtain objective results, starting from subjective evaluations, and to measure the relevance of the social, cultural and aesthetic benefits each ecosystem brings to human well-being.

References

1. R. HAINES-YOUNG, M. POTSCHIN: *The Links Between Biodiversity, Ecosystem Services and Human Wellbeing, in Ecosystem Ecology: A New Synthesis.* Cambridge University Press, 110–139 (2010).
2. G.C. DAILY: *Introduction: What Are Ecosystem Services.* Island Press, 1–10 (1997).
3. R. COSTANZA, R. DARGE, R. GROOT, S. FARBER, M. GRASSO, B. HANNON, K. LIMBURG, S. NAEEM, R. O'NEIL, J.V. PARUELO, G.R. RASKIN, P. SUTTON, M. BELT: *The Value of the World's Ecosystem Services and Natural Capital.* Nature, 387(6630), 253–260 (1997).
4. B. FISHER, R.K. TURNER, P. MORLING: *Defining and Classifying Ecosystem Services for Decision Making.* Ecological Economics, 68, 643–653 (2009).
5. M.A.(Millennium Ecosystem Assessment): *The Service of Nature, Living Beyond Our Means: Natural Assets and Human Wellbeing.* Island Press, 155 (2005).
6. C.R. HEARNE: *Managing Ecosystem Services: Tools and Theory for Understanding the Dynamics of Multiple Ecosystem Services on a Landscape.* PHD Thesis. McGill University, Department of Geography, 220 (2009).
7. B. FISHER, R.K. TURNER: *Ecosystem services: classification for valuation.* Biological Conservation 141, 1167–1169 (2008).
8. B. AMBROSE-OJİ, T. PAGELLA: *Spatial Analysis and Prioritisation of Cultural Ecosystem Services: A Review of Methods. Research Report.* Forest Research, 53 (2012).
9. TEEB (The Economics of Ecosystems and Biodiversity): *Ecological and Economic Foundation.* Earthscan, 456 (2010).
10. R. HAINES-YOUNG, M. POTSCHIN: *CICES V4.3 – Revised report prepared following consultation on CICES Version 4.* European Environment *Agency, 34 (2013).*
11. K. SCHMIDT: *Tourism Wellbeing and Ecosystem Services. (TObeWELL) Short Term Scientific Mission Report, 5 (2014).*
12. A. CHURCH, J. BURGESS, N.RAVENSCROFT: *Cultural Services. UK National Ecosystem Assessmen Technical Report, 60 0 (2013).*
13. B. BURKHARD, F. KROLL, F. MULLER, W. WINDHORST: *Landscapes Capacities to Provide Ecosystem Services – a Concept for Land-Cover Based Assessments.* Landscape Online, 15, 1–22 (2009).
14. C.W. ARMSTRONG, N. FOLEY, R. TINCH, S. van den HOVE: *Ecosystem Goods and Services of the Deep Sea.* HERMIONE, 68 (2012).

15. *I. KUBISEWSKI, R. COSTANZA: Sürdürülebilir Refah İçin Ekosistem Hizmetleri.* Worldwatch Enstitüsü Dünyanın Durumu 2012, *303–312 (2012). (in Turkish)*
16. C.D. BUTLER, W. OLUOCH-KOSURA: *Linking future ecosystem services and future human wellbeing.* Ecology and Society, 11(1), 30(2006)
17. M.A.(Millennium Ecosystem Assessment): *Ecosystem and Human Wellbeing-A framework for Assessment,* Island Press, 266 (2003).
18. DEFRA (Department for Environment Food and Rural Affairs): *An introductory guide to valuing ecosystem services.* DEFRA Publications, 68 (2007).
19. S. MAXWELL, D. HENDERSON, R. MCCLOY, G. HARPER: *Social Impacts and Wellbeing: multi-criteria analysis techniques for integrating nonmonetary evidence in valuation and appraisal.* Defra Evidence and Analysis Series, 107(2011).
20. ICSU, UNESCO, UNU: *Ecosystem Change and Human Wellbeing. Research and Monitoring. Report,* ICSU, UNESCO and UNU, 60(2008).
21. R. FISH, A. CHURCH, M. WINTER: *Conceptualising Cultural Ecosystem Services: A Novel Framework for Research and Critical Engagement.* Ecosystem Services, 21, 208–217 (2016).
22. J.ANGELOVSKA: *Wellness Tourism as Beneficiary of Cultural Ecosystem Services.* Linking Wellness Tourism to Ecosystem Services: Towards a Research Agenda Abstract Book, 6 (2013).
23. M.DUDLEY, P. COATES: *Cultural Ecosystem Services-A Keywords Manual.* UKNEAFO WP 5 Arts and Humanities Annex 3, 42 (2014).
24. D. NARAYAN, R. CHAMBERS, M.K. SHAH, P. PETESCH: *Voices of the Poor: Crying Out for Change.* Oxford University Press, 314 (2000).
25. P. CLOUGH: *The Value of Ecosystem Services for Recreation.* In Dymond JR ed. Ecosystem Services in New Zealand- Conditions and Trends, 330–342 (2013).
26. A. I. MILCU, J. HANSPACH, D. ABSON, J. FISCHER: *Cultural Ecosystem Services: A Literature Review and Prospects for Future Research.* Ecology and Society 18(3),44 (2013).
27. K. M. CHAN, M. RUCKELSHAUS: *Characterizing Changes in Marine Ecosystem Services.* F1000 Biology Reports 2, 54 (2010).

Elif Ebru Şişman[1], Pınar Gültürk[1]

Industrial Heritage Sites

[1]Namık Kemal University, Fine Arts, Design and Architecture Faculty Department of Landscape and Architecture 59030 Degirmenalti/Tekirdag esisman@nku.edu.tr, pgulturk@nku.edu.tr

Introduction

Industrialization, urbanization, and migration to big cities negatively affect the cities and city dwellers. Open green areas, which have gradually been reduced in urban areas, have become unable to meet the recreational needs of urban people. At this point, it becomes increasingly important to protect industrial areas, which have become idle in cities, by turning them into urban green spaces.

The fact that industrial areas have lost their economic value is the most important basic criterion leading to their transformation. The fact that industrial cities and their industrial function have broken down to an extent relative to the decrease of their importance and the loss of their value over time brings with it considerably serious social and economic problems. Some of these cities, which for the main part emerged as the products of the Industrial Revolution in the nineteenth century, largely lost their former functions and significance when it came to the mid-twentieth century, due to the rapid development and transformation of technology.

In general, it is possible to evaluate the transformation of industrial areas in two ways. The first of these is by looking at the transformations performed for the reclamation of industrial areas and the second is the transformation of industrial areas that have lost their economic value. While transformations performed for the reclamation of industrial areas are concerned with regulations in the industrial area, the transformation of the industrial areas that have lost their economic value falls within the field of urban transformation.

This study will examine the definitions related to the protection of industrial areas that have increasingly started to gain importance in the world, their history, their development in the world and in our country, and model applications in the process of reintegrating them to the city due to their features of being potential open areas.

Industry and Industrial Revolution

Industry is a word originating from the Latin word "industria" and is synonymous with the Arabic-based word sanayi (industry). Industry is "all the actions that are applied to give raw materials a structured form and the tools that are used to implement these actions" (1). Production is required for the existence and continuity of society, through development. Industry, in the most general sense, is a process involving the transition of human beings from a social organization based on hunting/gathering to a society based on industrial production. This process has shaped society and has been shaped by society, and involves century-long development (2). Along with industrialization, there have been economic developments, the concept of serial production with machinery was introduced by the industrial period, and there has been an increase in people's standard of living.

The Industrial Revolution is the second great social and economic transformation in the history of humanity after the agricultural revolution. The Industrial Revolution implied a significant transformation, with technological advances especially in machine production as a result of a series of developments taking place in the process of the industrialization of production.

The Industrial Revolution began with the use of steam engines in England. It continued to develop, growing with the establishment of factories operating with steam and electrical power instead of man and animal power. By the 1850s, the vast majority of English people had become workers who began to gather in industrial centers. This industrialization affected European countries in the second half of the 19th century and then quickly passed to America.

The first and most important transformation in the process of urbanization after the Industrial Revolution was rapid and intense population growth. Along with the Industrial Revolution, rural areas outside the cities began to be understood as a source of cheap labor, while colonies became a source of raw material for industrial production. As a result of the organization of industrial production patterns in urban areas and their surroundings as well as the manpower surplus caused by mechanization in agriculture and the general population increase, migration to cities increased and urban areas grew rapidly. Therefore the ratio of people living in urban areas increased very rapidly. This change also explains the formation of modern urbanization as we see it today (2).

The fact that machines became increasingly complex and the need for them to be collected in areas close to energy sources brought about large factories. Furthermore, industrialists settled in cities or established new cities. As a result of this, the difference between cities as opposed to villages and towns was further emphasized (3).

Transformation in the spatial sense is the change of the location selection criteria of settlements or cities depending on the location selected by industrial activity. The first public transportation movement emerged with the spread of cities to quite large areas by going beyond the walls of the previous 19th century. This was the end of that period, a time during which the sizes of the cities were determined by distances which could be covered on foot, and horses and carts were used. Cities were no longer pedestrianized spaces, due to developments in transportation. In general, the cities of this period are characterized as insecure, with increases in social crimes, and as unhealthy due to factors such as housing shortages, poor housing conditions and a lack of infrastructure. The city centers that were abandoned by the middle and upper classes due to these unhealthy conditions were transformed into the areas, and suburbanization activities were observed within the city perimeter for the first time. Despite the problems and negative conditions experienced in the cities of the period, Ebenezer Howard's Garden City approach and Tony Garnier's Industrial City approach emerged in the late 19th century and in the early 20th century respectively. And after the Industrial Revolution, the urbanization process moved to a new dimension with utopian and ideal urban approaches developed to overcome poor conditions.

The Industrial Revolution brought with it the use of building materials that would fundamentally change the landscape of the built environment as well as producing a change in the urban texture. Iron and steel were the primary new building materials of this period, and the first structures in which these materials were used were bridges (4). As a result of the use of new building materials and construction methods in the production of buildings and competition between countries in the field of technology, a new type of building emerged known as "exhibition buildings" in which the products of this competition were exhibited (2).

The Crystal Palace, which represented England at the first world exhibition in London in 1851, can stand as an example of these types of buildings. This structure became the pioneer of a new understanding of space with the removal of the thick walls providing a sharp separation between indoor and outdoor spaces, Among the fair and exhibition activities of the period, the most important world exhibition after the Crystal Palace was the 5th world exhibition opened in 1889 in Paris. At this exhibition, the Eiffel Tower built by engineer Gustave Eiffel to create a symbol of France and Paris and the Machinery Gallery (Galerie des Machines), to represent France, are examples of structures in which technology guided the architectural style (5).

The innovations in architectural style and construction techniques that emerged with the industrial revolution came into prominence with factory buildings, which were the new types of buildings that appeared in the late 18th century. These buildings were moved out of the cities along with the industrial developments. We can observe that efforts to reintroduce the ruined industrial structures remaining within cities to city life by refunctioning have increased since the mid-20th century (6).

Industrial Heritage

It is observed that, since the 1960s, industrial heritage has increasingly gained importance in Europe, especially in England; it has started to be perceived as a scientific discipline, and various other concepts related to it have come to the fore. These concepts can be listed as "industrial heritage", "industrial archeology", "industrial culture", "industrial landscape" and "brownfield". The aforementioned concepts are still under discussion, and they are addressed differently according to the current approaches to protection in various countries, the richness of the country's own cultural heritage, and the opportunities in the field of application (2, 7).

Brownfield is defined by the work group called CLARINET (Contaminated Land Rehabilitation Network for Environmental Technologies): "brownfields are sites that have been affected by the former uses of the site and surrounding land, are derelict and underused, may have real or perceived contamination problems, are mainly developed in urban areas, and require intervention to bring them back to beneficial use" (7, 8).

This definition was complemented by the European Expert Network CABERNET (Concerted Action on Brownfield and Economic Regeneration Network) according to which this term implies surfaces that have been affected by the former uses of the site and surrounding land (9).

Today, many world-famous non-governmental organizations, as well as institutions and other organizations, the majority of which are in Europe, are carrying out studies for the protection of industrial heritage. The most important of these are the TICCIH (The International Committee for the Conservation of Industrial Heritage) established between 1973–1978 and the ICOMOS (The International Council on Monuments and Sites) in the following periods (2). ICOMOS is the global non-governmental organization dedicated to the conservation of the world's historic monuments and sites. In particular, ICOMOS' network of experts counsels UNESCO on properties to be added to the World Heritage List. Therefore, TICCIH advises on historically significant industrial sites for the World Heritage List (10).

Some definitions are made regarding the industrial heritage and archeology. The most known of these is the definition included in the 'NIZHNY TAGIL' Regulation, which was prepared in 2003 by the members of the TICCIH at the congress gathered at the Nizhny Tagil City of the Russian Federation.

"Industrial heritage consists of the remains of industrial culture which are of historical, technological, social, architectural or scientific value. These remains consist of buildings and machinery, workshops, mills and factories, mines and sites for processing and refining, warehouse sand stores, places where energy is generated, transmitted and used, transport and all its infrastructure, as well as places used for social activities related to industry such as housing, religious worship or education" (11).

"Industrial archaeology is an interdisciplinary method of studying all the evidence, material and immaterial, of documents, artefacts, stratigraphy and structures, human settlements and natural and urban landscapes, created for or by industrial processes. It makes use of those methods of investigation that are most suitable to increase understanding of the industrial past and present"(11).

Industrial heritage has historical significance since it witnesses to people's daily lives and in a sense documents them. These areas are also important in terms of their location, the open spaces they have, their building stock, symbolic and monumental value, architectural features and public utility (12, 13).

World Industrial Heritage Sites

The areas that have suffered industrial deterioration and provided employment for thousands of people can be made productive again. These areas, rehabilitated and cleared of industrial waste, are potential areas that can developing technology and production processes (14).

The industrialization processes experienced during and after the Industrial Revolution in Europe are bound up with identity. Accordingly, the efforts to develop the criteria for the preservation and protection of the industrial heritage as a reflection of the effort to retain this identity in European culture are at quite an advanced level. The efforts to evaluate industrial heritage sites, which initially looked at the direct demonstration of concrete examples of industrial cultures, now touch upon dimensions integrated with regional and social life rather than individual structures and equipment. The Ruhr Basin and IBA Emscher Park, Kop van Zuid, London Port Areas: Docklands, Oslo Port Town, and Fjord City Project are some of these sites.

The largest industrial area in Germany and the world of that period, called the Ruhr Basin and located on both sides of the Emscher river, completed its life in the 1980s. It is a project of which the ecological and social rehabilitation was accomplished in 1999, by the integration of approximately one hundred different projects in an area of 800 km^2. A project that preserves social memory and does not alienate itself from the region was developed by means of revising the industrial history of the region instead of pushing for this to be forgotten, and preserving the existing topography and local natural vegetation. As a result of this application, the Ruhr Basin has become "a triggering factor for urban, economic, social and ecological change and also a symbol" (15).

The industrial buildings in the area, Power Plant, Gasometer, Blower House Complex, Casting House, Old Administration Building, Storeroom and Main Switching House, were restored and put into service.

The power plant built in 1902 started to be used for large-scale organizations by being modernized since 1997. The gasometer used to store the gas produced by the industrial plant has now been transformed into a man-made diving center. The fan house complex used as a part of the power plant is used for concerts and openings. While some parts of the Casting House are used for outdoor activities and as rope-climbing areas, some parts of it are preserved in their original form and are visited with guides. The former Administration Building has been used as a youth center with 140 beds since 2001. The storage area, one of the first buildings to be restored, was used as a food exhibition area for a period, but is now used for meetings and social activities. The main control house, which is the main control center of the plant, is used as the administration center where all kinds of information about the park can be obtained and restaurant and souvenir facilities are located (16).

New walkways and bicycle paths have been established to connect the new recreational areas introduced to the industrial plant. Different special sporting activities can be performed in the park areas. Vegetable elements serve as structural elements. While fruit trees and dense vegetation can be found in areas where the soil has a favorable quality, poplar and birch trees, raspberry and exotic plants that may increase soil quality over time are rarely located in areas where the ground does not allow it. Plane trees, parking trees and poplar trees along the channel and the black locust along the walking trail, among the plants in the park, add aesthetic and functional value. The flowers and effective ornamental plants at the entrance to the park have also become the symbol of the park. Some of the structures which remain from industrial plants have been used in children's playgrounds by being supported by play elements such as slides and a climbing wall. The channel in the area is a symbol for the main idea of the design. The channel transporting the sewage water in previous years has been purified and relocated, and is connected to the waterways and ponds where clean water is stored. This detail, which increases the aesthetic and ethical values of the design, also symbolizes the environmental improvements in the area.

The concept of industrial transformation is a new phenomenon for this moment in our country. There are no industrial areas that have been transformed, except for the re-functioning of several industrial structures. There is no institutional restructuring related to the subject other than the studies carried out at universities, but there are the representatives of the ICOMOS and TICCIH organizations. No industrial heritage in our country is present in the World Industrial Heritage list.

Recently, various scenarios have been produced for the transformation of industrial areas, and competition projects have called on studies carried out on industrial areas, and the transformation of industrial areas, which have lost their function into urban spaces. Zonguldak Washery Area, Uşak Old Tanyard Industrial Zone, Beykoz Shoe and Leather Factory and the Bursa Merinos Factory can be shown as examples of these competition projects (17). The first examples which come to mind of discussions in refunctioning or restoration works within the scope of industrial heritage in our country are the Lengerhane Building on the Golden Horn coast, which was transformed into the RahmiKoç Museum in Istanbul and the Hasköy Shipyard, TekelCibali Tobacco Plant, which was refunctioned as Kadir Has University, and the Silahtarağa Power Plant, the first power plant transformed into an international contemporary art museum and cultural center by the name of "Santral Istanbul" (18).

In our country, the refunctioning project addressed within the scope of concern about protection which is closest to its peers in Europe and the transformation of public space is the example of İzmitSekapark.

İzmit, Turkey's largest industrial city, has reserve areas that can allow it to serve as a pilot city in this sense. The Seka Paper Factory is the best qualified of these areas and is located in the central coastal strip region of İzmit. Its nature as a paper factory was removed after it was transferred to the Kocaeli Metropolitan Municipality, where it assumed the identity of a city park and its name was changed to Sekapark (17). The factory is one of the symbols of the industrial revolution in Turkey of the Republican period during which the first modern paper production was carried out in Turkey. The main idea of the project was the transformation of the region where the Seka Paper Plant is located by reconstructing with public functions, the establishment of strong bonds with the city and city dwellers and ensuring the continuity of green along the gulf. The design consists of 3 stages. Due to the size of the area and the structures within it, Stage I and Stage II Projects of Sekapark that were primarily regulated for recreational purposes, were completed in 2007 and opened for public use (19, 20).

The area between the coast, which is owned by Seka Plant, and the railway was designed as Stage I. All-purpose turf football fields, cafés and restaurants, a wooden pier into the sea, football, tennis and basketball fields, a mini golf course, children's playgrounds, a kiting hill, a bicycle road, running paths, fitness fields, parking areas, a stage on the sea, bridges over the pier, streamside bridges and synthetic turf mounds have been created to meet the city's recreational needs (20). The facilities which had lost their function in the project area were planned to create different usage areas at the second stage, and the aim was to structure the field to the west of the area as a park recreation site. In the third stage of the project, the plan was to modernize the buildings and structures in the industrial area and to use them as science and culture fields (21). When all stages of the project implementation are completed, the aim is that Sekapark shall become an exemplary industrial transformation park with an international dimension on the regional scale rather than the urban scale, and where memory, people, space and urban perceptions are strengthened, as well as sea, land, and railway urban transportation connections being provided (19).

Conclusion

Industrial plants contributing to the economic development of countries can be re-evaluated as traces of industrial history to be used for urban and public interest. Significant studies are carried out on this subject in many European countries. It is basically thought that the Industrial Revolution was not only a technological revolution but also a transformation with social, cultural and economic dimensions that changed people's thoughts and lifestyles. Therefore, technical monuments are addressed with their social environments in studies of industrial heritage preservation. It is clear that the "industrial archeology" that emerged in the 1950s and examines industrial heritage is related to many professional fields such as preservation, architectural history, sociological/technological history, design, urban, regional and landscape planning, machinery, construction, and electrical engineering and economics, leading it to require interdisciplinary study.

The preservation of an old factory and industrial plant is an important subject that should be addressed not only as a work of protecting architectural heritage but also in terms of increasing the city's cultural significance and quality of life, being able to convey the industrial heritage to future generations and bringing these buildings into the life of the city with appropriate functions. The refunctioning method, which is less costly than rebuilding the structures on these sites, depending on the state they are in, should be given preference in these areas.

References

1. D. HASOL: *Ansiklopedik Mimarlık Sözlüğü. İstanbul: Yapı Endüstri Merkezi Yayınları* (1975).
2. B. KAYA: *Endüstri Mirasımızın Korunmasında Planlama Yaklaşımı. Kültür ve Turizm Bakanlığı İzmir II Numaralı Kültür Varlıklarını Koruma Bölge Kurulu Müdürlüğü Uzmanlık Tezi, İzmir* (2012).
3. S.TANİLLİ: *Uygarlık Tarihi, Alkım Yayınevi, 600s., İstanbul* (2006).
4. B. KIRAÇ: *Endüstri Mirasının Korunması ve Yeniden Değerlendirilmesi, Özaslan, N. ve Özkut, D. (Ed), Mimari Korumada Güncel Konular, s.116–133, Eskişehir, Anadolu Üniversitesi Yayınları* (2010).
5. G. BİROL: *Modern Mimarlığın Ortaya Çıkışı ve Gelişimi, Megaron (Mimarlar Odası Balıkesir Şubesi Dergisi), Ekim 2006, s. 3–16* (2006)
6. H. KASAP ÖZSIRKINTI: *Endüstri Yapılarının Dönüşümü Sonucunda Ortaya Çıkan Loft Yapıları ve Estetik Artium Cilt 2, Sayı 2, 151–164.* (2014).
7. A. ALPAN: *Eski Sanayi Alanlarının Yazındaki Yerine ve Endüstri Arkeolojisinin Tarihçesine Kısa Bir Bakış. Planlama. 1–2: 21–28* (2012).
8. S. PEROVİC, N.K. FOLİĆ: *Brownfield regeneration – imperative for sustainable urban development. GRAĐEVINAR 64 (2012) 5, 373–383* (2012).
9. CABERNET: *Sustainable Brownfield Regeneration, CABERNET Network Report, University of Nottingham* (2006).
10. URL:http://ticcih.org/about/about-ticcih/)
11. ICOMOS: *The Nızhny Tagil Charter for the Industrial Heritage, July 2003,* http://www.icomos.org/18thapril/2006/nizhny-tagil-charter-e.pdf (2006).
12. Y.C. SEVERCAN, A. BARLAS: *The Conservation of Industrial Remains as a Source of Individuation and Socialization International Journal of Urban and Regional Research Volume 31.3, 675–82* (2007).
13. Y.C. SEVERCAN: *Endüstri Mirasının Korunması ve Yeniden İşlevlendirilmesine İlişkin Özelleştirme Yaklaşımları: Sorunlar ve Olanaklar. Planlama. 1–2: 40–46* (2012).
14. E. COLLATEN, C.BARTSCH: *Industrial Site Reuseand Urban Redevelopment An Overview,Cityscape: A Journal of Policy Development and Research, Vol:2, Number:3* (1996).
15. C.S. ERSİNE: *Doku Nakli" ve Yeni Kimlik Oluşturma Bağlamında Endüstri Mirasının Değerlendirilmesi. İTÜ Fen Bilimleri Enstitüsü, Yüksek Lisans Tezi* (2012).

16. URL: http://en.landschaftspark.de/the-park
17. S. KAYA, Ö. YERLİ, S. DÖNER: *Endüstriyel Alanların Endüstriyel Parklara Dönüşümü (Transformation of Industrial Areas to Industrial Parks) Düzce Üniversitesi Bilim ve Teknoloji Dergisi, 3:518–534* (2015).
18. A.A. ÖZÜDOĞRU: *Adana'da Dokuma Sanayi Yapılarının Endüstri Mirası Kapsamında İncelenmesi, Ç.Ü. Yayınlanmamış Yüksek Lisans Tezi, Adana* (2010).
19. D. OĞUZ, H. SAYGI, N. AKPINAR: *Kentiçi Endüstri Alanlarının Dönüşümüne Bir Model: İzmit/Sekapark, Coğrafi Bilimler Dergisi, CBD 8 (2), 157–167* (2010).
20. B. GEMİCİ: *Eski Endüstri Alanlarının Yeniden İşlevlendirilmesinde Kentsel Kalite Ve Değer Artışına İlişkin Kullanıcı Algısı : İzmit-Sekapark Örneği, İTÜ Fen Bilimleri Enstitüsü, Yüksek Lisans Tezi* (2011).
21. M. UZUN: *Endüstri Alanı Dönüşüm Projesi Sekapark'ın (Kocaeli-İzmit) Kıyı Kullanımına Etkisi, Marmara Coğrafya Dergisi 30: 154–179, ISSN:1303–2429 E-ISSN 2147–7825* (2014).

Assist. Prof. Dr. H. Candan Zülfikar[1],
Assist. Prof. Dr. Aysel Yavuz[2]

Imagine Accessible Cities for Everyone

[1]*Trakya University, Faculty of Architecture, Department of Landscape Architecture*
[2]*Karadeniz Technical University, Faculty of Forestry, Department of Landscape Architecture*

1. Introduction

Societies consist of individuals with very different characteristics and very different needs. In order to regard a group of individuals as a society, communication, common values and sharing between the members of this group as well as common spaces for these interactions are essential. In urban areas, where individuals of a society maintain their lives together, the common areas in which individuals can interrelate, gather, socialize, share common values and interact in various ways are streets, squares, waterfronts, parks, open spaces. All these areas in which community members can coexist are expected to offer individuals of a society equal opportunities, though they all come with different characteristics and needs. Through laws and regulations, it is stated that all individuals of a society have equal rights and that it is a social and legal requirement that individuals may benefit equally from social facilities and public spaces in communal life. Once urban areas become accessible and available to all individuals, they will have a characteristic of what is described as "a living organism" by various theorists and urban researchers. It is important to consider designs in a multidimensional manner, so that particularly disabled people can meet their social needs independently, without social exclusion, and benefit equally from public spaces without obstacles.

1.1. Social Features of Urban Spaces

The quality of urban life, which consists of physical, economic and social environmental components, remains as a pursuit for all societies. Modern and contemporary cities of the future would be formed in public environments where individuals of different cultures, ages and statuses could live independently and feel safe and comfortable. Urban spaces meet the social and cultural needs of individuals through activities that promote the quality of life in dimensions such as planning, design, implementation and management. Montgomery (1998) argues that cities are prosperous and decent when evaluated in terms of street life [1].

This, because all elements of urban life are united in these multi-purpose spaces: streets. Streets bring together people who do not know each other through social interactions such as social life, shopping, and sightseeing. In this context, accessible design of an urban space becomes a measure of its utility. It is considered that different utilization parameters exist for different ecological groups, for the utilization of the same space. Designers are expected to analyze and define these parameters, to make common utilization assessments, and to render the urban space an accessible public space which serves for utilization by each group equally. Public open spaces are designed as recreational or open green spaces for gatherings, for all individuals from all social categories. Frequently, disabled users were limited in the ways they could use these public spaces, for various reasons. [2].

On the other hand, research on the positive and restorative effects of contact with nature has indicated that viewing the natural environment contributes to reducing emotional stress and mental fatigue [3; 4; 5] and that areas qualified as urban which consist of natural components, also have restorative effects. Therefore, it is necessary to employ sensitivity in ensuring that public spaces that stand out for their social dimensions have restorative attributes for disabled individuals as well as for healthy individuals.

1.2. Public Space Design For Everyone

Designers recognized that designs should include everyone in a society from all areas of life, and the concept of designing for everyone, or universal design, are issues that have long beem under consideration. The term "design for everyone", was first used by the American architect Ronald L. Mace in 1985 [6]. Ronald L. Mace describes design for everyone as: "Design for everyone/all is simply designing to ensure that all products, buildings, open spaces could be used by as many individuals as possible" [7, p. 2]. Design for everyone is intended to satisfy all users in society, making no positive or negative discrimination for a certain group in society, and give labels to no group. A design that is intended particularly and only for disabled individuals is not a design for everyone [8]. The designers who adopt the concept of design for all pursue solutions with an integrative attitude in their designs [9]. Figures 1, 2, 3, 4, 5 present good examples of design for everyone in the world.

Figure 1. Good examples of playgrounds designed for all

a) Example of play equipment designed for everyone [10] *b) [11]*

Figure 2. Good examples of sidewalks and pedestrian spaces designed for everyone in the world

a) Japan *b) Japan (Photos by A. Yavuz)*

Figure 3. Good examples of stairs and ramps designed for all

a) Switzerland (Photo by C.Acar) *b) Spain(Photo by A. Yavuz)*

Figure 4. Good examples of stops designed for all
a) Bulgaria *b) Barcelona (Photos by A. Yavuz)*

Figure 5. Good examples of public transport designed for everyone in the world
a) Japan, (Photo by H.C.Zülfikar)

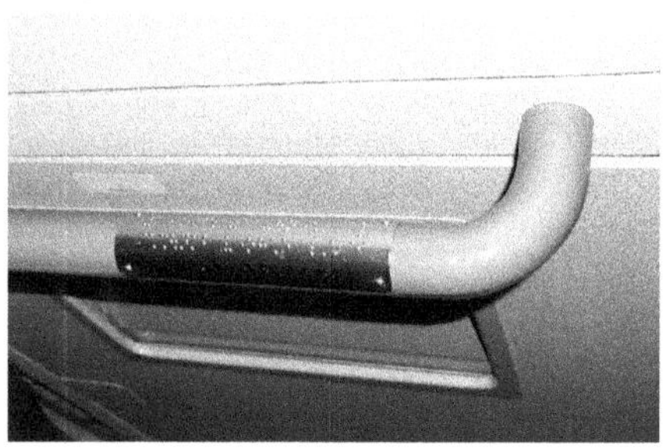

Design for everyone takes into account the public space needs of different users using universal design principles. As planners and designers who are a part of society and are trained with this consciousness increase, universal design will be adopted further with the support of local governments. In order to achieve this, the exercise of penal sanctions and a monitoring of the compliance of implementations to regulations should be supervised by different government institutions, in addition to the empathy developed by each responsible individual. It is acknowledged that the right to equal use for disabled individuals in public spaces be protected through standards and accessibility policies determined by legal legislations, which are audited by the Ministry of Family and Social Policy of T.C. in Turkey, and that implementations be carried out in order to eliminate the deficiencies that are determined due to audits [12]. According to Turkish Statistical Institute (TURKSTAT) data, 12% of Turkey's population consists of individuals with disabilities. In Turkey, where the percentage of individuals with disabilities is considerably high, there are many problems regarding the independent activities and free utilization of urban spaces for individuals with different types of disabilities. The principal barriers regarding the participation of disabled individuals relate to transportation and the physical environment (roads, pavements, public buildings, parks and gardens, urban facilities, residences, transportation). In order to propose solutions for these problems, it becomes necessary first to recognize the causes of these problems. The following Figure (Figure 6,7) presents negative examples of the design for all concepts in the world.

Figure 6. Negative examples of sidewalks and pedestrian spaces designed for all
a) Turkey[13] *b) Turkey (Photo by A.Yavuz)*

Figure 7. Negative examples of stairs and ramps designed for all
a) Trabzon, Turkey (Photo by U. Büyükkurt) *b) Edirne, Turkey (Photo by H.C.Zülfikar)*

2. Material and Method

In order that disabled people can exist within society without necessitating the help of other individuals, their opinions on urban open spaces should be recognized as first and foremost, especially regarding the residential zones they inhabit and the vicinity of these zones [14].

In this study, a questionnaire was conducted which evaluated the suitability of urban open spaces in terms of their functions, physical properties and utility for all, which contained different statements of awareness. The participants of the questionnaire were students of the Landscape Architecture departments of Karadeniz Technical University and Trakya University, as well as graduate students (master's degree) from Karadeniz Technical University, and architects, engineers and landscape architects working for the Trabzon Metropolitan Municipality. A total of 105 participants completed the questionnaire. 50 participants were undergraduate students at Karadeniz Technical University, 28 were undergraduate students at Trakya University, 11 were graduate students at Karadeniz Technical University, and 16 were architects, engineers and landscape architects working at the Trabzon Metropolitan Municipality.

2.1. The Questionnaire

34 expressions were given in five scales in order to measure the opinions of the participants, which were the verbal expressions of participants' personal experiences, in line with the Likert Attitude Scale developed by Rensis Likert in 1932. Following this the outcomes of the questionnaire from the 105 participants were evaluated.

3. Findings and Discussion

Demographic characteristics of participants such as gender, age, institution, occupation and country are presented in Table 1. 1.9% of participants answered the question "Do you have any disabilities?" with yes and 98.1% responded with no. The responses to the question "Do you have a disabled person in your family or in your close circle?" were yes by 19% and no by 81%. 29.5% of the participants answered the question "Did you ever deliver help to any association for disabled individuals?" as yes and 70.5% as no. The responses to the question "Were you ever involved in any disability-related organization?" were yes by 27.6% and no by 72.4%. Participants responded to the question "Do you think the studies regarding disabled individuals in Turkey are sufficient?" with 1% as yes and 99% as no.

Table 1. Frequency and percentage (%) distribution of the data on demographic characteristics

		TOTAL	
		Frequency	**Percent(%)**
CITY	Trabzon	**77**	**73.3**
	Edirne	**28**	**26.2**
JOB	Landscape architect	**6**	**5.7**
	Construction engineer	**6**	**5.7**
	Interior designer	**1**	**1.0**
	Architect	**1**	**1.0**
	Geological engineer	**1**	**1.0**
	Graphic artist	**1**	**1.0**
	Student	**78**	**74.3**
	Academician	**11**	**10.5**
İNSTİTUTION	Karadeniz Technical Univercity	**61**	**58.1**
	Trakya Univercity	**16**	**15.2**
	Trabzon Metropolitan Municipality	**28**	**26.7**
AGE	16–20	**32**	**30.5**
	21–25	**47**	**44.8**
	26–30	**10**	**9.5**
	31–35	**4**	**3.8**
	36–40	**4**	**3.8**
	41+	**8**	**7.6**
GENDER	Male	**26**	**24.8**
	Female	**79**	**75.2**

The data obtained were analyzed using the SPSS 20.0 software. The arithmetical means were calculated in order to exhibit the level of assessment for all of the statements in the questions which regarded urban open spaces. The participants' attitudes towards these statements were explained via factor analysis. Factor analysis allows several variables to be collected and interpreted in several groups. Cronbach's Alpha analysis was also used for an inquiry into the reliability of the factor groups.

Varimax rotation factor analysis was used in grouping the statements within the study. The purpose of using factor analysis in this phase was to moderate expressions via grouping and to name the generated factors. The Kaiser-Meyer-Olkin (KMO) Test was used to determine whether the data obtained within the scope of the present study were suitable for factor analysis and the value was found to be 0.740.

Four factor groups are defined as the result of factor analysis. The total variance of the 1st factor is 27.422%. The results indicate that the statements under this factor group (awareness) are more significant than the expressions under the other groups. A total of eleven statements were calculated under the awareness group and the average value for the group was calculated as 3.96. The maximum arithmetic mean, 4.55, corresponds to the expression "Vehicles parked on the roadway or in front of the ramp catch my attention". The minimum arithmetic mean value 3.46 corresponds to the statement "Audible impulsive sounds in public places, relief symbols or notes for informing the visually impaired catch my attention". These results are, in reality, a reflection of national practices.

The total variance of the second factor group (functional characteristics) was determined as 19.770%. Under the functional characteristics group, there were seven statements, and the average value of the group was found to be 2.63. The maximum arithmetical mean, 3.02, corresponded to the statement that "urban open spaces are feasible for social activities". The minimum arithmetical mean, 2.24, corresponds to the statement "Urban open spaces provide enough alternatives for our avocations".

The total variance of the third factor group (physical properties) was found to be 15.911%. Eleven statements were listed under the physical properties group and the average value of the group was 2.8. The maximum arithmetical mean, 3.88, corresponds to the statement "Urban open spaces support social life when they are accessible and available". The minimum arithmetical mean, 2.17, corresponds to the expression "There are audible warnings in public transport for the access to urban open spaces".

The total variance of the fourth factor group (perceptual properties) was found to be 5.302%, and the total explained variance as 68.405%. This value is above the desired level and is significant. Five statements within the perceptual characteristics group and the average value of the group were calculated as 2.57. The maximum arithmetical mean value, 2.99, corresponds to the statement "Urban open spaces facilitate the users' relationship with the natural environment". The minimum arithmetical mean value, 2.10, corresponds to the statement "Urban open spaces are adequate and accessible in the view by children".

Cronbach's Alpha α reliability was examined for the internal consistency of the statistical scale. A Cronbach's Alpha analysis was conducted to test the reliability of the Likert Attitude Scale and a α value of 0.893 was found for all statements. The reliability coefficient $\alpha = 0.893$ is well above the desired 80% level and this value indicates that there is a high level of internal consistency among the attitude statements that were generated.

4. Conclusion

Accessible cities for everyone should be planned and developed carefully, the communities who live in them should be well integrated, and all individuals should be happy, respectful and tolerant of each other's rights. All requirements should be met easily, and the infrastructure should be complete. It is important that urban open spaces are designed for all for integration within society. Designing an environment that is appropriate for all means equal, flexible, comprehensible, safe, simple and uncomplicated features in an environment that can accommodate all individuals [15].

In recent years, sociologists, along with urban planners, architects and landscape architects, have argued that a community could not achieve goals unless the administration encouraged people to participate in urban decision-making, urban planning and design processes. The healthy growth and development of society could be achieved only through such participation. In this context, "equality" means that society, which could be considered as a city, belongs to all. A good urban vision for every individual corresponds to a city where all possibilities and opportunities become accessible to provide a better life for all [16].

It is an agreeable result that the level of the awareness factor group is high among those who participated in the survey and who are practitioners. However, the fact that the problems have not been sufficiently resolved demonstrates that the level of awareness needs to attain even higher levels in order to better comprehend and solve the problems in practice. Besides, this awareness needs to be spread throughout society. Meanwhile, the number of technical staff (architects, landscape architects, engineers) is not sufficient in all local governments throughout the country. Several problems in some areas might be disregarded. Therefore, once the individuals who construct the society become conscious and aware of the problems, they can communicate the detected problematic issues to the practitioners and lead the way towards solutions to these problems.

Education and awareness activities could be carried out to increase awareness in society. Efforts should be made so that issues regarding individuals with disabilities become more prominent. However, even though individuals with disabilities constitute a higher population percentage than 12% in Turkey, these individuals are often not encountered in daily life. This is because the existing accessibility problems in urban open spaces and the viewpoint of society towards disabled people affect disabled people and their families negatively. The community therefore faces a vicious cycle. Creating urban open spaces primarily for everyone would help to integrate individuals with disabilities within society, and the level of awareness within society and for practitioners would shift towards a more positive perspective.

It is also important for practitioners to be aware of the fact that they could create an urban vision for all, have adequate technical knowledge and sensitivity, and have sufficient authorization.

Even though disabled people are still exposed to barriers in urban open spaces in Turkey, it is expected that the desired level will be attained, in the light of legal requirements and standards as well as sensitive individuals.

References

1. J. MONTGOMERY: *Making a City: Urbanity, Vitality and Urban Design, Journal of Urban Design*, 3(1), 93–116 (1998).
2. K. SEELAND, S. NICOLE`: *Public green space and disabled users. Urban Forestry & Urban Greening*, 5 29–34 (2006)
3. S. KAPLAN, & R. KAPLAN: *The experience of nature: A psychological perspective. New York: Cambridge University Press* (1989).
4. A. E.van den. BERG, H. STAATS, T. HARTİG: *Preference for Nature in Urbanized Societies: Stress, Restoration, and the Pursuit of Sustainability. Journal of Social Issues*, 63(1), pp. 79—96 (2007).
5. H. STAATS, T. HARTİG, *Alone or with a friend: a social context for psychological restoration and environmental preferences. Journal of Environmental Psychology*, 24 (2), pp. 199–211 (2004).
6. E. OSTROFF: *Universal Design: The New Paradigm, (Chapter 1) Universal Design Handbook, McGraw Hill, New York, USA*, (2001)
7. R.L. MACE, G.J. HARDIE, J.P. PLACE: *Accessible Environments: Towards Universal Design, The Center for Universal Design, North Carolina State University, Raleigh NC. USA*, ISBN:0279442 (1991)
8. A.N. EVCİL: *Herkes İçin Tasarım: Evrensel Tasarım. Boğaziçi Yayınları* ISBN:978–975-451-325-7 (2014).
9. N. DOSTOĞLU, vd.: *Herkes İçin Tasarım: Tanımlar, Hedefler, İlkeler, Mimarlık*, 347, 23–27 (2009)

10. https://s-media-cache-ak0.pinimg.com/originals/44/7a/35/447a35bb0f91c4966c7046d2baede1c8.jpg
11. https://s-media-cache-ak0.pinimg.com/564x/fb/f6/29/fbf6297843fbb727f1ab10db614db7fb.jpg
12. H. ACAR, İ. GÜVEN: *Trabzon kent merkezindeki açık yeşil alanların fiziksel engelli bireylerin kullanımı açısından değerlendirilmesi. TMMOB engelli Mühendis Mimar ve Şehir plancıları sempozyumu, Ankara* (2015).
13. http://www.wanhaber.com/belediyenin-gorme-engelliye-verdigi-degere-bakin-195970h.htm
14. A. YAVUZ: *Engellilerin yaşam kalitesi üzerine kentsel alan engelliliği ve erişilebilir-kullanılabilir tasarım yaklaşımları. TMMOB engelli Mühendis Mimar ve Şehir Plancıları Sempozyumu, Ankara.* (2015).
15. J. HANSON, 2004, *The Iclusive City: Delivering a more Accessible Urban Environment Through Inclusive Design, Conferance Paper, International Construction Conference Responding to Change, York,* (7–8 September 2004).
16. A. MAHMOUDİ, K. FANAEİ: *Revision of urban spaces to make it accessible for disabled people in order to achieve the aim of city for all, Cities for everyone. Livable, healthy, prosperous, Edited by Manfred Schrenk, Vasily V. Popovıch, Peter Zeıle real corp 2010 pProceedings/tTagungsband, Vienna,* (2010).

Suat Çabuk[1]

Conservation of Safranbolu – Bağlar Settlement in the Context of Sustainability

[1]*Department of Urban and Regional Planning, Faculty of Fine Arts, Design and Architecture, Namık Kemal University, Tekirdag, Turkey*

1. Introduction

Safranbolu is a city located on the edges of the Gümüş and Akçasu creeks, which flow into the Soğanlı River. It does this in the geographical region called Paphlagonia by Homer in his first famous epic, the Iliad, set during the 12th century BC. Today, Safranbolu is a living museum of architecture, history and culture with traditional houses, streets, bazaars, inns, baths and fountains etc. which date from the Ottoman Empire period. Due to these features, Safranbolu was included in the UNESCO World Heritage List in 1994.

The traditional city of Safranbolu consists of three regions, the *Bazaar* where the historical works, trade structures and quarters are situated, *Kıranköy* where non-Muslims lived until October 1924, and *Bağlar,* which was the region used by inhabitants as a resort during the summer months. These three regions have been taken under conservation within the boundaries of the urban site. Safranbolu is known as the "*Conservation Capital of Turkey*" because it has largely been preserved using local initiatives rather than conservation orders coming from government authorities and it has been the stage for pioneering works regarding its conservation from 1975 to the present day.

The Bağlar region of Safranbolu is one of the few resort sites that have been able to preserve their authenticity. In cities such as Ankara, Malatya and Diyarbakır, the vineyards and vineyard lifestyle have vanished. In cities such as Kayseri, Konya, Mersin and Muğla, the vineyards and vineyard lifestyle have come to the point of extinction.

Due to the development of transportation facilities post-World War II and Safranbolu becoming a "*satellite town*" of the Karabük iron and steel factories, with inhabitants starting to live throughout the year in their houses in the Bağlar region and selling their houses in the Bazaar region to people from the countryside, the way the region is used has also changed. In addition, the Has Tarla locality which lies between Bağlar and Kıranköy, and the Esentepe locality, have become the new development areas of Safranbolu, from 1945 until 2000 and from 2000 respectively.

Due to the pressure of urban development observed in the Bağlar region of Safranbolu, and to increase tourism activities in recent years, the Bağlar region has been transformed into a tourism area. In this study, the problem of conserving the Safranbolu Bağlar region is addressed in the context of sustainability.

2. Formation of Historical City of Safranbolu and the Bağlar Region

Throughout history, the Gasgas, Hittites, Phrygians, Persians, Pontus, Romans, Seljuks, Cobanogulları, Candarogulları and Ottomans respectively have dominated Safranbolu, which is situated in an area called Paphlagonia [1]. While architectural works dating from every period can be found around Safranbolu, those which date from the Ottoman Empire within the urban area are particularly eye-catching. For this reason, it can be said that the configuration of Safranbolu in the Turkish period took place during Ottoman times.

Safranbolu, like other Ottoman cities, consists of a combination of the Friday mosque, the bazaar and quarters or neighborhoods. The first important urban feature is the Old Mosque (the Gazi Suleyman Pasha Mosque, Cami-i Kebir), where Friday prayer is performed and which serves as a religious center.

The second basic feature of the city is Safranbolu Bazaar. In the past, Safranbolu served as the main center of a large area covering approximately 100 km in diameter and including the numerous villages and towns situated therein. Safranbolu was also an important stopping point on the caravan route extending from Gerede to Kastamonu and then to Sinop, both during the dominance of the Candarogulları and the Ottomans [2].

It is believed that the population of Safranbolu was less than 2000 before 1519, when it was thought that its social facilities were limited to two mosques, two bathhouses, one masjid, one caravanserai, one madrasah and a few dervish lodges, and also a bazaar which had the dimensions of a town bazaar. As the population of Safranbolu increased over time, new lines of business emerged at the bazaar.

Leatherworking was at the forefront of these new businesses. Paphlagonia was a leading livestock center from the Roman period, onward and, thanks to the Akhis, tanneries were established first of all in Kastamonu [3] and then in Safranbolu and Gerede.

The second phase in the configuration of the Safranbolu Bazaar took place in the 17th century. Karabashzade Huseyin Efendi, nicknamed Cinci Hodja, had an important role in this formation. Cinci Hodja earned fame, money and authority within a short period of time by causing the Ottoman Sultan, Ibrahim, to be healed between the years 1640–1647. During his tenure, he built the Cinci Inn and New Bath House in Safranbolu and established 50 shops between them.

In 1661, a large mosque was built in the northwest area of the Cinci Inn. A range of identical craft shops (Arasta) was also established in order to ensure revenue for the mosque [4]. The outlines of the "Upper Bazaar" part of the Ottoman Period – the Safranbolu Bazaar – thus emerged. As in other Ottoman city bazaars, every professional group in Safranbolu conducted their trade activities in their own identical craft shops, streets and bazaars.

The final stage in the formation of the Safranbolu Bazaar was realized with the mosque that Izzet Mehmet Pasha commissioned in his own name. Grand Vizier Izzet Mehmet Pasha commissioned a mosque, madrasah, library, clock tower and fountain to be built in Safranbolu.

The third element of the city was the Safranbolu neighborhoods. Detailed neighborhood information about Safranbolu can be obtained by referring to 16th century Ottoman records. It was determined that there were 392 houses in Safranbolu from the records of 1519 and 744 from 1568 [5]. The total number of registered houses in Safranbolu increased from 392 to 744 in about 50 years. This increase continued in the following centuries as well.

On the other hand, the existence of 10 neighborhoods in 1519 and 11 neighborhoods in 1568 was recorded in the city. The Gebran Neighborhood, mentioned in order no. 11, was the neighborhood where non-Muslims lived and its existence in the records of 1519 is notable. Its current name is Kıranköy, but it is named as Gebran in the records of that period.

In the light of all of these details, we can conclude that the current spatial view of the city (City-Bazaar, Kıranköy and Bağlar) was available with its outlines in the 16th century. In other words, it is known that there was a bazaar and neighborhoods on the outskirts of the city's castle, Gebran, namely the Kıranköy Neighborhood and the Bağlar region.

As for the Safranbolu Bağlar region, it was a vineyard settlement in the form of houses which were inhabited in the summer months, and where grapes were grown to meet the requirements for winter grapes, molasses and vinegar. The first document in which the expression "Bağlar" is mentioned in Safranbolu is the record of a foundation associated with the "Gazi Suleyman Pasha Foundation". In this document, it is stated that there was a locality called Cave situated near Safranbolu, the water from which was allocated to the vineyards of the borough and said mosque, madrasah and bath house [6]. Even if it is assumed that the foundation was founded by the family after the death of Suleyman Pasha, at the latest in the 1420s, it can be said that the Bağlar region was present when the Ottomans captured Safranbolu.

The Safranbolu Bağlar region emerged as a result of the desire of Safranbolu inhabitants to get closer to nature. Higher than the Bazaar region, the Bağlar region is located at an altitude of approximately 262 meters. Inhabitants of Safranbolu who wished to escape the hot and sweltering weather in the Bazaar region during the summer time benefited from the blessings of nature in the Bağlar region, which was more airy and cool, and refreshed their souls.

Other than the inhabitants of Safranbolu residing in the Akcasu Neighborhood in the early 20th century, many people owned a summer house in the Bağlar region [7]. Possession of a summer house was also considered to be an indicator of wealth. At present, there are magnificent residences in the Bağlar region of Safranbolu belonging to the 18th and 19th centuries and the first quarter of the 20th.. The Safranbolu locals moved to their summer houses in the Bağlar, which was about an hour's distance from the city, in the summer season.

The people of Safranbolu had a taste for nature due to their vineyards which were inherited from their fathers and grandfathers and were planted with vine and fruit trees. They also raised vegetables in this area where water and green areas were abundant.

The Bağlar region begins from the present-day Aslanlar area, which used to be a graveyard and extends to Değirmenbaşı in the east. While Bulak Creek forms the boundary in the west, it ended with the Ashagi Tokatli Village in the east.

Köyiçi is considered as the bazaar of the Bağlar region. Here, there are coffee shops and a few grocery and butchers' shops whose history dates back to the 18th century. Upon the construction of the Girls' Vocational High School at Köyiçi-Bağlarbaşı as an Idadi (high school) in 1919, the development of the area was initiated. According to in the State Archive of the Preme Minister's Office, the establishment of a separate municipality in Bağlar region was considered in the agenda in 1919 but was not deemed appropriate [8]. On the other hand, in October 1924, Orthodox residents of Safranbolu subject to trade arrangements were sent to Bartın to be deported. Since almost every Safranbolu Orthodox family, especially the priests, had a vineyard in Bağlar, many vineyards and gardens were left unoccupied. While some of these were granted to the Turkish families who came back through the exchange agreement, a great deal of them passed into the possession of other habitants of Safranbolu. In 1940, a total of 2851 houses were registered,1776 of which were in the Bazaar and Kıranköy regions and 1075 of which were in the Bağlar.

3. Sustainable Conservation of Safranbolu Bağlar Region

The cities in which we live in modern Turkey are usually the products of modern urban planning. Therefore, there are very few cities in Anatolia which have survived to the present day without losing traces of their past. Safranbolu comes first among the cities which have maintained traces of their past.

The first important document regarding the conservation of Safranbolu is the report prepared by the Provinces Bank which aimed to provide guidance to city planners who participated in the competition to develop a masterplan for Karabuk–Safranbolu. In the report regarding the conservation of Safranbolu, conservation of the exceptional character of the city was requested by stressing the fact that the old houses had original architecture [9].

The second important development regarding the conservation of Safranbolu was when the subject of the conservation of Safranbolu was introduced to the agenda of the academic world by thearchitect Yavuz Ince under the title of "Conservation of the Historical Environment Seminar", which was presented at the ITU Faculty of Architecture between 19–20 June 1975. The conservation of Safranbolu and its relationship with tourism was included in the agenda for the first time during these academic discussions.

The most important, concrete step in the conservation of Safranbolu was realized with an activity called "Safranbolu Architectural Values and Folklore Week" which was organized with participation from ministerial employees, consultants, artists and mainly academics from ITU's Faculty of Architecture under the leadership of the municipality, between 30 August and 5 September 1975. With thebroadcasting of a documentary film named "*Time in Safranbolu*" which was produced by Suha Arın in 1976, the historical city entered the country's agenda [10].

Finally, the Supreme Council of Real Estate, Antiquities and Monuments accepted Safranbolu as an urban site area on 8 October 1976. The Council specified the "Temporary Restitution Conditions" until the conservation plan for Safranbolu was prepared. Bazaar, Kiranköy and Bağlar were placed under conservation orders without discrimination (Figure 1).

The first "urban conservation plan" in Turkey was initiated for Safranbolu. The responsibility for the plan was entrusted to the ITU Faculty of Architecture. The conservation plan was to be prepared by İsmet Okyay. The plan was approved on 27 November 1990.

In the Okyay Plan, the minimum parcel size for Bağlar was specified as 2000 m^2 but the people of Safranbolu reacted unfavorably to this decision. Later, it was agreed that the size of parcels should be at least 1000 m^2 [7]. The reduction of the parcels in Bağlar led to a serious increase in building density and a gradual deterioration in the zonal texture.

Figure – 1 Urban Site and Interaction Areas in Safranbolu [11]

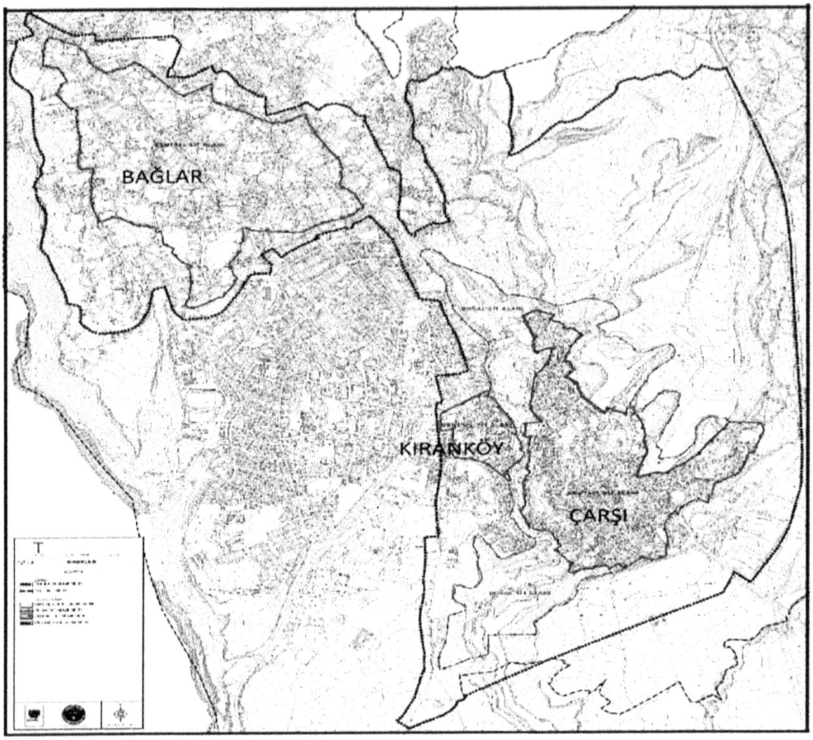

For the sustainability of the Bağlar region, it was decided that the following elements must absolutely be maintained:

- The identity, character, philosophy of the Bağlar region to be conserved as a whole
- Conservation of registered and unregistered mansions in Bağlar
- Conservation of the parcel order in the Bağlar
- Protection of the green spaces, vineyards and gardens in the Bağlar
- Conservation of street order in the Bağlar
- Conservation of the water channels (arklar) in the Bağlar
- Conservation of the walls of vineyards and gardens

- Conservation of the social structures in the Bağlar region like the historic mosque, masjid etc.
- Conservation of the names of localities and street names used in the Bağlar
- Prevention of new building restructuring within the Bağlar urban site and also new buildings in the interaction transition area, which are to be kept at set heights and dimensions in compliance with the character of Safranbolu vineyard resorts
- The social and technical infrastructure requirements to be provided in accordance with the characteristic structure of the Bağlar.

The Safranbolu Conservation (Master and Implementation) Development Plan (SCDP), which is also applicable today, was approved in 2010. The SCDP consists of three different sites, namely the urban site, archaeological site and natural site. The three regions (Bazaar, Kıranköy and Bağlar) observed throughout the Ottoman and Republic periods in Safranbolu constitute the main zoning approach in the SCDP.

Since the Bağlar region was reviewed within the scope of the study, the decisions of the SCDP concerning the region (Figure 2) were analyzed.

Figure – 2 Bağlar Urban Site Area and Interaction Area in the Safranbolu Conservation Development Plan [11].

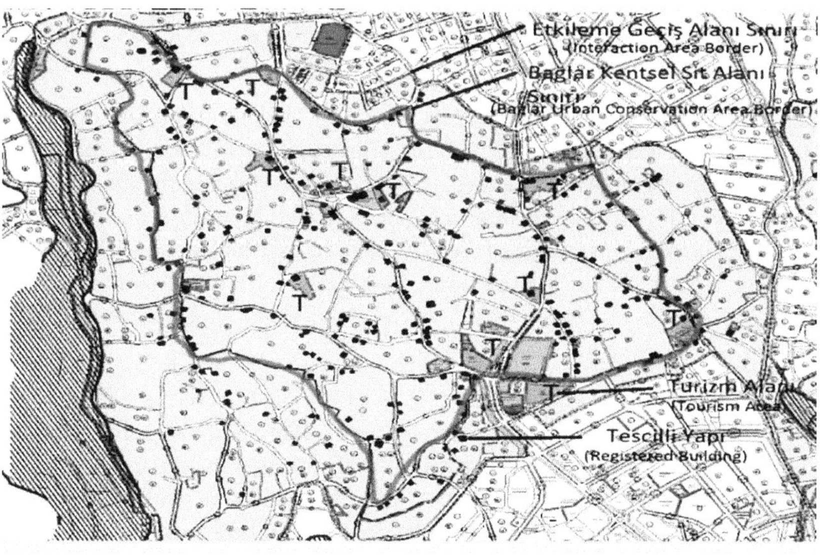

The decisions of the SCDP concerning the Bağlar region can be addressed under the following headings

MANAGEMENT: As in all urban site areas, the conservation legislation is also implemented in the site areas in the Bağlar region, staying within the scope of the SCDP and in keeping with the decisions of the conservation board. The Karabuk Cultural Heritage Conservation Board makes its decisions relating to the site through direction from the explanatory report of the SCDP, implementation regulation and applicable laws, regulations and principles. Also the municipality is an institution which provides notifying opinions on the site areas. The municipality is more effective in the interaction transition area. As emphasized in the SCDP in 2010, the "Area Management Plan of Safranbolu", which is included in the UNESCO World Heritage List, needs to be prepared. In addition, the "Water Channels Project", which was planned for the Bağlar region in the SCDP, has not yet been completed by the municipality.

BOUNDARIES: The first section which requires checking when the SCDP is scrutinized is whether or not the boundaries of the urban site area and the interaction transition area in the Bağlar region have been correctly determined. The northern and southern borders pass through the correct geographical spots, but the western and eastern borders did not pass through the wrong points. Due to mistakes in the western and eastern boundaries, some of the greenery and 59 registered buildings remain within the interaction transition area, outside the boundaries of the urban site.

PLANNED POPULATION OF BAĞLAR REGION: The total residential area in the SCDP reserved within the urban site area of the Bağlar is 86.9 hectares. The gross density of the region was calculated as 56.34 persons / ha and the total population was found to be 4896. In addition, in the calculations made according to the floor area difference of building (E = 0.20, 0.40, etc.) in the interaction transition area, it was foreseen that 5168 persons would reside within this area. As long as tourism continues to develop within this area, it does not appear likely that 5000 persons could live within the site area.

CONSERVATION OF THE GREENERY IN BAĞLAR: One of the fundamental targets specified in the SCDP is the conservation of the natural texture of parcels, but the flexibility provided in the conditions of parceling (at least 2,000 m^2 with 20 m frontal width) is in contradiction with this. Therefore, with an approach based only on the conservation of registered construction and the bringing in of parceling which conforms to this, the conservation of greenery will be limited in the Bağlar. Also, the recognition of the right to construct new buildings with a floor area up to 170 m^2 on empty parcels in the Bağlar causes damage to the greenery. More restrictive provisions are required in zoning. In addition to this, the conservation of irrigation ditches which allow the enhancement of the greenery, request projects for garden arrangement and the prohibition on using gardens as depots are positive decisions by the SCDP.

CONSERVATION OF TRADITIONAL VINEYARD RESORTS: A total of 293 registered buildings exist in the Bağlar region; 234 of these are within the urban site area and 59 are in the interaction transition area. Prohibiting new buildings within the registered parcel of the plan is favorable. Within the interaction transition area, it is permissible for a new building in parcels up to 1700 m^2 and two buildings within parcels larger than 1700 m^2 (a 300 m^2 total floor area) to be built. As with any structure within the site area, it is a positive step that the buildings to be built on the neighboring parcels of registered parcels within the interaction transition area must be subject to board approval, and that criteria for the altitude and urban silhouette are introduced.

VINEYARD AND ORCHARD WALLS: The SCDP has declared that the original vineyard and garden walls in the Bağlar area should be conserved; that permission should be obtained from the board prior to any changes; that local materials should be used and style compliance be followed when new walls are built. A positive decision in terms of conservation and development was made whereby street perspectives should also be conserved and developed.

DESIGNATION OF NEW FUNCTIONS: Despite the fact that residential use of registered buildings was taken as a basis, no limit has been set down for the number of boarding houses. In case of function changes, authorization should be granted by the conservation board. Therefore, 293 buildings registered in the Bağlar region could possibly be transformed for tourism activities in the future. As a matter of fact, transformations in the Bazaar region have clearly revealed that the authenticity of historic buildings has been lost.

TRANSPORTATION: There are no major interventions in the SCDP. However, three registered buildings for the several cul-de-sac streets opened in the Bağlar region were relocated. The most striking decision is a suggestion for a one-way traffic system and a new road with a 15 m width which is proposed within the greenery to the west of the Bağlar region. Despite the fact it would ease the flow of traffic, the integrity of the Bulak Valley and Bağlar greenery would suffer. The lack of decision related to an auto parking lot in Bağlar has resulted in road parking bays. In the interaction transition area, it was decided that a space for one vehicle per residence and two vehicles per 50 m^2 – a business place and one vehicle space per room – should be reserved from the tourism facility areas.

SOCIAL AND TECHNICAL INFRASTRUCTURE: It is observed that the current provision for social facilities in the SCDP is ongoing within the Bağlar region site area and only a few places for children's playgrounds are reserved in order not to spoil the greenery. The social and technical infrastructural areas are provided in the interaction transition area. The fact that all kinds of infrastructure activities are subject to the approval of the conservation board is also a favorable decision.

4. Conclusion

Forty-two years have passed since the commencement of the formal conservation process in Safranbolu. In this process, more than 250 residences and numerous monumental works and fountains have been restored. We can see that a significant portion of the restored residences were left for tourism purposes in the 1990s. Today, 37 accommodation facilities in the Bazaar region, 18 in the Kıranköy region and 21 in the Bağlar region continue to operate.

In 2016, the number of people who stayed in Safranbolu reached 205 144. The municipality hopes to increase this to 1 000 000. Despite the favorable effects of tourism, unfavorable effects are also being experienced in Safranbolu, such as a deterioration of the originality of historical buildings due to their being converted into hotels, a decrease in the number of residences situated in the historical fabric of the place due to their having been transformed into hotels, and the spreading of hotels to the Bağlar region which is one of the most unique places in Safranbolu. The Safranbolu Bağlar region is one of the rare places that could have preserved its authenticity.

In this context, the Bağlar region needs to be more radically preserved than is outlined in the SCDP. An approach which preserves only the vineyards, garden walls and water channels should be extended to one which also conserves the textures of parcels. Thereafter, calculations can be made in terms of its capacity for tourism. In the light of these calculations, clarity might be achieved as to the level of accommodation required.

References

[1] A. GÖKOĞLU: *Paphlagonia, Kastamonu, Sinop, Safranbolu, Bartın, Bolu, Gerede, Mudurnu, İskilip, Bafra, Alaçam ve Civarı Gayrimenkul Eski Eserleri ve Arkeolojisi*. Doğrusöz Matbaası, Kastamonu, (1952).

[2] İ. CANBULAT: City of Safranbolu. *Cultural and Natural Heritage of Turkey*. (https://gulevisafranbolu.files. wordpress.com/2013/07/safranbolu.pdf), (2012).
[3] ANONYM: *İslam Ansiklopedisi*. Cilt 24, Türkiye Diyanet Vakfı Yayınları, İstanbul, (2012).
[4] S. AKTÜRE, T. ŞENYAPILI: Safranbolu'da Mekânsal Yapının Gösterdiği Nitelikler ve 'Koruma' Önerilerinin Düşündürdükleri. *ODTÜ Mimarlık Fakültesi Dergisi*, 2 (1): 61–96, (1976).
[5]. K. Z. TAŞ: 16. Yüzyılda Safranbolu İdari Yapısı ve Vakıfları. *Tarih Araştırmaları Dergisi*, 18 (29), 1–8, (1996).
[6]. H. YAZICIOĞLU: *Küçük Osmanlı'nın Öyküsü Safranbolu Tarihi*. Şa-To Yayınları, İstanbul, (2001).
[7]. K. ULUKAVAK: *Bir Safranbolulunun Penceresinden Safranbolu (Gözlemler-Derlemeler-Denemeler)*. Bizim Büro Basımevi, Ankara, (2007).
[8] A. KUŞ: *Belediye Başkanı Gadartalıoğlu Osman Akın'ın Özel Arşivinden: Bir Zamanlar Safranbolu (1931–1946)*. Safranbolu Belediyesi Kültür Hizmeti, Safranbolu, (2009).
[9]. Y. DÖNMEZ, S. ÇABUK, M. ÖZTÜRK, E. GÖKYER: Safranbolu Kentsel Sit Alanında Otopark Sorunu ve Çözüm Alternatifleri. *Bartın Orman Fakültesi Dergisi*, 18 (2): 137–145, (2016).
[10]. Ü. ÖZDEMİR: Safranbolu'nun Kültürel Miras Kaynakları ve Korunması. *Doğu Coğrafya Dergisi*. 26, 129–143, Konya, (2011).
[11] Safranbolu Conservation Plan and Analyses, (2010).

Aysel Gürkan[1]

Biotope Mapping in Landscape Planning

[1]*Mustafa Kemal University, Faculty of Architecture, Department of Landscape Architecture, 31040 Antakya, Hatay, Turkey*

Introduction

The term "biotope" was introduced by the German scientist Dahl in 1908, as an addition to the concept of "biocenosis" earlier formulated by Möbius (1877). Initially it determined the physical-chemical conditions of existence of a biocenosis (the biotope of a biocenosis). Furthermore, both biotope and biocenosis were respectively considered as abiotic and biotic parts of an ecosystem (Olenin and Doctory, 2006). The conceptual basis is the classification of biotopes mainly based on land use and habitat type, as most habitats in urban areas have important social and economic functions, e.g. parks and school grounds (Sukopp and Weiler, 1988; Lövenhaft et al., 2002).

Biotope mapping in turn is the process of identification and specification of existing biotopes and landscape units (Sukopp and Weiler, 1988; Freeman and Buck, 2003; Löfvenhaft et al., 2002). Biotope mapping should be based on a well-structured, fundamental concept that responds to different demands, especially maintenance and development of biodiversity and nature-like experiences for urban dwellers (Werner, 1999).

Initially, biotope maps produced for the protection of native species and habitats usually focused on the protection of endangered plants and animals. However, nowadays biotope maps have wider and more comprehensive uses, such as nature conservation management in the urban ecosystem (Byun et al. 2002; Mun et al. 2005; Hong et al, 2005). They are also implemented in landscape plans (Bastian, 1996). Landscape planning is defined as planning for the conservation and development of the landscape as a spatial pattern of ecosystems with the aim of maintaining a persistent functioning of natural balance and quality for the experience of the landscape by people (Kiemsted and Gustedt, 1991). It is the practice of planning for the sustainable use of physical, biological, and cultural resources. It seeks to protect unique, scarce, and rare resources, avoid hazards, protect limited resources for controlled use, and accommodate development-appropriate locations (Fabos, 1985) Landscape planning is most fundamentally linked with the latter, the spatial dimension, and predominantly at the scale of the landscape. The planning explicitly addresses multiple abiotic-biotic-cultural goals and resources. It was conceived to be transdisciplinary, as it includes knowledge from science, planning and stakeholders, and citizens. It is based largely on landscape-ecological theory and concepts, as understood and applied through spatial assessments and spatial concepts (Ahern, 1999).

Biotope Mapping Approaches

Two major approaches, the selective and the comprehensive, have been used for biotope mapping in recent years (Qio et al., 2010). Selective biotope mapping is a 'quick method for assessing the importance of open spaces in towns for nature conservation', developed by Wittig and Schreiber (1983). Biotopes worthy or potentially worthy of protection are characterized (Hong et al., 2005), which means that only green spaces are mapped and built areas are not taken into account. These valuable biotopes should have at least one of the following characteristics: (1) a richness in species with at least some rare species, (2) high structural diversity, and (3) refuges for animal and plant species that have been displaced by intensive land use (Jarvis and Young, 2005). In selective biotope mapping, green spaces are evaluated according to several criteria, such as the age of each biotope, size of the site, shortest distance to a similar habitat of type as a measure of habitat rarity, and biotope structure (Starfinger and Sukopp, 1994).

Comprehensive biotope mapping means that all biotopes in an area are investigated, including open green spaces as well as built-up areas (Wachter, 1999). There are some similarities between these methods and each method has some advantages and disadvantages compared with others. The advantage of selective mapping carried out to detect biotopes worth protecting or potentially worth protecting is that it can be carried out in a very short time and with the lowest expenditure. Its disadvantage is the fact that only the areas which have special value are taken into consideration for the protection of species in residential areas by selective mapping, and some biotopes which cover large areas and which can be important elements of the biotope systems network may be overlooked. Other methods developed in parallel with selective mapping aim at the mapping of all biotopes in the study areas. In comprehensive mapping, the areas contained by all land use types are examined separately (Yılmaz et al., 2013).

Mapping Process and Evaluation System

There are different methods for biotope mapping. Recent studies focus on a method for systematic biotope mapping based on the analysis of surveys, categorization, and evaluation criteria (Hong et al., 2005).

Survey method

Surveying is the first stage of biotope mapping. The collection of basic materials like city master plans, topographic maps, satellite images, vegetation maps, etc., generate this stage. After that all data should be analyzed for biotope mapping.

Categorization

A standardized and uniform biotope category which reflects specific regional characteristics needs to be developed for the purpose of landscape planning and development. This can be done by examining and analyzing existing categories and characterizing non-biological elements and biological features as well (Hong et al., 2005). Gürkan, (2016) identified biotope types under three levels (Table 1). The first level is very general and includes just three landscape types: (1) "Built up", (2) "Non-built-up", and (3) "Agricultural area". The three main categories were divided into 7 main categories under the second level; "Urban", "Industry", "Forest", Wetland", "Shrubland", "Bare soil" and "Agriculture". The third level divided 19 sub-biotopes; "Cemetery", "Open pit-mines", "Educational institutions", "Recreational institutions", "Parks", "Settlement (garden houses)", "Settlement (multi-story)", "Settlement (site)", "Settlement (with low density)", "Other public institutions", "Commercial institutions", "Forest", "Maquis", "River bank", "River and banks", "Garrigue", "Open area", "Agriculture (annual)" and "Horticulture". Most biotope classifications are primarily divided into land use types and habitat types (Bastian, 1999).

Table 1. Biotope types identified under 3 level in the research area.

Level 1	Level 2	Level 3
Built-up	Urban	Cemetery
		Open pit-mines
		Educational institutions
		Recreational institutions
		Parks
		Settlement (garden houses)
		Settlement (multi-story)
		Settlement (site)
		Settlement (low density)
		Other Public Inst
	Industry	Commercial institutions

Level 1	Level 2	Level 3
Non Built-up	Forest	Forest
		Maquis
	Wetland	River bank
		River and Banks
	Shrubland	Garrigue
	Bare	Open areas
Agriculture	Agriculture	Agriculture (annual)
		Horticulture

Biotope types are mapped acording to Level 2 in the study; agriculture, urban, forest, shrubland, bare soil, wetland and industry (Figure 1). The study area is situated in the city of Antakya, in the Mediterranean region of Turkey. This meant that most of the study area included urban areas. After that, the classification was supported by the field survey, which was a very important step.

Figure 1. Map of biotope types acording to Level 2 (Agriculture, urban area, forest, shrubland, bare soil, wetland, industry) (Gürkan, 2016).

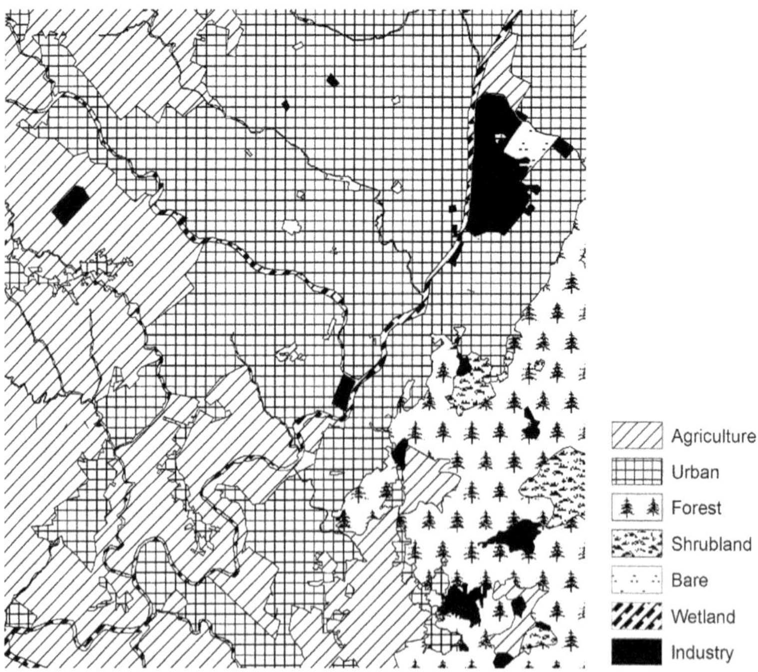

Evaluation

Biotope mapping itself is not sufficient for landscape planning. It is necessary to assess the importance of biotopes for the aims of nature conservation and landscape planning (Bastian, 1996). So there are many criteria for different studies.

For example, Altan (1997), Guzelmansur and Altan (2001), and Mansuroğlu et al. (2006), used two parameters to assess biotope sensitivity to human disturbance: 1) the rarity of biotopes and 2) the re-establishment ability (Table 2).

Table 2. Sensitivity matrix (Altan, 1997; Güzelmansur and Altan 2001; Mansuroğlu et al, 2006).

Rarity	Re-establishment			
	Impossible (RI)	**Very difficult (VRD)**	**Partly possible (RPP)**	**Possible (RP)**
Very rare (VR)	I	I	II	II
Rare (R)	I	I	II	II
Partly present (PP)	I	II	II	III
Sufficiently present (SP)	II	II	III	IV

I. Very sensitive, II. Sensitive, III. Less sensitive, IV. Not sensitive

On the other hand Bastian (1996) used seven parameters to assess biotopes: 1) degree of naturalness of vegetation; 2) regenerative capacity, age, duration of development; 3) diversity; 4) spatial criteria (biotope size, isolation, connectivity); 5) representancy; 6) rarity, endangering; 7) complex biotope values.

Degree of Naturalness of Vegetation: This used human-influenced vegetation types. These were divided into: I) completely natural, II) natural, III) medium value of changes, IV) little value in the changes, V) complete changes.

Regenerative capacity, age, duration of development: Biotope types can be differentiated with regard to their age/regenerative capacity. I) These biotopes need about 5 years for re-establishment II) These biotopes need about 5–25 years for re-establishment, III) These biotopes need about 25–50 years for re-establishment, IV) These biotopes need about 50–200 years for re-establishment, V) These biotopes need more than about 200 years for re-establishment (Bastian, 1996).

Diversity: This concerns the richness of landscape biotopes, landscape elements or land use types (Bastian, 1996).

Spatial criteria: The larger the biotope, the more species can settle there (Bastian, 1996).

Representancy: I) Extremly high, II) Very high, III) High, IV) Low, V) Very low

Rarity/endangering: Some species and biotopes occur seldom from natural causes but they are not endangered automatically.This concerns (Altan, 1997): 1) the rarity of biotopes I) Very rare: Those biotopes standing alone or distributed in a very restricted area; II) Rare: Those biotopes limited in numbers or distributed in a restricted area; III) Present: these biotopes seen sufficiently; IV) Partly present: Those biotopes seen at intervals in large areas sufficiently; V) Sufficiently present: Those biotopes regularly seen in large areas.

Complex biotope values: Complex biotope values are very relaxed for planning purposes because they are easier to handle for the authorities.

Five of these criteria were used in the study for evaluation biotopes; naturalness, regeneration, diversity, representancy, and rarity (Table 3).

Table 3. Evaluation of biotope types (modified from Bastian, 1996).

	Evaluation of Biotope Types				
Criteria	I –Very High Value	II –High	III– Medium	IV– Little	V–Lowest Value
Naturalness	Completely natural	Natural	Medium value of changes	Little value of changes	Completely changes
Regeneration	Very difficult	Difficult	Partly possible	Possible	Possible
Diversity	Very high	High	Middle	Little	No diversity
Representancy	Extrimly high	Very high	High	Low	Very low
Rarity	Very rare	Rare	Present	Partly present	Sufficiently present

Biotope types are evaluated acording to Table 3. As a result of the evaluation of biotopes, built-up areas have a low value while natural areas like forest and wetland have the highest value (Figure 2).

Biotope Mapping in Landscape Planning

Biotope mapping has been adopted by an increasing number of countries in order to develop a view on sustainable development in which biodiversity is an important component (Lee et al., 2005). For example, much of the work of the habitat survey in the UK has been repeated with small-size thresholds and no quality cut-off point, thus increasing the coverage of the investigation (Handley et al., 2003). In accordance with the Swedish Planning and Building Act Amendments from January 1996, the maintenance of biodiversity in urban

Figure 2. Evaluation of biotope types (I: Very high, II: High, III. Medium, IV. Little, V: Lowest value (The value 5 is not exist in the study area).

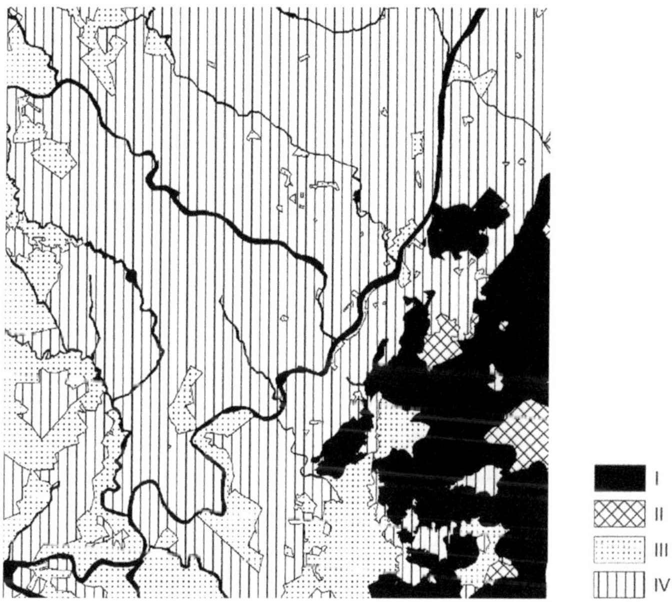

areas is one of the Swedish national environmental objectives (Regeringens Proposition (Government Bill), 1997/1998). Stockholm City Planning Administration developed a biotope mapping model for the spatial planning of biodiversity issues in Stockholm. The model designated the core areas, connectivity zones, buffer zones and green development areas according to the identification of different values of target biotopes in green and built-up areas. The relative strategies adapted to priorities in the spatial planning of biodiversity were presented (Lofvenhaft et al., 2002). An investigation among the administrations of 120 towns in Germany conducted by Werner (2002) showed that urban biotope mapping was used in very diverse ways, primarily for: (1) habitat protection; (2) green space planning; (3) landscape planning and (4) measures for species protection. All applications were primarily focused on reinforcing urban biodiversity as a component of human daily life (Qio et al., 2010).

Conclusions

Biotope mapping was also conducted to provide an ecological basis for sustainable development. Thus, the aim of the biotope maps can be specified to allow lands to be used within a reflection of their environmental states (Letio and Ahern, 2002). The study comprised three stages: introducing types of biotope mapping, the mapping process and evaluation, and the discussion of biotope mapping in landscape planning. The survey method was the main stage of the mapping, so all data should have been collected at this stage. Categorization methodology differed for each piece of research, which divided under Level 3 in the study. Biotopes evaluation methodology, developed by Bastian (1996), was used in the contribution. The following criteria were used: degree of naturalness of vegetation; regeneration capacity; diversity; representancy; and rarity. After that, biotope mapping and landscape planning was discussed in the final stage. Results of the evaluation showed very useful information about biotopes. The highest values of biotopes had very natural areas which were very important for landscape planning, so this kind of biotope should be protected under sustainable landscape planning. On the other hand, the lowest value for biotopes should be recommended for urban development areas in urban ecosystems. Biotope mapping and assessment play an important role in nature conservation and landscape planning (Bastian, 1996). Thus biotope mapping provides the following benefits for landscape planning: 1) It helps to get basic information for landscape planning which is important in defining priorities for nature conservation; 2) It helps to develop programs for endangered species; 3) It helps to get some recommendations for nature management; 4) It helps to get basic information for green areas; 5) It helps to provide an ecological basis for urban planning activities. Therefore, biotope mapping is very necessary for sustainable development and for sustainability. The results presented in the paper show that the mapping and evaluation of biotopes is a very important tool for landscape planning. It can be used for the management of sustainable land uses. Just one disadvantage of the evaluation is that it includes very different criteria, and it takes time to collect and evaluate these.

References

Altan, T. (1997) Participation of ecological planning in urban area and importance of biotope mapping- *II. Coastal problems and environmental problems*, Kusadasi, Turkey, pp. 1–16.

Ahern, J. (1999) Spatial concepts, planning strategies, and future scenarios: a framework method for integrating landscape ecology and landscape planning. In *Landscape Ecological Analysis* (pp. 175–201). Springer New York.

Bastian, O. (1996). Biotope mapping and evaluation as a base of nature conservation and landscape planning. *Ekologia(Bratislava)/Ecology(Bratislava), 15*(1), 5–17.

Byun, B.S., Song, I.J., Hong, S.K., Yoo, S. (2002) Biotope mapping for ecological management and urban planning in Korea. In: INTECOL (ed) Proceedings of the 8th INTECOL international congress of ecology. INTECOL, Seoul

Fabos, J.G. (1985) Land-Use Planning: From Global to Local Challenge. Chapman & Hall, New York.

Freeman, C., Buck, O., (2003) Development of an ecological mapping methodology for urban areas in New Zealand. Landscape and Urban Planning, 63(3): 161–173.

Guzelmansur, A. and Altan, T., (2000) Analysis of coast land use and alternative land use proposals for Erzin Sarımazı-Burnaz Beaches and some holiday houses in coastal zone of Erzin. – Çukurova University, *Journal of Agricultural Faculty* 2001, 16 (3): 85–94.

Gürkan, A. (2016) Biotope mapping in an urban environment for sustainable urban development–a case study in the southern part of Turkey. *Applied ecology and environmental research, 14*(4), 493–504.

Handley, J., Pauleit, S., Slinn, P., Barber, A., Baker, M., Jones, C., and Lindley, S. (2003) Accessible natural green space standards in towns and cities: A review and toolkit for their implementation. *English nature research reports, 526.*

Hong, S. K., Song, I. J., Byun, B., Yoo, S., and Nakagoshi, N. (2005) Applications of biotope mapping for spatial environmental planning and policy: case studies in urban ecosystems in Korea. *Landscape and Ecological Engineering, 1*(2), 101–112

Jarvis, P. J., and Young, C. H. (2005) The mapping of urban habitat and its evaluation. *Wolverhampton: University of Wolverhampton.*

Kiemstedt, H., Gustedt, E., (1991) Landschaftsplanung in der Bundesrepublik Deutchland. Inter. Conference: Landschaftsplanung als Instrument umfassender Umweltvorsorge, Hannover.

Lee, K.J., Han, B.H., Hong, S.H., Choi, J.W., (2005) A study on the characteristics of urban ecosystems and plans for the environment and ecosystem in Gangnam- gu, Seoul, Korea. Landscape and Ecological Engineering 1, 207–219.

Leitao, A. B., and Ahern, J. (2002) Applying landscape ecological concepts and metrics in sustainable landscape planning. *Landscape and urban planning, 59*(2), 65–93.

Lövenhaft, K., Bjorn, C., Ihse, M. (2002) Biotope patterns in urban areas: a conceptual model integrating biodiversity issues in spatial planning. – Landscape and Urban Planning 58 (2–4): 223–240.

Mansuroglu, S., Ortacesme, V., Karaguzel, O. (2006) Biotope mapping in an urban environment and its implications for urban management in Turkey. – Journal of Environmental Management 81:175–187.

Mun, S.K., Seong, H.C., Ku, B.H., Byun, B.S., Yu, H.S., Lee, D.K., Lee, S.M., Lee, E.Y., Lee, E.H., Lee, J.J., Jeon, S.W., Jeon, Y.O. (2005) Environmental planning (in Korean). Bomoondang, Seoul

Olenin, S., and Ducrotoy, J. P. (2006). The concept of biotope in marine ecology and coastal management. *Marine Pollution Bulletin*, *53*(1), 20–29.

Qiu, L., Gao, T., Gunnarsson, A., Hammer, M., and von Bothmer, R. (2010). A methodological study of biotope mapping in nature conservation. *Urban forestry & urban greening*, *9*(2), 161–166.

Sukopp, H., Weiler, S. (1988) Biotope mapping and nature conservation strategies in urban areas of the Federal Republic of Germany. – Landscape and Urban Planning 15 (1–2): 39–58.

Starfinger, U., and Sukopp, H. (1994). Assessment of urban biotopes for nature conservation. *Landscape planning and ecological networks. Elsevier, Amsterdam*, 89–115.

Wachter, M. (1999). Comprehensive biotope mapping in Germany: The example of Leipzig. *Deinsea*, *5*, 67–76.

Werner, P. (1999). Why biotope mapping in populated areas. *DEINSEA-Annual of the Natural History Museum Rotterdam*, *5*, 9–26.

Werner, P. (2002). Urban biotope mapping—The use of data and maps for urban management and planning—Experiences from Germany. *Institute for Housing and Environment, Darmstadt.*

Wittig, R., Schreiber, K.F. (1983) A quick method for assessing the importance of open spaces in towns for urban nature conservation Biological Conservation, 26, pp. 54–64

Yılmaz, B., Gülez, S., Kaya, L.G.(2013) Mapping of biotopes in urban areas: A case study of the city of Bartın and its environs, Turkey. *Scientific Research and Essays*, *5*(4), 352–365.

A. Esra Cengiz, Çiğdem Kaptan Ayhan,
Aybike Ayfer Karadağ, Demet Demiroğlu

Rural Tourism and Traditional Turkish Villages in the Development Process

1. Introduction

Today tourism is both of crucial importance to developing countries as a sector and described as an essential economic source for many countries. This is due to the requirement for no additional cost to the triangle of nature, culture, and history – the most important raw materials of the sector – and the extent of the added value which is ultimately created. With the negative impacts of tourism activities on natural and cultural resources, and with understanding that these impacts endangered the respective future of tourism, the concept of sustainability has also been reflected in tourism, and the issue of 'sustainable tourism' has emerged. Today, all advanced countries of the world are considering the types of tourism which are more respectful of environmental resources and which aim to ensure the sustainability of the environment while using it, and plan their tourism in this vein. For this purpose, types of tourism such as alternative tourism, ecotourism, nature tourism, conscious tourism, responsible tourism, and green tourism have been defined in all developed and developing countries. In this way, the dimension of environmentalism has been introduced to tourism, at least in discourse. Alternative tourism aims to create an environment of good quality by preserving the stocks of natural resources, and gaining economic benefits to this end by analyzing the tourism-related activities of the local people. There is therefore a close relationship between the fundamentals of sustainable development and the concept of alternative tourism. As the understanding of local development has become more prominent, rural tourism, which is included in the alternative tourism sector, has begun to become more important to all those countries undertaking development, and action has begun to be taken to make this development sustainable. Ultimately, rural tourism has qualities whereby it can both serve the understanding of local development and play an effective role in making development sustainable [1; 2; 3; 4; 5]. Some tourism planning which is to be implemented by adhering to the principles of sustainability will both reduce the negative impacts of tourism on the environment and bring about economic, social, and cultural contributions [6].

The beginnings of rural tourism may also be ascribed to the diversification of alternative tourism, the increase of the importance attached to cultural events, and the negative outcomes of uncontrolled development in cities [7]. Furthermore, attention shifted to rural tourism so as to create new economies relating to the decrease in the agricultural and forestry sectors in rural areas. Rural tourism is a valuable source of alternative tourism which supports the protection of the rural environment and the continuity of small-sized enterprises, which offers a source of income and job opportunities, and which combines or allows the responsibilities of tourism and housework for women. In addition, it may also be evaluated as a new source of income for rural settlements [8].

With their landscape characteristics, local lifestyles, and originality, rural areas are attractive points for tourism activities. Rural tourism is a type of tourism which is both intertwined with rural areas and based on natural resources. On the one hand the concentration of tourism particularly in rural areas in recent years may pose the danger of extinction of local cultures and originality, but on the other, rural tourism activities have significant and positive impacts on the conservation and development of local cultures. Thanks to its positive impacts, its importance is being understood better with every passing day as an alternative to, or complement to, the existing types of tourism in countries which are developed or developing in this trade [9; 10; 11].

Even though there are considerably varied definitions of rural tourism, it is predominantly defined as a type of tourism which is performed in relation to farms and nature and in which tourists participate in activities such as touring farms, fishing, riding a horse, and "trekking" (a long and challenging walk), whereas in some definitions it is evaluated as being all tourism activities in rural areas [12]. In fact, rural tourism generally involves many types of tourism. It is carried out in natural areas, rural settlements with low population density, and areas where tourism is associated with agriculture, forestry, and traditional occupations [8].

Rural tourism has a quality which also overlaps the policies of the tourism that have taken shape in our country in recent years regarding its spread over geographical space and time [13]. Turkey has many destinations where rural tourism can be carried out, and rural tourism may provide great support to the tourism of our country given its innumerable benefits [9; 11]. As also stated by Soykan [9], the use of rural settlements for tourism has become widespread in our country in recent years. Nevertheless, we can observe that few villages have been opened to rural tourism in the true sense to date, and there is an awareness of great potential. This study firstly aims to reveal the scope of rural tourism as a concept which embraces all of village tourism, farm tourism, green tourism, ecotourism, and agritourism, together with the related concepts, and then looks at how it originated and developed abroad and in Turkey. In this context, the relationship of rural tourism with the concept of rural development, its impact on the development process of villages and the utilization of traditional Turkish villages within the scope of rural tourism are the other subjects which are addressed.

2. The Concept of the Rural Area

Before defining the concept of rural tourism, it is useful to briefly explain the concept of a 'rural area'. The concepts of rural areas and agricultural areas are mostly interchangeable concepts. However, rural areas are multifunctional and dynamic systems. They encompass different land uses and activities such as settlement, transportation, industry, forest, tourism, and recreation. Moreover, one of the most evident impacts on the restructuring process of the economy in rural areas is the requirement for creating job opportunities which are alternatives to agriculture [8]. Rural areas are defined differently by various researchers who consider different criteria such as population status, geographical location, and administrative form. According to the definition formed by taking the administrative form into consideration, the spatial terrain which consists consisting of the districts and villages outside the metropolitan settlements is called the 'rural area', and the subdistricts and villages outside the municipal boundaries of provincial/district centers are also acknowledged as the rural area [14]. According to Soykan [15], rural areas are *"generally settlement areas with a population of fewer than 10 thousand people."* However, this definition does not form a structure which applies to all countries; furthermore, it is a slightly relative definition. Therefore, rural areas should be considered *"the places where agricultural activities are generally carried out and a traditional life is sustained"*. On the other hand, the OECD[1] [16] defines rural areas as *"those places where the population density per km² is 150 at the maximum"*.

In the Report by the Specialization Commission for Rural Industry in the Seventh Five-Year Development Plan by the State Planning Organization of the Prime Ministry, a 'rural area' is defined as *"the environments where the impacts of economic quality are based heavily on the utilization of natural resources; where face-to-face relationships are relatively more common; where the rules of living are shaped substantially according to traditions and customs; and where technical and technological developments as well as economic, social, and cultural developments*

1 OECD: Organisation for Economic Co-Operation and Development.

take place more slowly" [17]. Pezikoğlu [18] stated that the overwhelming majority of the natural resources were in rural areas.

Rural areas mainly have four important functions [17], namely (a) they are places of food production; (b) they lay the groundwork for village handicrafts and industry; (c) they are the settlement areas of people working in agriculture and other economic branches; and (d) they are resting-places for people in cities and industrial settlements.

Different classifications are available both abroad and in this country as regards the classification of rural areas. Besides, a wide variety of criteria is used to classify rural settlements. This is primarily because rural settlements have a wide range of different features. Rural settlements may be places of production for some sections but places of residence and consumption for some others [19].

According to Village Law No. 442, which entered into force on March 19 1924, the settlement areas with a population of 2000 to 20000 people are called towns, while the settlement areas with a population of more than 20000 people are called cities. In this case, given the criterion of population, the settlement areas with a population of fewer than 20000 people are at the same time within the scope of the 'rural area' (URL-1).

Turkey is a country which comprises seven geographical regions with vast territories and different characteristics. Those rural settlements in which the natural geography shapes cultural landscapes and their functions by playing a significant role are very sparsely populated in some places but have seasonally used (temporary) qualities in some others. The forms of settlement offer innumerable images ranging from *scattered* to *collective*. The rural settlement units comprise village, neighborhood, *oba, kom, mezra,* and *divan*. Nevertheless, the basic rural settlement unit of Turkey is the 'village' [20; 17]. Even though what a village means was defined with Village Law No. 442, it is true that this definition is inadequate. According to the law, "*the spatial units whose population is below 2000 people and which consist of those people who are engaged in agriculture and live in collective or scattered houses are defined as villages*"(URL-1). The qualification of a village is a concept with two different meanings. One of them is the 'administrative' meaning of a village, whereas the other one is its 'settlement' meaning. As a settlement term, villages are defined as "*the settlements whose economic activities are mostly agriculture, animal husbandry, and handicrafts; where the division of labor, the community spirit and a traditionalistic understanding of life dominate; which have some unique social organization in which traditional relationships are common in the family; where the process of social change takes place relatively slowly; which comprise the social and physical environments that go on existing in a rural space; and which differ from cities in all these aspects of theirs*" [21].

Human beings adopted a sedentary life on earth upon the commencement of agricultural activities. However, people behaved selectively when choosing the places of establishment of settlements and preferred those areas in which they could live easily. People established the early settlements in those areas where water sources and fertile agricultural lands were available and which were easily accessible in moderately temperate climatic fields. Forming the basis for the early villages, these settlements underwent change over time and made up today's village settlement forms [22]. Forming the basis for rural settlements, villages are generally the settlements in which basic production (e.g. agriculture and animal husbandry) is carried out in Turkey. Throughout history, villages have displayed a dynamic structure, sometimes developed and grown, sometimes disappeared, and sometimes changed places. Resulting from the impacts of natural, historical, and social factors, this change did not cause villages to lose their significance in any period [21].

The villages in our country have significant potential for, and a privileged place in, rural tourism. When considered in terms of the development of rural tourism, villages are extremely convenient for being utilized due to both their settlement characteristics and sociocultural structures.

3. Rural Tourism

3.1. Its Origination and Development Abroad

Rural tourism is a type of tourism which develops depending on cultural tourism and which goes on existing as a sub-branch of culture and, in this context, of cultural tourism [23]. Cultural tourism is one of the alternative tourism types which emerged as an alternative to mass tourism. It aims to inculcate one with the consciousness of protecting/conserving natural and cultural resources, and deal with these assets within the scope of sustainable tourism and rural tourism, which is integrated within it.[13]. The first coinage of the concept of rural tourism dates back to the Romanticism movement, which commenced in the late 18th century and developed in the 19th century. This movement originated as a reaction to industrialized cities and to the Industrial Revolution. Rural tourism flourished with the construction of railway links to rural areas, together with the themes of natural landscapes, popular traditions, and nature in the 19th century.

The concept of holidays based in rural areas started in this same century. However, this understanding differs from the understanding of rural tourism accepted today, and encompasses an approach which emerged upon the development of the railway network and which was based on the hotels and resorts in rural areas. On the other hand, modern rural tourism as we understand it today began in the post-World War II era [24]. Rural tourism after the 1970s differed slightly, and there was a profound increase in the number of tourists who participated in rural tourism. Today 70% of the American population participates in rural tourism [12].

Different approaches and definitions are available for the concept of rural tourism, which has changed and evolved since its beginning [16; 25; 26; 27; 9; 28; 29; 15; 10; 30; 1; 31; 23; 2; 3; 12; 13; 18; 11; 32; 8; 7; 4; 33; 34; 5; 35; 36]. Furthermore, no consensus on the definition of rural tourism has been reached yet in the relevant literature. This is because there has not yet been any common accepted definition on the concept of 'rural area'. The definition of rural area is different in every country [26].

The tourism sector has also been the scene of some changes due to the phenomenon of globalization, as with many sectors worldwide. The new trends in tourism in the world are changing and evolving in terms of types of tourism such as historical, health, nature, cultural, and rural tourism, depending on demand. The countries or regions with potential for such different types of tourism can realize their rural development through the new tourism policies they develop, depending on the demand for tourism [30].

Rural development encompasses the multidirectional processes, activities, and organizations which aim to develop the production, income, and welfare levels to change the structure of the human communities that reside in rural areas in social, economic, and cultural terms; to eliminate the imbalances in human-soil relationships; to create in rural areas the physical and social infrastructures which exist in urban areas; and to make better use of the agricultural products [37]. Rural development and the relationship of rural development to rural tourism will be elaborated upon later.

However, although there are many different definitions of rural tourism, there is a consensus about the fact that this type of tourism has a significant place in terms of rural development. As also stated by Sharpley and Telfer [33], many countries also include rural tourism in their regional and national development strategies in order to create economic diversity. According to Ryan et al. (2009), it is thought that the development of tourism, which is supported so as to increase the incomes and create inter-societal harmony in underdeveloped rural regions, is an important driving force for the elimination of poverty as well as for sustainable development and environmental protection [38]. To speak of a novel concept of rural development, agriculture in rural areas will be replaced by the activities which are important to create rural sources of income, alongside potential rural and non-rural activities. Rural tourism is one of them [25].

The OECD [16] very basically defines rural tourism as *"all tourism activities carried out in the rural area"*. According to Lane (1994), however, rural tourism is a type of tourism which takes place in the rural area; which is functionally rural and small-sized; which includes tailor-made products and services; which has substantially traditional qualities; and whose growth should be sustainable, gradual, and controlled [34]. According to Fennel [31], rural tourism is equivalent to farm tourism and the majority of the activities carried out by tourists are related to farms and their close vicinities. Both these are approaches which confine rural tourism only to farm tourism, as in this definition, and definitions which evaluate it within a broader framework are also available. For instance, Nair et al. [35] define rural tourism as a multidirectional type of tourism which also encompasses the experiencing of activities concerning farms and nature as well as the cultural features and traditional life habits of various rural localities (Figure 1).

Figure 1. Examples of rural tourism practices abroad-I (URL-2; URL-3)

Rural tourism takes place as tourists go to, and stay in, rural settlements and watch or attend the activities unique to that rural settlement in order to spend their holidays in natural environments and be together with different cultures [12]. Besides, it is a type of tourism which is able to attract people to rural areas for holidays in many countries of the world; which integrates them with the rural culture, the natural environment, and agriculture; and which is also able to integrate them easily with other types of tourism [39] (Figure 2).

Figure 2. Examples of rural tourism practices abroad-II (URL-4; URL-5)

The development of rural tourism is encouraged in the places expressed as rural areas in agreement with the economic, socio-political, cultural, historical, geographical, and human structures of the locality concerned. Even though rural tourism is intertwined with concepts such as farm tourism, green tourism, and tableland tourism, it contributes to the rural area, the rural activity, rural heritage, and rural life (Figure 3). It is a must that rural tourism is performed in rural areas. Nevertheless, there are different approaches regarding whether every type of tourism performed in rural areas is rural tourism [32; 4; 12; 40]. At a difference to all the above-mentioned forms of tourism, rural tourism is not a movement which merely has economic impacts. Along with its economic dimension, it is an activity which has social, cultural, and environmental aspects and which influences the social structure [11].

Figure 3. The concept of rural tourism/components of rural tourism [28; 29]

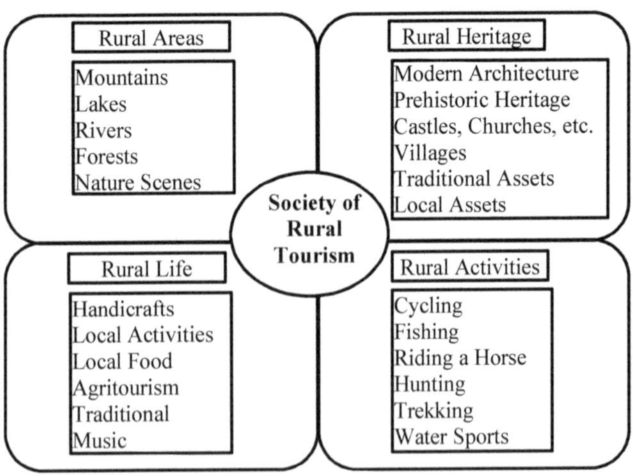

It is possible to summarize the general characteristics of rural tourism as follows [9]: (a) Rural tourism is a type of tourism which can be performed every season; (b) Rural tourism is an element of geographical balance in the distribution of tourism; (c) Rural tourism can integrate with many types of tourism; (c) The activities carried out in rural tourism are very diverse and original; (d) The tourist profile for rural tourism is different; (e) Rural tourism contributes to the protection/conservation of the natural environment and cultural heritage; (f) Rural tourism serves the understanding of sustainable tourism; and (g) Rural tourism is an essential instrument for the promotion of a country.Thanks to all these characteristics, rural tourism is the type of tourism which is the most harmonious with the environment and culture and which has the fewest negative effects. The provision of accommodation in rural areas, mostly the available houses of rural settlements, does not lead to such environmental problems as concretion and land consumption. Rural tourism does not cause any seasonal concentration due to its feature of being carried out in all four seasons of the year, and eliminates the pressure of tourism [9]. Rural tourism also has significant power as it revives the potential of the people in a rural region for entrepreneurship, activates local initiatives and the economy, supports rural development at the national and regional levels, as well as the employment of local women [36]. Local people and their support of and participation in local governments provide adequate financial sources for the development of tourism. On top of this, technical assistance and knowledge for the development and promotion of tourism and good strategic planning are necessary for the successful development of tourism in rural areas [27].

3.2. Rural Tourism in Turkey

When the issue is examined in terms of Turkey, it is a fact that 35% of the population lives in rural areas and that agricultural activities are the most important sources of livelihood in rural areas, as also stated by Gülçubuk (2007). In the sociocultural and economic development processes of Turkey in recent years, the differences in development between cities and rural areas have become clear as a result of the endeavors of modernization, industrialization, and socio-economic transformations. Owing to their socio-economic structures, rural areas have unfortunately failed to catch up with the rate of development displayed by cities. Even though there have been some developments in Turkey regarding the development of rural areas and the enhancement of the quality of life of inhabitants of rural areas by putting into practice policies to minimize the differences in development nationwide, these developments have not yet been able to reach the desired level.

In fact, the most important way of eliminating this difference in development is rural tourism – one of the leading sectors of rural development [30].

Turkey not only has quite a strong potential for alternative tourism but also, with both its natural beauty and possession of a rich extant culture, is an ideal example of those countries in which alternative tourism can be developed [41]. Today Turkey is one of the countries where rural tourism, which has a significant place among the alternative tourism types, can be performed in a substantial way. Now that the meaning 'rural' means that the concept of rural tourism is perceived directly as being in "rural areas", in our country tableland tourism, hunting tourism, cave tourism, ecotourism and outdoor nature sports are considered to be the same as rural tourism (Figure 4). Although this is not incorrect, it is a fact that between them there are some differences in goals and activities, since the actual purpose of rural tourism is to spend a holiday by staying in a village, on a farm, in a chalet, etc. and by meeting/interacting with the rural culture [9].

Figure 4. Activities from nature tourism and ecotourism carried out within the scope of rural tourism in Turkey [42; 43]

In Turkey, rural tourism is developed under the headings of tableland tourism, ecotourism, and adventure tourism by the public and private sectors [12]. Nevertheless, it is known that the early formations of rural tourism in Turkey started with the concept of 'tableland'. Within the framework of 'the Project of Tableland Tourism' [44], launched in 1990 by the Ministry of Culture and Tourism in parallel with the change in the worldwide tourism demand, tourism activities were encouraged on a total of 26 tablelands in Antalya, Gaziantep, Sinop, Ordu, Giresun, Trabzon, Rize, and Artvin provinces. These areas were determined with the name 'the Center for Tableland Tourism' by the Ministry according to the following criteria (The Ministry of Culture and Tourism, 2015): (a) Having rich potential for natural assets; (b) Having a civil architectural character; (c) Having easy access to transportation to the city center; (d) The social structure of the local people that is close to tourism; (e) Sustainability of an activity in economic terms; and (f) The possibility of ensuring its integration with the other types of tourism.

The studies on this issue are still ongoing nationwide. In tableland tourism, the Ministry aims to utilize the tablelands with the unique natural and cultural assets of our country and incorporate the tableland life within tourism activities by taking the conservation-use balance into consideration [12].With the planning period, rural development projects were implemented to reduce poverty and eliminate the regional imbalance in various provinces in Turkey in the 1960s. One of the goals of each rural development project implemented was to prevent migration from the rural area to the urban area. The development of rural tourism, one of the components of rural development, also appears to be an indispensable component in preventing migration [32].As the endeavors of rural development, particularly of rural tourism, gained momentum in Turkey in the early 21st century, the studies by the Study Group on Rural Development created by the Ministry of Agriculture and Village Affairs of the Republic of Turkey helped by arousing the initial interest in the issue, in the true sense. Turkey found an opportunity to closely meet the regional development policies within the framework of the process towards full membership of the European Union. To provide sustainability for rural areas, the lack of which was one of the important reasons for regional imbalance, rural tourism is acknowledged as an essential instrument for rural development [4].

Different rural tourism projects for the development of rural areas are being implemented in the projects with domestic and foreign sources in Turkey. When the implementation of these projects is considered, examples of successful projects in which sustainability has been prioritized can be seen. There were subsidies via the IPARD (The Instrument for Pre-Accession Assistance for Rural Development) Program for rural tourism for the first time between 2007 and 2013, given by the Ministry of Agriculture and Village Affairs. A probable development is that it will be an important income-providing activity, particularly for those people who live in mountainous areas with geographical conditions that are inconvenient for carrying out any economic activities and that are likely to be regarded as disadvantaged [32].

As indicated in Article 1586 of the 8th Five-Year Development Plan, it is very important to "*direct to alternative tourism activities such as winter tourism, mountain tourism, thermal tourism, health tourism, yacht tourism, congress tourism, and ecotourism in order to create new alternative areas by providing the balanced distribution of tourism seasonally, and to the geographical regions and also by con-*

sidering the change in tourists' preferences in foreign markets" [46].The development of different tourism products, the spreading of tourism activities over the twelve months of the year and the minimization of the impact of the available mass tourism activities are essential to Turkish tourism policies and planning. Action Plan 2023 of the Turkish Tourism Strategy (2007) puts emphasis on the requirement for diversifying tourism products so as to spread the tourism season over the whole year. Development of tourism types such as health tourism and thermal tourism, winter tourism, golf tourism, sea tourism, ecotourism and rural/tableland tourism, and congress and fair tourism out of the alternative tourism types is placed at the forefront in these diversification studies [47].

Various projects supported by the public, local governments, and nongovernmental organizations have been conducted in rural settlements which, due to their natural and cultural assets, stand under the headings of rural tourism, tableland tourism and ecotourism in Turkey in recent years. However, it is also known that they have not been put into practice very widely. It is also the case that, in Turkey, rural tourism may give great support to the tourism in our country, given its innumerable benefits. Firstly, it may be effective in reducing the concentration of tourism in specific months of the year, and contribute to the elimination of the unequal distribution of tourism throughout the country. Furthermore, it may enable our market to expand by creating demands from different and conscious tourist groups. The power of rural tourism to easily integrate with other types of tourism may dynamize these. Rural tourism is the most harmonious type of tourism which can be understood as sustainable tourism. Rural tourism has an essential function in the conservation of our natural and cultural heritage and in introducing it to the world [9].

4. Place of Rural Tourism in the Rural Development Process of Villages

As also stated by Kaypak [48], speaking of development in the true sense in the 21st century is now possible only if sustainable development can be ensured. This will not happen with urban development alone, ruling out rural areas.

Rural development refers to a form of development which is included in the general development programs of countries and which has a significant share in regional development. The main goal of countries, particularly developing countries, is general development. And yet it is certain that the profound impact to be created by the added value obtained through rural development will add a potential value firstly to regional development and then to general development [4]. There are different ideas and conceptual confusions about the concepts and definitions of rural development, as there are for rural tourism. Nevertheless, as also stated by Ellis and Biggs [25], the attempts to describe the ideas which developed in rural development in the last half century also contain a risk of oversimplification.

Rural development refers to the processes, activities, and organizations which concern the development of production, income, and welfare levels in order to alter the structures surrounding the people living in rural areas in socio-economic and cultural terms; the elimination of imbalances; the creation of physical and social infrastructures in rural areas such as exist in urban areas; and the better utilization of agricultural products [32].

According to Pretty and Koohafkan (2002), rural development was one of the issues for which the United Nations laid down serious criteria in the Conference on Environment and Development held in Rio de Janeiro in 1992. The term 'rural development' was defined as an *"improvement of the rural life"* in Agenda 21 [18]. Rural development is an important instrument for enhancing the contribution of rural areas to the national and local economies; the development of the socio-economic status and quality of life of rural society; reducing the differences in development between rural and urban areas; ensuring balanced and sustainable development; and the protection, development, and sustainable use of natural resources and ecosystems [49]. Considering these definitions, it is possible to state that rural development aims to minimize the difference in development between urban and rural areas; utilize the local potential and resources; protect/conserve natural and cultural properties; develop the rural section in economic and sociocultural terms; and make it sustainable [50; 51].

No matter how rural development is defined, the most significant building block in this type of development is agriculture. Rural communities to be developed through rural development are largely villages. All in all, rural development programs are implemented for those villagers who make the overwhelming majority of their living from agricultural activities.

One of the different methods of applying agriculture developed in economic terms and for the sustainability of employment in rural areas is rural tourism. Thus, having rural tourism activities adopted substantially by the communities residing in villages is the prerequisite for achieving rural development through rural tourism [18; 4].

The gradual deepening of the differences in income levels and living conditions between rural areas and cities causes new models and approaches which accelerate the development of rural areas so that they are continually on the agenda. Today rural tourism is evaluated as an alternative development strategy for the restructuring of rural areas in economic and social terms in many western countries, for such reasons as the decrease in economic efficiency in rural areas, the restructuring process that the agricultural sector is undergoing, the gradual shrinkage of rural industry, and the decrease in the population of rural areas through migration. Furthermore, rural tourism is a supported sector in developed and developing countries, for it develops economic and social activities. Rural tourism is encouraged particularly in those places where traditional agriculture has decreased, as it can bring a reduction in unemployment, economic diversity, the provision of a source of income, and a contribution to the infrastructure and the superstructure of the places where it is carried out. In summary, the impact of rural tourism on rural development is an undeniable reality [52; 12; 50]. It is considered necessary to put into practice rural development programs which will slow down migration from villages to cities but meanwhile raise the living standards of the people living in the rural area to meet those of the inhabitants of cities. It is a known fact that tourism activities have a significant place in the development of rural areas. Therefore, there is a general opinion that the use of rural tourism in rural development will cause significant changes which will increase the living standards of the people living in rural areas [4].

It is evident that different approaches have been developed and implemented in the process since the foundation of the Republic of Turkey, so as to protect and develop villagers in rural areas, minimize the differences between rural areas and cities, and to develop rural areas [53]. When the historical perspective on the endeavours in rural development carried out by Turkey is scrutinized, it is possible to state that the rural development models or projects which present themselves are products of periodic and different cause-and-effect relationships, in the context of rural development experiments which have not had any widespread effects [49]. Moreover, the results obtained from the policies and projects implemented clearly reveal that there was a failure to attain the majority of the objectives and that the increase in rural-urban problems could not be halted. The basic reasons for this are generally the mistakes made in the selection and implementation of policies and projects, as well as their follow-up stages, along with an inattention to sociological realities [53].

In fact, one of the basic problems experienced by present-day Turkish society is the problem of rural development. Although more than one-third of the country's population still lives in rural areas, and in spite of the rapid social change experienced in our country in recent years, we can see that no noteworthy developments have emerged concerning the establishment of socio-economic problems in the rural areas of Turkey in the recent past [54].

Economic development is the primary issue on which countries put too much emphasis. Every country around the world wants its own society to live with good welfare and therefore implements various programs for itself. The fact that the population living in rural areas outnumbers the urban population is an element which complicates development for most countries. Rural development is strongly emphasized in these countries. Rural development is the increase in production and income per capita in rural regions as well as the changes in, and development of, economic, social, and cultural structures. One of the leading and most important sectors in rural development is rural tourism. Furthermore, one of the ways of attaining rural development is to develop the sources of tourism in a rural region. Today rural development is considered quite important and not only agricultural activities but also opportunities for rural tourism are utilized in development. By forming the relationship between rural tourism and culture, the intention is to obtain sustainable economic and social benefits. Thus, various projects to yield positive outcomes in terms of the development of rural tourism have been presented and tried to be realized to date [32; 55; 7].

Thus, the concepts of rural tourism and rural development define a process in which they are absolutely directly interrelated [56] and also involve the relationships between agriculture, tourism, and the rural area (Figure 5). It is possible to make use of different components by accurately utilizing the potential of an area when ensuring the rural development of a locality or a region. A development approach based fundamentally on agricultural activities might mostly be adopted and such economic activities as animal husbandry, forestry, tourism, and rural industry might also stand out depending on the geographical conditions of the locality. Rural tourism is of special importance in this respect as well [57].

Agriculture is the cornerstone of economic capital in rural areas, and this capital is considerably consumable. Thus, different practices are being carried out in many countries today so as to ensure the sustainability of employment in rural areas. Rural tourism is the symbol of success for nonagricultural diversification in rural regions. In addition, one of the leading and most important sectors in rural development is rural tourism. The richness and diversity of natural and cultural properties in rural areas offer significant potential for the development of tourism and recreational activities. Within this scope, many activities such as ecotourism, agrotourism, agro-eco-tourism, and agritourism contribute to the economic, ecological, and social development and sustainability of villages within the scope of rural tourism [18; 58].

Figure 5. The interaction between rural tourism and rural development [4].

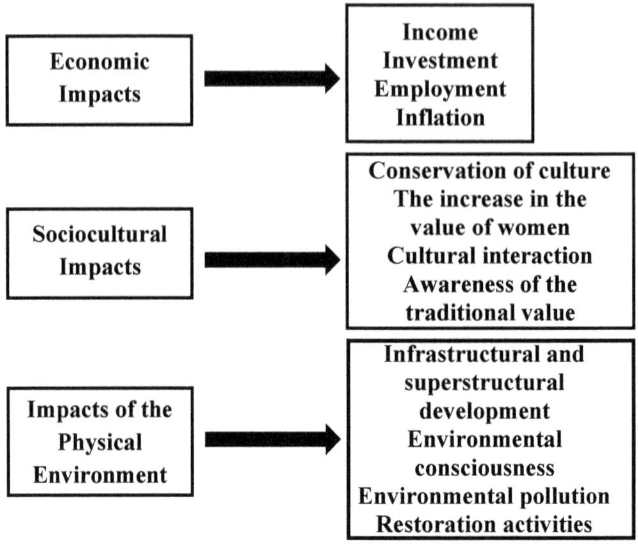

Besides referring to farm tourism, green tourism or tableland tourism, rural tourism is a type of tourism which is indexed to nature holidays and particularly ecotourism, shopping, skiing, nature tours on bicycles or horseback, adventure, rafting, sport, thermal tourism, hunting and fishing, art, history, and the ethnic structure [32].

The impact of rural tourism on rural development is predominantly proportional to the efficient use of the historical, natural, cultural, and agricultural resources in the locality. Today development based on rural tourism works better with the presence of rural entrepreneurs and provides people who make a living from agriculture in rural areas with a secondary source of income. Furthermore, rural areas offer tourists mystical attractiveness resulting from their ethnic and geographical characters, history, different cultures, and rural nature [30].

Additionally, rural tourism contributes to socio-economic rural development by providing rural women with employment in the local dimension, reviving local economies, and preventing migration. Moreover, it prevents the rural culture from becoming extinct, supports the protection of natural environments, and can easily integrate with other types of tourism. Carrying out tourism in different areas allows the implementation of long-term activities over the year, attracts tourists in almost every period of the year, and is also important in that it provides justice by reducing interregional differences in the distribution of incomes [59].

Having become increasingly important in recent years, rural tourism is one of the basic instruments for rural development in terms of providing an opportunity for utilizing the local potential and resources, increasing the welfare level of people at the locality by creating additional sources of income and employment, preventing migration from villages to cities, contributing to cultural interaction, encouraging natural and cultural properties, and developing physical infrastructure in rural areas. Thus, rural tourism is a quite essential activity which has a multidirectional impact on the development of the areas with potential for development [57].

5. Place/Role of Traditional Turkish Villages in Rural Tourism

As the primary issues which have begun to predominate on the tourism agenda in our country in recent years, protection of the environment and the use of natural and cultural resources with an understanding of sustainable tourism have begun to be adopted as concepts. The spaces untouched by human beings are further decreasing with every passing day and the importance of the areas with this quality is gradually increasing. At this point, rural areas, particularly villages, are of quite profound importance [60; 46].

The neglected phenomenon of villages and the overlooked problems of the rural Turkish people underlie many problems experienced by Turkish society today. In other words, the reality of the failure to satisfy people and enable them to live in human dignity in the places where they were born underlies many social problems such as migration from rural areas to urban areas, squatting, the failure of urbanization, urban unemployment, radicalization, and a rapid increase in crime rates and socio-psychological problems [61].

Villages, which represent rural settlements in Turkey, have a rather complicated profile in terms of their socio-economic and cultural characteristics. Scattered settlements and topographical locations have a major effect on the occurrence of this situation. The difficult topography in Turkey has led to the occurrence of scattered rural settlements. The rural population, whose livelihood is based on agriculture and animal husbandry, has practised agriculture in areas which are, for the most part, inconvenient for this activity. This was in order to meet the need for grain in unevenly mountainous areas. There was an irregular and excessive use of forests in order to meet the need for firewood, and forests and pastures were often overgrazed by animals. The villages whose socio-economic sides have failed to develop for these reasons have remained economically dependent on the forests and protected areas in their close vicinity [62; 63]. Rural areas require different activities so as to be able to meet economic expectations. One of them is rural tourism.

The types of tourism and the economies in rural areas should be developed by creating tourism areas within them. A maximization of the tourism supply, balancing the tourism supply and demand and ensuring the use of tourism as an instrument for regional development takes place by means of the villages where rural tourism is implemented. Therefore, it should firstly be ensured that the comfort and welfare levels of the local people in many villages of Turkey are increased through rural tourism, so that the natural, cultural and economic assets and traditional social life can also be preserved in villages [64; 65].

In fact, rural tourism, in its true sense, is that people desiring to experience village life by leaving behind the busy lifestyles of the city center participate in village life fully, for example by staying in the buildings in those areas in their natural state, hoeing the soil, and milking. Furthermore, a person who has become part of village life pays a fee in return for this. Thus, these actions are called rural tourism. This activity both provide earnings for the region where it takes place and stops the emigration of people from that region, thereby enabling them to actively participate in production and the sector [8]. Unfortunately, a rural tourism movement in this sense has not been available in the villages of our country yet. Besides all that was mentioned above, in rural areas where tourism is hardly seen, the local people do not consider it likely that their traditional life habits, which have persisted for a considerable time and which are ordinary for them, might have any potential for tourism. Therefore, the local people should be informed about the fact that such activities as milking, collecting crops from the field/garden or making cheese, etc. might each become an element of tourism and that, if applied correctly, rural tourism might become an important source of income in addition to agricultural activities. In addition, raising awareness among the local people regarding the fact that rural tourism will continue only if original rural assets are conserved is of great importance in terms of sustainability. As Akça (2004) also stated, what comes to mind when rural tourism is considered in our country is predominantly tableland tourism or outdoor and recreational activities, including treks and nature sports performed in rural areas, and daily village visits which encompass shopping and daily eating & drinking [12].

Indeed, Turkey has a great potential for rural tourism, perhaps more than any other country. At this point, the Ministry of Culture and Tourism in our country has begun to pursue a policy for the spread of tourism from coasts to inner parts. It should be considered normal that there has not been a demand for rural tourism within the scope of domestic tourism in our country. Our large rural population, until recent years, the rural origin of the majority of those living in our cities and the strong kinship relationships in our cultural structure have not left much necessity to go to rural areas for tourism, as the main motivation for participating in rural tourism is the longing for rural geographical environments and for the traditional culture. This is the main reason why rural areas mostly connote the concept of holidays in the western countries which have become urban societies [9].

Natural and cultural landscape characteristics are among the most important features of rural settlements [66]. This is one of the most important factors which play a role in the development of rural tourism in that locality. There are various natural and cultural assets in rural areas which can facilitate tourism there. Rural areas are perhaps the most important building blocks in the region where rural tourism will be practised. Especially the tourists who come from urban living areas prefer this type of tourism, as they long for nature. In such circumstances, any natural and cultural assets in the rural areas have special meanings for rural tourism [4]. The overwhelming majority of natural and cultural assets of our country are in villages in rural areas. In this context, traditional villages bring great potential for rural tourism.

Even though the primary expectation of people who engage in a touristic activity is to see new places and get to know new cultures, discovering new and different tastes and getting to know local handicrafts can also feature in their expectations. Tourists setting off with the purpose of practicing rural tourism hope that this expectation of theirs will be met [67]. At this point, the roles of local tastes and local handicrafts are great. The traditional Turkish villages have considerable potential in this respect. It is possible to encounter the existence of a distinct food culture and different types of traditional handicrafts in almost every village from west to east.

Evaluations of the fact that settlements have diverged from their original identities, become copies of each other, and then standardized, have become widespread in recent years. As a contrary initiative to this standardization, studies into conserving settlements together with their original fabric as well as with examples of traditional civil architecture are not new [68]. The conservation of traditional settlements, one of the sources of rural tourism, gains importance at this point since a tourist who arrives in the locality in order to practise rural tourism at the same time wants to get to know the traditional settlement of that place and to breathe that air. Also, with their traditional and relatively unimpaired settlement fabric, the villages of our country qualify as sources of rural tourism.

Villages around the big cities and on the roadside, as well as the rural settlements in the vicinity of ancient cities in Turkey, have not been foreign to tourism in recent years. Rural restaurants, shopping exhibitions, fish farms and rural roads have been visited by tourists or excursionists. Daily tours to such places or the rural areas in which they are located are organized by travel agencies, along with individual tours. Furthermore, there are intentions to revive the ancient village remains through various projects [32]. Putting such developments into practice for traditional Turkish villages as well will be effective for the spread of rural tourism over the entire country, rather than it existing merely for specific regions of the country.

The villages of our country perform various activities likely to become sources of rural tourism (e.g. village weddings, charity events, local handicrafts, traditional weaving, and local dishes). As also stated by Kadanalı and Yazgan [52] as well as Karamehmet et al. [36], the presence of such activities as village weddings, local handicrafts, traditions unique to a locality, local dishes, folk dances, clothes, and lifestyles in the rural tourism activity is also of great importance in promoting the region as well as developing the rural economy. Thanks to this, cultural tourism will be supported and rural tourism will be strengthened.

To obtain the economic, sociocultural, and environmental benefits expected from rural tourism, first it is essential that the local people are willing to be volunteers [52]. In other words, tolerance and hospitality are among the most important priorities in the realization of rural tourism for traditional Turkish villages because it is the villages where rural tourism should indeed take place. At this point, as also stated by Karamehmet et al. [36], opening the settlements of the families living in rural areas to provide accommodation, also using the houses owned by the local people as guesthouses, and in this way enabling the local people to participate in tourism activities in person are the most important conditions for the development of rural tourism.

In order for tourism to provide sustainable development in rural areas, first the local people should be willing. The stay of incoming visitors in the houses of the local people, and the involvement of tourists in traditional tasks (e.g. agricultural activities; production of cheese, bread, and tomato paste; or handicrafts) together underlie rural tourism. Hence, willingness of the local people to involve the visitors with a different culture, religion, and language in their own houses and everyday lives is of profound importance. Additionally, we can see that with an increase in tourism activities in any place, the younger population in particular is influenced on a sociocultural level (e.g. a change in the clothing habits and the occurrence of the desire to go to big cities or other countries) by the incoming visitors. The local people should also be able to access such impacts, which are likely to occur together with rural tourism, if it is carried out with tolerance.

6. Conclusion and Recommendations

In most countries worldwide, be they developed or developing, the important achievements obtained by rural tourism enable it to be seen, in addition to agriculture, as an alternative approach to the development of rural localities. When the issue is addressed in terms of Turkey, we see that Turkey is a country which is rich in agricultural potential. In those regions where the agricultural sector is concentrated, rural developement will be easier to carry out, through rural tourism. What is important here is the effective use of the potential for rural tourism in the locality and the establishment of coordination with the leading sectors in the area. [30].

It is known that our country has the power to provide opportunities for the development of almost all types of tourism, given its natural and cultural appeal, in line with the increased demand for alternative tourism types worldwide as of the 1990s, and studies to diversify the types of tourism [69]. Many existing cultural, geographical, historical, social and human assets of Turkey have qualities likely to create economic added value for rural tourism [4]. Rural areas are areas where natural and cultural beauties come together. The villages of Turkey are also quite rich in cultural and natural beauties [8]. Furthermore, this richness means that there is great potential for the realization of rural tourism activities in villages.

In particular, the low education level in rural areas, inefficient practices by local governments and the impacts of political and economic instabilities may be mentioned as Turkey's weaknesses in terms of rural tourism. To get the desired positive economic results from rural tourism, it is necessary to enhance social awareness, enable the central government to transfer its power to the local government in a controlled fashion, and make society conscious of the European Union's financial subsidies for rural tourism projects. For this purpose, it is considered a must to receive help from Nongovernmental Organizations and universities [4].

In Turkey, the current clientele for rural tourism are foreign tourists. In other words, rural tourism should take its place within foreign tourism. That's why the villages to be opened to rural tourism should be selected meticulously, as they will serve as sorts of representatives of traditional village culture. Village people should take part in every task, from the planning stage to all initiatives, and they should certainly be enabled to get organized and to undergo training on this issue. Now that every village has unique qualities and local resources, the requests of the local people and their creativity should be given prominence [9]. The only thing required to ensure the sustainability of rural and agricultural areas is some rural tourism planning which will be made in agreement, and accurately, with the existing natural and cultural features of each village.

Currently, rural tourism continues to have shortcomings and mistakes are made in some villages in our country. To make it more productive, accommodation should first be included in villages. The foundation of a village association, which would deal with booking transactions for guests, and limits on excursions, could be introduced as they have been in western rural tourism. It is not too difficult to do so in our country. What is currently difficult is to use the richness our country has in rural tourism culture in a sustainable fashion and not allow it to degenerate. [9]. At this point, training people living in rural areas on the issue is of prime importance.

The unique structure of the rural area may complicate adaptation to a service sector like tourism [70]. Even though rural tourism involves coexistence and sharing some tasks, it is also unavoidable that visitors demand some services (e.g. cleanliness of rooms or preparation of food). At this point, it is of extreme importance that the local people are also resigned to fulfilling the requirements of the service sector and undergo some training on this issue. The greatest and most important role belongs to women in the rural area, by the nature of rural tourism. In this respect, women's undergoing of training on this issue especially will be effective to the success of tourism activities in the area.

In conclusion, tourism activities carried out in rural areas are regarded as solutions to the problems experienced by the rural population. This also applies to the rural settlements in our country, particularly the villages. In the event of an increase in migration from villages to cities due to the attractiveness of urban life, migration may be reduced or prevented by encouraging rural tourism. With rural tourism, there will be an increase in services in rural areas and in their quality. Rural society can be enabled to expand outwards via the visits to rural areas by tourists and the exchange of cultures. With the encouragement of rural tourism, local people will act more meticulously as regards the protection of the natural resources in the region [52].

For all these reasons, it might be stated that rural tourism, which is in harmony with sustainable tourism, brings significant contributions to tourism in the country in both micro and macro terms. When considered from this perspective, it is appropriate to determine, develop, and ensure the sustainability of regions undertaking rural tourism – one of the alternative tourism types with a positive impact on the national economy – by diversifying tourism and spreading it over four seasons [13].

As also stated by Soykan [9], representatives from the tourism sector, public institutions, nongovernmental organizations, local governors, local people and tourism experts at the university level should work with the plans, projects, and programs to be carried out, to initiate rural tourism in the true sense, conduct it in the spirit of sustainable understanding, and carry it into the future in Turkey.

References

1. P. KISA OVALI: *Kitle Turizmi ve Ekolojik Turizmin Kavram, Mimari ve Çevresel Etkiler Bakımından Karşılaştırılması. Megaron/Yıldız Teknik Üniversitesi Mimarlık Fakültesi Dergisi,* 2 (2), 64–79 (2007).
2. E. AKPINAR, Y. BULUT: *Ülkemizde Alternatif Bir Turizm Dalı Olan Ekoturizm Çeşitlerinin Bölgelere Göre Dağılımı ve Uygulama Alanları. III. Ulusal Karadeniz Ormancılık Kongresi,* 20–22 Mayıs, Cilt: IV, 1575–1594 (2010).
3. N. ERDOĞAN: *Eko-turizm ve Doğa Temelli Turizmin Geldikleri Noktada Kavramsal Çerçeve. Orman Ekosistemlerinde Eko-turizm Çalıştayı ve 10. Yılında TODEG,* TODEG-Ankara, 42–51 (2010).
4. N. KUŞAT: *Sürdürülebilir Kırsal Kalkınma için Bir Alternatif Olarak Kırsal Turizm ve Türkiye'de Uygulanabilirliği. Ekonomik ve Sosyal Araştırmalar Dergisi,* 10 (2), 65–87 (2014).
5. M. A. KASALAK: *Dünyada Ekoturizm Pazarı ve Ekoturizmin Ülke Gelirlerine Katkıları. Journal of Recreation and Tourism Research,* 2 (2), 20–26 (2015).
6. N. ERDOĞAN: *Ekoturizm: Kavramsal ve Eleştirel Yaklaşımlar.* Available from: http://researchturkey.org/tr/ecotourism-conceptual-and-critical-approaches/, Date of Access: 21.04.2017 (2015).
7. E. KUT: *Kültür Köy Turizm Projesi İle Sürdürülebilir Kırsal Turizm Kalkınmasının Sağlanması, Gökçe Çerkez Kültür Köyü Politika Önerileri. Uluslararası Sosyal ve Ekonomik Bilimler Dergisi,* 3 (1), 33–41 (2013).
8. D. D. GÖKALP, M. E. YAZGAN: *Kırsal Peyzaj Planlamada Agroturizm ve Agriturizm. KMÜ Sosyal ve Ekonomik Araştırmalar Dergisi,* 15 (24), 25–29 (2013).
9. F. SOYKAN: *Kırsal Turizm ve Türkiye Turizmi için Önemi. Ege Coğrafya Dergisi,* 12, 1–11 (2003).
10. A. USLU, T. KİPER: *Turizmin Kültürel Miras Üzerine Etkileri: Beypazarı/Ankara Örneğinde Yerel Halkın Farkındalığı. Tekirdağ Ziraat Fakültesi Dergisi,* 3 (3), 305–314 (2006).
11. M. UÇAR, H. UÇAR, A. KURNAZ, H. AKYURT KURNAZ: *Kırsal Turizmin Sosyo-Ekonomik Yapıya Etkisi ve Fethiye Örneği. KMÜ Sosyal ve Ekonomik Araştırmalar Dergisi,* 14 (22), 75–82 (2012).

12. O. AYDIN: *AB'de Kırsal Turizmde İlk 5 Ülke ve Türkiye'de Kırsal Turizm. KMÜ Sosyal ve Ekonomik Araştırmalar Dergisi,* 14 (23), 39–46 (2012).
13. D. KODAŞ, S. SÜ ERÖZ: *Kırsal Turizm İle Kültürel Turizmin Bütünleşmesi. KMÜ Sosyal ve Ekonomik Araştırmalar Dergisi,* 14 (22), 169–174 (2012).
14. Y. AKÇA, R. ÖZDEMİR: *Turizm Yoluyla Kırsal Kalkınma: Amasra ve Köyleri Örneği. Bartın Üniversitesi İktisadi ve İdari Bilimler Fakültesi Dergisi,* 5 (9), 85–98 (2014).
15. F. SOYKAN: *Avrupa'da Kırsal Turizme Bakış Açısı ve Kazanılan Deneyim. II. Balıkesir Ulusal Turizm Kongresi,* 20–22 Nisan, Balıkesir, 71–87 (2006).
16. ORGANISATION FOR ECONOMIC CO-OPERATION AND DEVELOPMENT (OECD): *Tourism Strategies and Rural Development,* Paris, OCDE/GD (94), 49 (1994).
17. T. KİPER: Safranbolu Yörükköyü Peyzaj Potansiyelinin Kırsal Turizm Açısından Değerlendirilmesi. Ankara Üniveristesi Fen Bilimleri Enstitüsü Peyzaj Mimarlığı Anabilim Dalı, Basılmamış Doktora Tezi, 367 s., Ankara (2006).
18. F. PEZİKOĞLU: *Sürdürülebilir Tarım ve Kırsal Kalkınma Kavramı İçinde Tarım-Turizm-Kırsal Alan İlişkisi ve Sonuçları. KMÜ Sosyal ve Ekonomik Araştırmalar Dergisi,* 14 (22), 83–92 (2012).
19. E. TÜMERTEKİN: *Kırsal Alanlar ve Coğrafya. İstanbul: İTÜ Mimarlık ve Şehircilik Enstitü Dergisi,* (17), 51–63 (1980).
20. N. TUNÇDİLEK: *Türkiye'nin Kır Potansiyeli ve Sorunları. İstanbul Üniversitesi Yayınları,* No: 2364, İstanbul (1978).
21. B. TAŞ: *Türkiye'de Coğrafi Bakış Açısıyla Köy Kavramı. TÜCAUM Uluslararası Coğrafya Sempozyumu,* 13–14 Ekim, 292–302 (2016).
22. M. SARIBAŞ, A. PINAR: *Silifke Göksu Vadisi'nde Yer Alan Köylerin Beşeri Coğrafya Özellikleri. Selçuk Üniversitesi Sosyal Bilimler Enstitüsü Dergisi,* 29, 207–221 (2013).
23. F. AKLANOĞLU: *Geleneksel Yerleşmelerde Kültür Turizmi: Beypazarı Örneği. Kastamonu Üniversitesi Orman Fakültesi Dergisi,* 10 (2), 125–136 (2010).
24. B. LANE: *Rural Tourism: An Overview, Tourism Studies.* Ed. by T. Jamal and M. Robinson, SAGE Publications Ltd., London, 354–370 (2009).
25. F. ELLIS, S. BIGGS: *Evolving Themes in Rural Development 1950s-2000s. Development Policy Review,* 19 (4), 437–448 (2001).
26. L. ROBERTS, D. HALL: *Rural Tourism and Recreation: Principles to Practice.* Ed. by L. Roberts, D. Hall, CABI Publishing, London UK, 1–23, (2001).
27. S. WILSON, D. R. FESENMAIER, J. FESENMAIER, J. C. VAN ES: *Factors for Success in Rural Tourism Development. J. of Travel Research,* 132–138 (2001).

28. WTO: *World Trade Organization.* Available from: https://www.wto.org/english/res_e/statis_e/its2004_e/its2004_e.pdf, Date of Access: 28.04.2017

29. J. C. HOLLOWAY, N. TAYLOR: *The Business of Tourism (Seventh Edition).* Prentice Hall, ISBN-10: 0–273-70161-4, 697p., England (2006).

30. H. ÇEKEN, L. KARADAĞ, T. DALGIN: *Kırsal Kalkınmada Yeni Bir Yaklaşım: Kırsal Turizm ve Türkiye'ye Yönelik Teorik Bir Çalışma. Artvin Çoruh Üniversitesi Orman Fakültesi Dergisi,* 8 (1), 1–14 (2007).

31. D. FENNEL: *Ecotourism.* Third Edition, London and New York: Roudledge, Taylor & Francis Group (2008).

32. E. ÜN, F. TUTAR, E. TUTAR, Ç. ERKAN: *Ekonomik Kalkınmada Kırsal Turizmin Rolü: Türkiye Örneği. International Conference on Eurasian Economies (SESSION 4C: Turizm),* 345–350 (2012).

33. R. SHARPLEY, D. J. TELFER: *Conclusion: Tourism and Development. Tourism and Development Concepts and Issues (2nd Edition).* Ed. by R. Sharpley and D. J. Telfer, Aspects of Tourism: 63, Channel View Publications, Bristol, UK., 453–468 (2014).

34. M. CARNEIRO, J. MARIA, J. LIMA, A. L. SILVA: *Landscape and the Rural Tourism Experience: Identifying Key Elements, Addressing Potential, and Implications for the Future. J. of Sustainable Tourism,* 23 (8–9), 1217–1235 (2015).

35. V. NAIR, U. T. MUNIKRISHNAN, S. D. RAJARATNAM, N. KING: *Redefining Rural Tourism in Malaysia: A Conceptual Perspective. Asia Pacific J. of Tourism Research,* 20 (3), 314–337 (2015).

36. B. KARAMEHMET, G. AYDIN, C. U. ÖZGÖKER: *Türkiye'de Kırsal Turizm Destinasyonlarının Rekabetçi Gücünün Geliştirilmesi için Stratejik Öneriler. Balkan Sosyal Bilimler Dergisi,* 5 (10), 106–116 (2016).

37. C. GERAY: *Kırsal Kalkınma Yöneltileri, İlçe Yerel Yönetimi ve İlçe Köy Birlikleri Önerisi. Çağdaş Yerel Yönetimler Dergisi,* 8 (2), 11–42 (1999).

38. B. SU: *Rural Tourism in China. Tourism Management,* 32 (6), 1438–1441 (2011).

39. F. SOYKAN: *Doğal Çevre ve Kırsal Kültürle Bütünleşen Bir Turizm Türü: Kırsal Turizm. Anatolia Turizm Arastırmaları Dergisi,* Yıl: 10, Mart-Haziran, 67–75 (1999).

40. C. DURAN: *Türkiye de Dağlık Alanların Kırsal Turizm Açısından Önemi. KMÜ Sosyal ve Ekonomik Araştırmalar Dergisi, 14 (22), 45–52 (2012).*

41. C. AKYOL: *Kırsal Turizmde Ev Pansiyonculuğu Modeli ve Karadeniz Örneklemesi-Artvin. Uluslararası Sosyal ve Ekonomik Bilimler Dergisi,* 2 (2), 79–83 (2012).

42. G. GÜNEŞ: *Türkiye Kırsalı'nda Kadın.* Available from: http://e-bulten.library.atilim.edu.tr/sayilar/2013-03/makale.html, Date of Access: 01.06.2017 (2008).

43. G. GÜNEŞ: *Sürdürülebilir Kırsal Planlamada Doğa Turizmi ve Yerellik.* Available from: http://docplayer.biz.tr/8716618-Surdurulebilir-kirsal-planlamada-doga-turizmi-ve-yerellik.html, Date of Access: 01.06.2017 (2013).
44. S. ERDOĞAN: *Kırsal Kalkınmada Kırsal Turizmin Rolü. Türk İdare Dergisi,* 435, 197–205 (2002).
45. THE MINISTRY OF CULTURE AND TOURISM: *Yayla Turizmi.* Available from: http://yigm.kulturturizm.gov.tr/TR,11532/yayla-turizmi.html, Date of Access: 21.04.2017 (2015).
46. S. CEYLAN, H. DEMİRKAYA: *Kış Turizmine Bağlı Olarak Gelişen Bir Kırsal Yerleşme: Çobanisa Köyü (Isparta). Doğu Coğrafya Dergisi,* 14 (21), 79–94 (2009).
47. G. ÇETİNKAYA: *Bir Macera Turizmi Etkinliği Olarak Kaya Tırmanış Sporu Ve Antalya-Geyikbayırı'nın Potansiyeli. Doğu Coğrafya Dergisi,* 19 (31), 83–100 (2014).
48. Ş. KAYPAK: *Ekolojik Turizm ve Sürdürülebilir Kırsal Kalkınma. KMÜ Sosyal ve Ekonomik Araştırmalar Dergisi,* 14 (22), 11–29 (2012).
49. K. ÖZCAN, A. AKCİ: *Kırsal Kalkınma için Model Önerisi: Köy Kümeleri. TEAD,* 2 (1), 1–11 (2016).
50. M. CİVELEK, T. DALGIN, H. ÇEKEN: *Agro-Turizm ve Kırsal Kalkınma İlişkisi: Muğla Yöresindeki Agro-Turizm Alanlarında Bir Araştırma. Turizm Akademik Dergisi,* 1 (1), 15–28 (2014).
51. DPT (STATE PLANNING ORGANIZATION): *Ulusal Kalkınma Stratejisi. Devlet Planlama Teşkilatı, Ankara* (2006).
52. E. KADANALI, Ş. YAZGAN: *Kırsal Turizmin Ekonomik- Sosyal ve Çevresel Etkileri. KMÜ Sosyal ve Ekonomik Araştırmalar Dergisi,* 14 (23), 97–100 (2012).
53. E. ÇOLAKOĞLU: *Kırsal Kalkınma Problemine Bir Arayış Olarak Köy-Kent Projesi. ZKÜ Sosyal Bilimler Dergisi,* 3 (6), 187–202 (2007).
54. A. ARSLAN: *Bir Köy Sosyolojisi Çalışması: Kavaközü Köyü'nün Sosyo-Ekonomik Yapısı ve Sorunları. Osmangazi Üniversitesi Sosyal Bilimler Dergisi,* 4 (1), 1–26 (2003).
55. M. BOZ, M. Y. ERYAMAN, Ç. ÖZKAN: *Kırsal Kalkınmada Turizmin Önemi: Nusratlı Köyü Örneği. JED / GKD,* 9 (2), 293–308 (2014).
56. S. ÖZDEMİR: *Kırsal Kalkınmada Kırsal Turizmden Yararlanma Olanakları: Gökçeada Örneği. KMÜ Sosyal ve Ekonomik Araştırmalar Dergisi,* 14 (23), 19–21 (2012).
57. C. AKBULAK: *Ardahan İlinde Kırsal Turizm Potansiyelinin Sayısallaştırılmış Swot Analizi ile Değerlendirilmesi. Humanitas,* 4 (7), 1–30 (2016).

58. A. KARAMAN, S. BALTACIOĞLU, Y. E. GÜRHAN: *Kırsal Kalkınma Açısından Kırsal Turizm ve Halkapınar için Proje Önerileri. Sosyal Bilimler Meslek Yüksekokulu Dergisi,* 18 (2), 165–185 (2015).
59. E. TORUN: *Kırsal Turizmin Bölge İnsanına Katkıları. KMÜ Sosyal ve Ekonomik Araştırmalar Dergisi,* 15 (24), 31–37 (2013).
60. S. YILMAZ: *Serçeme Vadisi'nin Rekreasyonel Kullanım Potansiyelinin Belirlenmesi. Ekoloji Dergisi,* 13 (51), 1–6 (2004).
61. A. ARSLAN: *Bir Ankara Köyü (Kavaközü)'nün Sosyolojik İncelemesi. Sosyal Bilimler Enstitüsü Dergisi,* 17 (2), 53–70 (2004).
62. İ. ATALAY: *Türkiye' de Kır Yerleşmelerinin Arazi Degredasyonu Üzerindeki Etkileri (Türkiye'de Kır Yerleşmeleri). Coğrafya Araştırmaları Dergisi,* 1 (1), 91–103 (1989).
63. S. GÖRMÜŞ, D. OĞUZ: *Kırsal Yerleşim ve Korunan Alan Arasındaki Etkileşimin Değerlendirilmesinde Peyzaj Karakter Analizinin Rolü: Kapısuyu Havzası Örneği. Tarım Bilimleri Dergisi,* 19, 310–322 (2013).
64. E. POSTALCI ALTINKAYA, Ç. TEKİN, Ö. EREN: Kırsal Yerleşim Bölgelerinde Sürdürülebilirlik: Çanakkale Bölgesi Ayvacık İlçesi Adatepe ve Demirciköy. Politeknik Dergisi, 14 (2), 109–113 (2011).
65. Ö. TUĞUN, A. KARAMAN: *Çekirdek Köylerin Eko Turizme Kazandırılması İçin Sürdurulebilirlik Kavramı Çerçevesinde Bir Model. MEGARON/Yıldız Teknik Üniversitesi Mimarlık Fakültesi Dergisi,* 9 (4), 321–337 (2014).
66. B. YILMAZ, G. ATİK: *Doğal Peyzaj Özelliklerinin Kırsal Yerleşimler Üzerindeki Etkileri: Bartın Örneği. ZKÜ Bartın Orman Fakültesi Dergisi,* 8 (10), 1–9 (2006).
67. S. ETİKAN, T. ÇUKUR: *Kırsal Turizm Faaliyetlerinin Çomakdağ-Kızılağaç Köyü El Sanatları Üzerine Etkisi. Süleyman Demirel Üniversitesi Güzel Sanatlar Fakültesi Hakemli Dergisi,* 8, 1–15 (2011).
68. P. ŞENOL, A. ER AKAN: *Kırsal Yaşam / Kırsal Konut: Bir Yaşam Biçimi Olarak Geleneksel Kırsal Konut Üretiminde Kızılcık Köyü Örneği. SDÜ Fen Edebiyat Fakültesi Sosyal Bilimler Dergisi,* 24, 143–160 (2011).
69. G. EMEKLİ: *Avrupa Birliği'nde Turizm Politikaları Ve Türkiye'de Kültürel Turizm. Ege Coğrafya Dergisi,* 14, 99–107 (2005).
70. R. SHARPLEY: *Rural Tourism and the Challenge of Tourism Diversification: The Case of Cyprus. Tourism Management,* 23, 233–244 (2002).

URL-1: Village Law No. 442, Available from: http://www.mevzuat.gov.tr/MevzuatMetin/1.3.442.pdf, Date of Access: 01.06.2017 (1924).

URL-2: *Agritourism.* Available from: http://www.johnstoncountync.org/agritourism, Date of Access: 12.05.2017 (2016).

URL-3: *Louisiana Agritourism.* Available from: http://louisianaagritourism.blogspot.com.tr/2010/08/wesmar-farms-in-moreauville-la.html, Date of Access: 12.05.2017 (2010).

URL-4: *Weekend in agritourism between nature and relax.* Available from: http://www.in-lombardia.com/visiting-lombardy/nature-vacation-lombardy/weekend-in-agritourism-between-nature-and-relax, Date of Access: 12.05.2017 (2015).

URL-5: *Agritourism.* Available from: http://www.massvacation.com/blog/2016/04/massachusetts-home-grown-and-delicious/, Date of Access: 12.05.2017 (2016).

Melih Öztürk[1]

Driver and Passenger Perception Along the Roadside Landscapes: Safety and Recreation

[1]*Bartın University, Faculty of Forestry, Department of Landscape Architecture, 74100, Bartın, Turkey, melihozturk@bartin.edu.tr*

1. Introduction

The landscape involves the total environment within the field of vision, not only that which is perceived by the naked eye but is also the part observed through certain instruments and remote sensing techniques. Roads are connections between communities, cultures and economies [1, 2]. Hence, roadside landscapes consist of road-integrated environments, parts of which are included in the travelers' field of vision. Here, `travelers' refers both to the driver and the passenger. Thereby, the driver and passenger perceive only part of the roadside landscape which they pay attention to along their journey. The travelers' perception and attention are influenced by many factors associated with the roadside landscape. In general, roadside buildings, objects, and vegetation, as well as road and landscape topography, landscape climate and roadside fauna are some of those factors. Those factors either enable or impair visibility, either promote or disturb tranquility, either supply or disrupt aesthetics, either concentrate or distract attention, and ultimately either add comfort to or complicate travel. In brief, they favor or interrupt safety and recreation. Here, the driver is particularly concerned with safety while recreation is especially relevant to the passenger. The quantity and quality of these factors vary, based on the spatial and temporal characteristics of road and roadside. Hence, this chapter focuses on the travelers' perception of the roadside landscape considering the spatial and temporal dynamics of the existing factors. Eventually, the seasonal and daily variations in urban and rural roadside landscapes are discussed around the fundamental concepts of safety and recreation. Consequently, sustainable roadside landscape management objectives are also proposed in order to prevent accidents or at least mitigate their impact, and provide more recreation both for drivers and for passengers. Ultimately, sustainable roadside landscape management not only deals with the travelers' perception but also considers the conservation of that roadside landscape ecology.
Keywords: Safety and recreation; Roadside vegetation; Traveler perception; To-

pography and climate; Environmental aesthetics and behavior; Visibility and tranquility; Seasonal-daily urban and rural landscape.

2. Urban Roadside Landscapes

Urban roads involve not only inner-city roads but also inter-city roads including urban highways, freeways and bypasses (Figure 1; *Left*). Therefore, buildings, parked vehicles, trees, communities, stable or mobile objects and equipment such as bus stops, billboards, garbage containers and traffic signs in particular often accompany urban roadsides and constitute average urban roadside landscapes (Figure 1; *Right*).

The components of urban roadside landscapes are several and variable compared with rural roadside landscapes. These multiple landscape components either distract the perception of the drivers and passengers or contribute to the concentration and relaxation of those travelers. Urban roadside landscapes therefore constitute more complex factors, associated with the safety and recreation of drivers and passengers. Based on a study of drivers' perspectives, drivers favored the open urban expressways rather than urban tunnel expressways [3].

*Figure 1. Buildings, slopes and bypasses within the roadside landscape along an urban highway (**Left**). Buildings, trees, shrubs and traffic signs within the roadside landscape of an inner-city road (**Right**).*

2.1. Safety Risks and Benefits of Urban Roadside Landscapes

Although urban roads carry more accident risks compared to rural roads because of their heavy traffic potential, the severity and fatality of accidents are relatively higher for rural roads because of higher velocity and longer distances travelled along these roads. In fact, no kind of road can be declared absolutely safe. Roadside landscapes play a significant role in both the occurrence and prevention of accidents along urban roads. It has been suggested that the more relaxing the roadside landscape, the less likely it is that car accidents will occur [4].

Firstly, intensive and tall buildings in the roadside landscape are a source of negative pressure, first on the drivers and then on the passengers. According to a video-taped simulated drive test study, participants who drove with an artifact-dominated view suffered more stress than those with a nature-dominated view [5]. In addition to the heavy traffic, intensive and tall buildings stimulated anxiety and stress amongst drivers and passengers (Figure 2; *Left*). Besides this, alternating colors and sizes of roadside buildings, objects, and equipment such as dynamic advertisement billboards also distracted the vision of the driver who was barely able to focus on the road and traffic flow (Figure 2; *Left*). On the other hand, well-coordinated roadside buildings which are proportionate with each other particularly in terms of their configurations, textures, colors and sizes, allow drivers to concentrate on the road and traffic flow, enabling them to easily follow the road and traffic signs (Figure 2; *Right*).

Similar safety benefits are ensured by the beneficial effect of appropriately designed roadside trees, objects and equipment, such as bus stops located in harmony with the road. As a matter of fact, the roadside landscape improvement in Texas (USA) led to a pronounced contribution to traffic safety, apparently diminishing the crash rate [6].

*Figure 2. Buildings, traffic signs and monuments along the urban roadside landscape; an image of heavy traffic on the road (**Left**). An example of a relatively well-coordinated roadside of buildings and trees; an image from a section of an inner-city roadside landscape (**Right**).*

Impacts of day/night and seasonality

Urban traffic is heavier in the course of daytime hours than it is in the night-time hours. During daytime hours, sunlight directly penetrating through the tall buildings or reflected from those buildings dazzle the driver if it is not absorbed by the roadside trees, or at least by the sun visors. Depending on the angle of the penetrated or reflected sunlight, the intensity of the light to which the driver is exposed can change. Correspondingly, the dazzled driver's field of vision is restricted; posing a risk of accident within heavy traffic in particular. Consequently, the organization and configuration of the roadside trees not only provide a relatively regulatory impact on the traffic, but also serve as a roadside buffer in order to absorb direct or reflected sunlight, ultimately contributing to accident risk reduction. Indeed, the statistics-based research and driving simulator studies have shown a preference for tree-dominated aesthetic landscapes, and proposed them in place of built-up landscapes, as elements working against driver stress and reducing the risk of accidents along the urban roadsides [7]. The roadside vegetation within the central refuge partially contributes to the obstruction of the headlight beams from vehicles in the opposite lane, mitigating any dazzling for the driver.

During nighttime hours, the electrical elements within the roadside landscape contribute to the road lighting for the vehicles (Figure 3; *Left*). This artificial light is therefore intentionally directed and oriented towards the road and supplies concentrated lighting upon it, particularly concealing and shading the roadside landscape farther back (Figure 3; *Left*). This means that throughout these nighttime hours, the drivers pay more attention to the road and the traffic rather than being influenced by alternating city lights and accordingly, by the relatively distracting urban silhouette and landscape. To be specific, elements such as the roadside trees, the lighting equipment design, and, in particular, the organization and configuration of the lampposts and the orientation of the lamps are significant from the point of view of lighting intensity and therefore road safety.

Seasonality also affects urban traffic safety, principally in terms of the combined effects of topography and precipitation, mainly in the form of snow, within the roadside landscape [8]. Indeed, when the roadside landscape is covered with snow in winter, due to the higher snow albedo – known as the dimensionless measure of solar radiation reflection potential of the surfaces [9], the sunlight reflected from the snow will almost temporally blind the driver.

Although urban heat induces the rapid melting of snow, winter cities that have emerged mostly at high altitudes and at high northern latitudes experience long-term snow cover within their roadside landscape (Figure 3; *Right*). In these winter cities, the heavy snow cover is ploughed from the road surface and drifts onto the pavements within the roadside landscape, threatening the traffic by hiding these pavements (Figure 3; *Right*). After snow, fog and mist are also effective for traveler and traffic safety because they hide and blur the roadside landscape visibility, which then possibly leads to crashes of the vehicles with each other or with roadside objects and components. Frost causes serious problems, particularly during highway construction. [10].

*Figure 3. Alignment of lighting equipment along the urban road within the seaside landscape; an image during the nighttime hours (**Left**). Ploughed and drifted snow cover on the roadside pavements; an image from a winter city (**Right**).*

2.2. Recreation Potential of Urban Roadside Landscapes

Given the relative complexity of urban roadside landscapes, they often perturb and confuse the passenger who is endeavoring to prepare and gather his/her thoughts, usually before school or work, and who is struggling to strip his/her mind of these heavy burdens. Nevertheless, harmonized roadside landscape components also offer recreation for those travelers.

The recreation potential of urban roadside landscapes is relatively restricted compared to that of rural roadside landscapes. This recreation potential of urban roadside landscape manifests itself through the remarkable and aesthetic characteristics of roadside vegetation (Figure 4; *Left*). Nonetheless, variety in roadside vegetation is preferred to a uniform composition, as mentioned in a study conducted in Tehran, Iran [11]. In addition, there are suggestions that urban forestry be considered during the planning and management proposals for urban transportation [12].

Moreover, spectacular urban built environments within archeological, old or heritage cities including, in particular, historical settlements involving antique and old buildings, monuments and objects, fascinate passengers, particularly those who visit these cities as tourists (Figure 4; *Right*). However, the impact of objects of heritage on the drivers is relatively contradictory as it places stress, or did on some drivers during the driving simulator test [13]. On the other hand, these historical urban roadside landscapes may be relatively normal for native travelers, rather than recreational.

*Figure 4. Deciduous trees along an urban road within a seaside landscape; an image during nighttime hours (**Left**). Historical houses and trees along the urban roadside landscape; an image from a heritage city (**Right**).*

Impacts of day/night and seasonality

The nighttime hours of an over-urbanized and built-up city are comparatively preferable to the daytime hours, as regards various aspects of the roadside landscape. In fact, even the complicated, irregular and messy scenery of that roadside landscape can be concealed and shaded or favorably altered and modified by means of convenient lighting techniques and equipment during nighttime hours. Moreover, planned and well-organized roadside landscapes with proportionally dimensioned and configured buildings, properly designed and positioned trees or shrubs, carefully selected and placed components and objects together with appropriately architected and located lighting techniques and equipment satisfy both drivers and passengers. Provided that such an urban roadside landscape occurs close to the sea, recreational satisfaction is ensured for travelers during the nighttime hours (Figure 4; *Left*). Furthermore, ancient urban castles, historic sights and bridges thoroughly accessorized with colorful lighting equipment, and/or cafés planted with decorative trees, shrubs or flowers, please and comfort travelers not only during the daytime hours but also during the nighttime hours.

Seasons make a difference in determining the visual preferences and quality of the urban roadside vegetation [14]. Along an urban roadside landscape, seasons manifest themselves primarily in the form of the foliation, blooming and fading, and defoliation processes which are respective indicators of spring and autumn (Figure 4; *Left*). These fading and defoliation processes are particularly associated with the ecophysiological characteristics of urban vegetation [15]. These changing colors of the urban roadside vegetation during the spring, summer and autumn, together with the snow cover in winter, offer recreation for travelers.

3. Rural Roadside Landscapes

Rural roads refer particularly to the inter-city roads that appear locally in the form of highways, freeways and bypasses, and that usually diverge into the side-roads leading towards and through towns and villages. Thereby, distinct or merged land use consisting frequently of forests, pastures and agricultural areas in which scattered villages (Figure 5; *Left*) and cattle occur are comprised within the rural roadside landscapes. These land uses and ecosystems are connected via the landscape corridors or dissolved into landscape fragments with alternating topography along these rural roads. Nevertheless, sometimes the road itself either constitutes a corridor [16, 17] or leads to such fragmentation [18]. This inevitable fragmentation is to some extent amended by constructing ecological corridors [19, 20] and tunnels in order to join isolated habitats (Figure 5; *Right*).

*Figure 5. Rural roadside landscape with various land uses including forest, agricultural land and pasture (**Left**). Ecological tunnel application to connect isolated habitats along the rural highway (**Right**).*

Occasionally rivers and seas escort these roads in patches and become part of these roadside landscapes. Travelers track these straight or meandering rivers along the plane and prairie landscapes (Figure 6; *Left*). Travelers seldom catch a glimpse of these rivers that are hiding behind the hills and trees within the forested mountainous landscapes. Highways partly assisted with viaducts and tunnels extend along the rural landscapes or serve as an alternative for those primary roads (Figure 6; *Right*).

Even though rural roads are often associated with scenic landscapes and are frequently reminiscent of recreation for the travelers, they may cause frustration and anxiety, risking safety here and there. However, the study that investigated the effects of roadside vegetation on the driver's anger and frustration revealed the positive impact of roadside vegetation in order to mitigate frustration [21].

*Figure 6. Meandering river and village along the rural roadside landscape; an image from a plateau (**Left**). Viaduct, side-road and topical bus stop along the rural roadside landscape (**Right**).*

3.1. Safety Risks and Benefits of Rural Roadside Landscapes

During the construction of rural roads based particularly on the cut-fill techniques, retaining walls are typically applied in order to compensate for emerging topographical alterations. The retaining walls of the curving rural roads, higher than the drivers' upper vision field and constraining their visibility distance, trigger drivers' anxiety, worrying them about the state of the traffic ahead of those curves (Figure 7; *Left*).

Accordingly, drivers suffer accidents throughvehicles, rocks or animals suddenly appearing behind those high retaining walls along rural roads. Landslides, sometimes including the collapse of the retaining walls, pose serious problems for the vehicles' safety, travelers and traffic [22, 23]. On the contrary, with reference to a driving simulator study, open roadside landscapes led the drivers tested to experience relatively less stress [24].

Parked vehicles ready for departure perturb drivers, particularly inexperienced ones who drive close to the roadside. This situation holds especially for the single lane rural roads where vehicles which are suddenly introduced can confuse the driver. On such occasions, the agitated driver is involuntarily forced or led to change lane, putting the vehicle in the opposite lane in jeopardy. The severity of relevant accidental risk is relatively higher on rural highways and freeways on which traffic flows with higher velocities and vehicles moving on the byway suddenly enter these highways and freeways. Regarding a driving simulator study, the vegetation density increment did not lead to any driver speed reduction [25].

Rural seaside roads, due to the refreshing, relaxing, calming, inspirational and pleasant impression of marine blueness, remove a spiritual burden from wayworn drivers and passengers (Figure 7; *Right*). Therefore, such drivers are, as it were, purified of anxiety and stress, and therefore avoid sudden motions of passing and overtaking. Indeed, drivers equipped with tolerance and patience pay more attention to the road and traffic rather than concentrating on the motion of individual vehicles and sticking with the behavior of particular drivers, and in turn minimize the possible risk of accidents. However, rural roads near the sea but at higher altitudes above sea level may excite and threaten them, especially the inexperienced drivers who pose considerable risks for the travelers.

*Figure 7. High retaining wall along the rural roadside landscape (**Left**). Marine landscape along the rural inter-city road (**Right**).*

Impacts of day/night and seasonality

Safety risks occur mostly on meandering rural roads, notably during the nighttime hours, in particular where the inadequate roadside lighting elements and barriers associated with sharp slopes and cliffs are involved. Driver failures along with the obscurity of the roads increase the severity of those safety risks. The daytime hours, however, mitigate these risks to lower levels, providing relative contributions to the relaxation and recreation of the drivers. According to a recent study along a Chinese freeway, the brighter colors of the roadside landscape reduced drivers' fatigue levels [26].

Precipitation has significant effects on drivers' perceptions of rural roads and the roadside landscape. Heavy rainfall narrows the drivers' field of vision, orienting them towards their own vehicle and leading them to concentrate on their own route rather than controlling the traffic around them. This situation is relatively dangerous in terms of increasing the possibility of traffic collision.

Principally the nighttime hours and then also the early daytime hours of the winter season pose considerable safety risks because of possible snow cover, ice and frost on the rural roads (Figure 8; *Left*). Sunlight either directly penetrating through the mountains (Figure 8; *Right*) or intensively reflected from the snow cover within the rural roadside landscapes dazzles the drivers, leading them to lose their steerability.

*Figure 8. Snow cover within the rural roadside landscape (**Left**). Sunlight penetrating through the mountains along the rural roadside landscape (**Right**).*

3.2. Recreation Potential of Rural Roadside Landscapes

Rural roadside landscapes have many aesthetic characteristics which already and intrinsically offer recreation for both drivers and passengers. Rural houses and cottages, either individual or aggregated in the form of villages, and rural houses and cottages embedded within agricultural areas and/or forests, display harmonious pastoral landscapes along these rural roads. Additionally, further natural components such as birds, rocks, creeks and reliefs, together with cultural components such as bus stops, old bridges and industrial mills. may play part of a supplementary role for the sake of the travelers' perception along these rural roadside landscapes. Consequently, beyond the transportation requirements of the travelers, rural roadside landscape elements and components are created and portrayed in order to satisfy their spiritual demands and to obtain recreation.

Variety in colors, textures, forms, dimensions and topography emphasize the spatial dynamics along roadside landscapes. Depending on the travelers' perceptions of all these variations, the degree of recreational content they add to the spatial dynamism of the landscape is revealed. Based on expert scenic evaluation of semi-rural highway landscapes, vividness appears often as a descriptor [27].

The topography is also determinant of the travelers' field of vision, either allowing wider and longer vision or restricting the vision, even down to solely that of the road corridor. The extension ranges of these vision fields fluctuate, based on the geomorphological features and attributes of the rural roads and roadside landscapes.

Impacts of day/night and seasonality

Usually the visibility of rural roadside landscapes is relatively limited during the nighttime hours, due to the only reciprocally localized lighting elements on the roadside, unless rural settlements, industries or tourist facilities exist within those rural roadside landscapes. Hence, daytime hours when the field of vision expands are more advantageous for drivers and passengers in terms of recreation.

According to the questionnaire survey amongst roadside users from northern England, the roadside vegetation is an important component of the roadside landscape [28]. The phenology and ecophysiology of the rural roadside vegetation just next to the road [2, 29] and within the rural roadside landscapes [30] are significant in terms of the recreation potential of these rural roadside landscapes for both drivers and passengers. The phenological and ecophysiological characteristics of the roadside vegetation are directly associated with seasonality and climate.

After the first onset of fresh and glossy leaves, the deciduous roadside vegetation inspires travelers, indicating and welcoming the coming of spring (Figure 9; *Left*). Therefore, the drivers and passengers experience peace and bliss, and enjoy comfortable and recreational travel along roadside forest landscapes. On the other hand, in autumn, despite its reputation as being a "season of sorrow", the colors of the deciduous roadside vegetation array sequentially turn into pale green, yellowish green, straw yellow, reddish brown and pale brown [30]. These color patterns are also seen in roadside agricultural landscapes as well as forest landscapes (Figure 9; *Right*). This riot of autumn colors and the combination of these colors, along with other roadside landscape elements, fulfill the recreational needs of drivers and passengers to a certain extent along their journey.

*Figure 9. Leaf onset of deciduous trees along rural roadside landscapes; an image during spring (**Left**). Leaf fading and defoliation along the rural roadside landscape; an image during autumn (**Right**).*

References

1. J. D. DELGADO, N. L. ARROYO, J. R. ARÉVALO, J. M. FERNÁNDEZ-PALACIOS: *Edge effects of roads on temperature, light, canopy cover, and canopy height in laurel and pine forests (Tenerife, Canary Islands). Landscape and Urban Planning,* 81, 328–340 (2007).

2. M. ÖZTÜRK, E. GÖKYER: *Seasonal variation in light transmission and canopy gaps of deciduous roadside vegetation: Assessment within forest landscape. Eurasian Journal of Forest Science,* 3 (2), 1–11 (2015).
3. J. S. YEUNG, Y. D. WONG, H. XU: *Driver perspectives of open and tunnel expressways. Journal of Environmental Psychology, 36, 248–256.* (2013).
4. I. MATIJOŠAITIENĖ, K. NAVICKAITĖ: *Aesthetics and safety of road landscape: are they related? Journal of Sustainable Architecture and Civil Engineering, 1 (1), 20–25.* (2012).
5. R. PARSONS, L. G. TASSINARY, R. S. ULRICH, M. R. HEBL, M. GROSSMAN-ALEXANDER: *The view from the road: Implications for stress recovery and immunization. Journal of Environmental Psychology, 18, 113–140.* (1998).
6. J-H. MOK, H. C. LANDPHAIR, J. R. NADERI: *Landscape improvement impacts on roadside safety in Texas. Landscape and Urban Planning, 78, 263–274.* (2006).
7. K. K. DIXON, K. L. WOLF: *Benefits and Risks of Urban Roadside Landscape: Finding a Balanced Response. Proceedings of the 3rd Urban Street Symposium (June 24–27, 2007, Seattle, Washington, USA).* (2007).
8. M. ÖZTÜRK, E. GÖKYER, A. ÇETİN, Y. DÖNMEZ: *Topography based transportation risk classification for negative winter conditions within Amasra-Bartın-Safranbolu route. 10–12 February 2016. Erzurum, Turkey. pp. 1257–1267.* (2016).
9. M. TÜRKEŞ: *Climatology and Meteorology (in Turkish). Kriter Publishing. İstanbul, Turkey.* (2010).
10. W. M. MARSH: *Landscape Planning, Environmental Applications, Fifth Edition. John Wiley and Sons, Inc. USA.* (2010).
11. M. FATHI, M. R. MASNAVI: *Assessing environmental aesthetics of roadside vegetation and scenic beauty of highway landscape: Perspectives and perception of motorists. International Journal of Environmental Research, 8 (4), 941–652.* (2014).
12. K. L. WOLF: *Freeway roadside management: The urban forest beyond the white line. Journal of Arboriculture, 29 (3), 127–136.* (2003).
13. H. ANTONSON, C. AHLSTRÖM, S. MÅRDH, G. BLOMQVIST, M. WIKLUND: *Landscape heritage objects' effect on driving: A combined driving simulator and questionnaire study. Accident Analysis and Prevention, 62, 168–177.* (2014).
14. E. EROĞLU, Z. DEMİR: *Phenological and visual evaluations of some roadside deciduous trees in urban area. Biological Diversity and Conservation, 9 (1), 143–153.* (2016).

15. M. ÖZTÜRK, E. GÖKYER, A. ÇETİN, Y. DÖNMEZ: *Urban aesthetics oriented evaluation of autumn-winter ecophysiological dynamics of forest landscape within Bartın City and Vicinity. International Winter Cities Symposium. 10–12 February 2016. Erzurum, Turkey. pp. 1268–1280.* (2016).
16. R. T. T. FORMAN, M. GODRON: *Landscape Ecology. John Wiley & Sons, Inc. USA.* (1986).
17. R. T. T. FORMAN, L. E. ALEXANDER: *Roads and their major ecological effects. Annual Review of Ecology and Systematics,* 29, 207–231+C2 (1998).
18. R. T. T. FORMAN, R. D. DEBLINGER: *The ecological road-effect zone of a Massachusetts (U.S.A.) suburban highway. Conservation Biology,* 14, 36–46 (2000).
19. M. G. TURNER, R. H. GARDNER: *Landscape Ecology in Theory and Practice; Pattern and Process, Second Edition. Springer-Verlag, New York, USA.* (2015).
20. V. INGEGNOLI: *Landscape Bionomics, Biological-Integrated Landscape Ecology. First Edition. Springer-Verlag, Milan, Italy.* (2015).
21. J. M. CACKOWSKI, J. L. NASAR: *The restorative effects of roadside vegetation; Implications for automobile driver anger and frustration. Environment and Behavior, 35 (6), 736–751.* (2003).
22. S. FLÜGEL, L. I. RIZZI, K. VEISTEN, R. ELVIK, J. D. DIOS ORTÚZAR: *Car drivers' valuation of landslide risk reductions. Safety Science, 77, 1–9.* (2015).
23. M. ÖZTÜRK, E. GÖKYER, M. SEZGİN: *Evaluation of the landslides that occurred along the Amasra-Bartın-Safranbolu road slopes and landscape reclamation applications against those landslides (in Turkish). National Landslide Symposium, 27–29 April 2016. Ankara, Turkey. pp. 635–646.* (2016).
24. H. ANTONSON, S. MÅRDH, M. WIKLUND, G. BLOMQVIST: *Effect of surrounding landscape on driving behavior: A driving simulator study. Journal of Environmental Psychology, 29, 493–502.* (2009).
25. C. D. FITZPATRICK, S. SAMUEL, M. A. KNODLER JR: *Evaluating the effect of vegetation and clear zone width on driver behaviour using a driving simulator. Transportation Research Part F, 42, 80–89.* (2016).
26. L. WANG, Y. BIE, S. LI: *The impact of roadside landscape colors on driver's mean heart rate considering driving time. Transportation Research Part F, 42, 151–161.* (2016).
27. G. R. CLAY, R. K. SMIDT: *Assessing the validity and reliability of descriptor variables used in scenic highway analysis. Landscape and Urban Planning, 66, 239–255.* (2004).

28. K. F. AKBAR, W. H. G. HALE, A. D. HEADLEY: *Assessment of scenic beauty of the roadside vegetation in northern England. Landscape and Urban Planning, 63, 139–144.* (2003).
29. M. ÖZTÜRK: *Complete intra-annual cycle of leaf area index in a Platanus orientalis L. stand. Plant Biosystems-An International Journal Dealing with All Aspects of Plant Biology, 150 (6), 1296–1305.* (2016).
30. M. ÖZTÜRK, E. GÖKYER: *Spatio-temporal Characteristics of Roadside Landscape Components Along the Coastal Zone Between Amasra and Kurucaşile Cities of Bartın Province, Turkey. Landscapes and Greenways of Resilience (Eds.: Jombach, S., Valánszki, I., Filep-Kovács, K., Fábos, J. Gy., Ryan, R. L., Lindhult, M. S., Kollányi, L.) "Proceedings of 5th Fábos Conference on Landscape and Greenway Planning (30 June-03 July; 2016)". Budapest, Hungary. pp. 293–302.* (2016).

Ayça Yeşim Çağlayan[1], Betül Atakan Öznam[2]

Sustainable Space-Sustainable Living: Physical Activity Friendly Urban Open and Green Spaces

[1]ayesim@stanbul.edu.tr
[2]pmbetul@istanbul.edu.tr

1. Urban Life Quality, Livability and Recreation

Pursuant to the definition from the World Health Organization, quality of life is a complex and broad concept which is shaped in relation to a wide range of factors such as the physical health of individuals, their psychological state, their economic status, their level of freedom, their social relations and their interactions with the main characteristics of the environment in which they live [1], [2], [3], (Figure 1).

In order to be able to talk about the concept of "livability" or "quality of life" in a space, it is also necessary to make allowances for subjective evaluations that determine the level of people's satisfaction as well as physical environment standards that are objectively measurable [4].

Physical environmental quality can be defined by features such as the availability of open and green spaces, transportation networks (accessibility), infrastructure and municipal services, communication (accessibility), socio-cultural activities, the conservation of natural and historical values, planned housing and living environments, housing types and their quality, the reduction of the environmental effects of work areas, and the presence of recreational areas [3], [5].

1. The Place and Importance of Urban Open and Green Spaces for Quality of Life

Architectural constructions, open green spaces, their relation to each other and so their integrity shape the character of a city. Urban open green spaces have an important position in balancing the disrupted relationship between humans and nature, and improving urban living conditions. Accordingly, the qualities and quantities of open green areas are regarded as a sign of civilization and quality of life in developed countries [6].

Figure 1. Basic Elements of Life Quality (RIVM, 2000) [3].

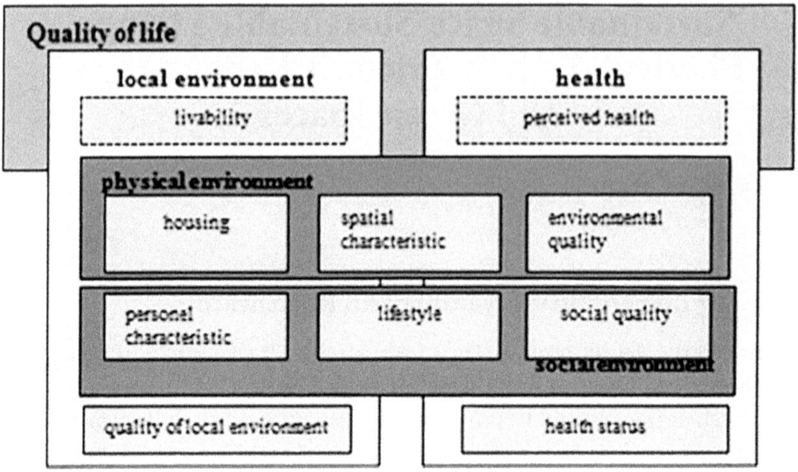

The quality of the urban space also determines the quality of life of the individual. What is important for ensuring the quality of urban life is establishing a healthy urban structure. It is therefore necessary to meet economic, cultural, environmental, social and political needs and to regard each of those as individual rights, and so to apply, protect and develop them [7].

Urban green spaces are public spaces that determine the quality of the physical and social environment within the city space, that allow for educational, cultural and recreational uses, and that are free for every individual in the community [8].

At the same time, open and green spaces are focal points where cultural accumulation is reflected and social communication is realized. These areas are divided into various classes according to their functions, service areas and locations. Made both ecologically and socially, this classification facilitates the planning of the urban distribution of green areas, and the quality of services, equipment, landscape value, accessibility and aesthetic value specified according to these categories also determines the contribution of green areas to the quality of urban life [9].

The spaces which provide a high quality of life are the ones where people can relate, hold on, feel belonging, and identify with themselves, and which they remember and miss. In other words, establishing links between people and spaces is an important theme in creating a high-quality urban life [10].

Planned treatment of active green spaces is one of the most important indicators for the quality of life, as it would enhance the quality of life [11].

In order for the open and green spaces to be functional, the amount of space per person, the equipment elements and the quality of each green area in terms of service within its boundaries, and its visual activity are important (Figure 2) [12].

Figure 2. The Impact of Open and Green Spaces on Quality of Life [9].

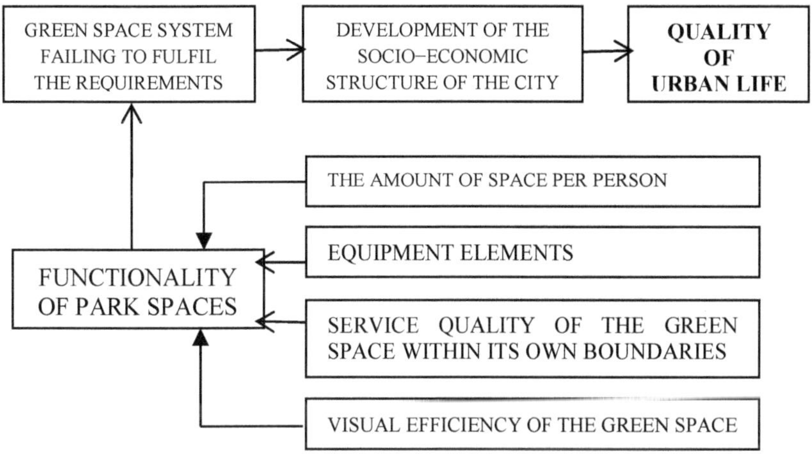

Along with the rapid growth of the urban population in our country, the change in social, economic, political and cultural conditions has lead to a gradual decrease of open green areas, especially in city centres. Cities that take such a form move people away from their natural environment, make them monotonous, adversely affect them physically and mentally, and eventually reduce the quality of life of the town-dwellers [2].

Green areas in urban spaces play a crucial role in finding a solution to environmental problems, fulfilling people's longing for greenery, and meeting their recreational needs [13]. In cities burdened with physical, social, biological and hygienic problems, there is a need for open and green spaces with characteristics that will give people vitality, be able to infuse them with new energy, and will enable recreational activities in particular [14].

Recreation can be defined as leisure time activities voluntarily chosen by human beings to repair their physical strength and enrich their mental capacity, regardless of forced labour [15].

Also, one of the inevitable consequences of rapid urbanization in Turkey is the lack of valid public open and green spaces where people can play sports. The inclusion of sports and recreation areas in contemporary city planning is becoming more of an issue in terms of the creation of livable, healthy environments [16].

3. Sports Recreation and Physical Activity

Sports activities or, in other words, "recreational sports" constitute a large part of the recreation as a leisure activity that is increasingly becoming popular and widespread in our time [17].

Of all recreational activities, those based on physical exercise or the recreational implementation of various branches of sport, and that form a large part of recreational activities are called **sporty recreation**. Physical activities create a foundation for sporty recreation that participants perform in their free time without expectation of financial gain, and that provide satisfaction, fun and socialization [17]. For example, recreational hikes and walks carried out towards achieving everyday goals form a significant part of the daily physical activity of the individuals. Scientific research reveals the cause and effect relationship between physical activity and well-being. Regular physical activity removes the risk of developing numerous chronic illnesses such as heart and vascular diseases, bowel cancer, diabetes, and tension disorders. Besides, it is also known that a balanced diet and regular activity result in weight loss, thus helping reduce the overall obesity rates across the community [11]. Especially in developing countries, government and health organizations have expressed great concern about the levels of chronic illnesses caused by a lack of physical activity (PA) in the community. This means that the associated health costs gradually increase, and given this, collaborative research is being developed with experts in different disciplines to find ways to improve physical activity [18].

4. Physical Activity (PA) in Urban Open and Green Spaces

The relationship between health and physical space has become one of the most widely researched topics of our time. Examining the relationship between qualities in the physical environment such as the presence of walking areas, people's PA behaviors (walking habits, etc.) and health, researchers refer to the importance of creating public green spaces within urban areas [11].

It has been observed that a variety of factors play a role in PA behaviors; factors such as individual / intrapersonal (biological, psychological), interindividual / interpersonal (social, cultural), organizational, social, physical environmental conditions and politics come into question. Being developed with a socioecological approach, these models underline the need for a comprehensive study of the understanding of multiple and interactive determinants of PA behaviors. Sallis (2015) summarizes the matter in terms of multilevel intervention approaches as follows [19]:

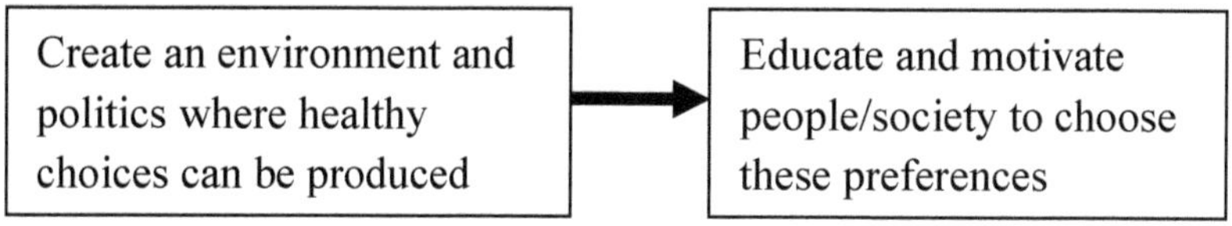

The built environment is the most important factor that either facilitates or prevents participation in PA. Public open spaces such as parks and green spaces are deemed to be the most basic environments that give a chance for a variety of physical activities such as recreational hiking and sports. For example, as mentioned previously, walking is one of the most frequent medium-intensity physical activities performed in urban open spaces where everyday life goes on. In this context, understanding the effects of the planning and design features of the structured environment on walking activity, and developing policies and strategies to guide the population into walking is of importance in raising the quality of life in the society as a whole [11].

The article concerning the "development of habitats" is emphasized as the 5th component within the framework of the "Multi–Stakeholder Health Responsibility Development Program" (2013–2023) by the Republic of Turkey's Ministry of Health. In that article, with the help of human and environment oriented regulations the aim is to provide the necessities of healthy living in and around houses; and preserve and improve the physical, mental and social well-being of human beings, with contemporary arrangements to be held in public places and places within social spaces [20].

Active life researchers usually describe that built environment as parks and green spaces, whereas public health experts express that this setting is not necessarily a green area but that what is important is that the intentional purpose of the space is recreational objectives. Accordingly, questions must be asked as to whether it is an active or passive space, and in terms of either size, design, or physical characteristics, if it meets the aims of a PA amenity, intended for recreation; such spaces being defined as urban environment areas that are available at all times for the use of the community andare freely accessible.

Baykan (2003) categorized open and green spaces according to the recreational function they undertake as parks (neighbourhood, city, district park etc.), special-purpose parks and gardens (roof gardens, hobby gardens, cultural gardens etc.), sports areas (golf, basketball, etc.), children's playgrounds, natural and semi-natural areas, corridors (urban forests, woodland etc.), other open and green spaces in the city (cemeteries, residential gardens, etc.) [21].

Open and green spaces are expected to provide three basic principles in order to be able to actively take on the task of "recreational space" [22]:

- Must be accessible, appropriately positioned for the purpose (time and money costs are taken into account in accessing costs); accessible by public transport on the urban and regional scale, and preferably within walking distance for pedestrians in the neighbourhood and sub-scales. It should be accessible to all segments of the community.
- Its size should be suitable for the purpose of use; the size of the area and the variety of actions it allows should increase as the distance from the user increases.
- Design must be appropriate for purpose; compliance with principles such as sensitivity to demand, ergonomics, security, aesthetics, etc. and a consideration of multi-purposefulness of use.

It is rather difficult to set a standard for recreational green spaces. However, it is both a very laborious and costly process to calculate recreational habits clearly, such as the frequency and duration of use of green spaces and the recreational demands of the town-dweller [21]. Since the changes that take place in the socio–economic structure of society over time could also affect the use of green space and recreational demands of potential users, the standards set out might be insufficient.

The service area of public open and green space may vary according to factors such as community welfare level (car ownership, income that can be allocated for transportation, time, etc.) ***for transportation by car***; topography, climate, and social preferences, also ***for pedestrian transportation***. Besides all of these, the concentration of settlements in city and urban sub-units and the type of residence affect the recreational choices of the residents. When the spatial norm is determined, housing properties in the service area should also be taken into account [21].

Some of the work carried out redefines the potential of parks to cover plazas, pocket parks, greenways, nature conservation areas, and other public open spaces, in order to create cities which are sustainable socially and environmentally [23].

Concordantly, it is emphasized that parks are one of the most important elements of the built environment that enable PA in their neighbourhood and vicinity [24].

5. Physical Activity Based Planning and Design of Public Open and Green Areas/Park Open Spaces

Scientific studies on the evaluation of free time in the past 40 years, especially emphasizing the physical, psychological, social, economic and environmental benefits of parks, have revealed the relationship between public open and green areas and PA [25].

In terms of PA, 4 issues were looked at:

- "frequency of participation in PA as leisure",
- "duration of participation in PA",
- "type of PA performed" and
- "the intensity of PA performed in the area (sedentary, moderate and vigorous activity).

Within this context, it was determined that the planning and design of public open spaces and green spaces have an effect on PA; and there was an attempt to establish the qualities that are effective in increasing the level of PA [25,26].

Factors such as the amount of free time the users possessed, their financial possibilities, personal health, level of education, possibilities for transportation and therefore access, and perception of security, as well as the continuity of maintenance, the adequacy of facilities in the area, social cohesion (presence of a companion during the activity) were the main criteria influencing the use of the area for PA [25].

In an attempt to increase the effect of parks on PA, during the planning and design process, each of the data pertaining to issues of "existing facilities", "maintenance conditions", "accessibility", "aesthetics", "security" or "management policy" had to be individually obtained, assessed, planned and designed for the four main zones that constituted the physical environment of the parks ("park activity area", "park supporting area", "general park environment " and "surrounding neighbourhood ") [25].

Usually used for PA or specifically designed for this purpose, **park activity areas** are the sections, zones and opportunity areas within the park. Examples are sports fields, swimming pools, routes and paths, playgrounds, open green areas, and other areas where PA is directly carried out [25]. Also, **park supporting areas** include facilities and equipment that make PA more interesting and safe for various users in parks. Such areas, even if they do not encompass the facilities that directly apply to the PA, are an integral part of the experience of the visit to the park. Picnic areas, eating and drinking places, and parking lots and so on can be can be given as examples of these areas. Those areas are the ones that directly affect "the frequency of the use of the park", "the duration of the stay in the park" and "the level of people's being active in the park" [25]. The park area is not just a region of the sum of certain parts. **The overall park environment** comprises the emphasis and meaning of the whole area. The aesthetic appearance of the park in general, its size, and the range of programs in the area are among the hallmarks of the overall environment of the park. Within this framework, especially "general park use", "size" and "accessibility" are the data types that need to be examined. The various features of **the surrounding neighbourhood** also affect people's perception and use of the park. "Traffic volume", "presence of neglected or derelict houses", "aesthetic qualities", "crime rate" and "demographic structure of the region" are the criteria that affect the use of the park [25].

In a piece of scientific research recently conducted, it was reported that facilitating access to the parks and similar spaces and adequate information increase the "PA frequency" by 48.4% [27]. For example, Kaczyski et al. (2014) found that regulation of lower traffic speeds supports the use of public open spaces [24]. Another study shows that with the perception of crime-free, safe, light in traffic but aesthetic, public open spaces around the neighbourhood are more used for walking purposes. At the same time, it was determined that the density and surface area of green spaces and parks in a region are related to how often that area is used for walking and cycling as means of transport. [27].

"Also, subjects that emphasize the types of data required in the planning and design of parks" form the basis for assessing the spatial and strategic characteristics of the park in relation to its levels of PA.

The most important issues affecting the level of use of the area for PA purposes are **"the park features"** that are located in the area. They can range from a sports ground used for football matches to a natural environment that allows you to experience calm and solitude.

Park features consist of the following elements [25]:

- **Facilities** that serve as basic environments for PA such as paths, routes, playgrounds, basketball fields, etc., and amenities which are defined as support factors such as picnic tables, lighting, parking lots, signs and warning signs for the realization of PA [28].
- Recreation organizations/programs practised within the area (e.g. neighbourhood matches or leagues after school etc.),
- Diversity composed of "park facilities", "organizations/programs" "locations where they are realized", and "combinations of users".

In the research aimed at the use of parks for PA purposes, it has been revealed that the number of park features in particular is one of the most important criteria. The parks whose average number of park features was 12.43 (5.6 facility structures and 6.57 amenities) were preferred for PA purposes, nevertheless, the ones with an average of 6.74 park features (2.74 facility structures and 4.00 amenities) were not. Among the facilities, the ones with paving materials, unpaved paths, and wooded areas are the most effective for increasing the level of PA. The parks where those facilities are available are used seven times more often for PA than those without. Due to enabling different physical activities (cycling, running, etc.), in other words, due to functionality, the presence of the path covered with paving material is the most preferred feature [26].

While park areas where facilities and amenities are regularly maintained are mostly preferred by the users, the ones with **degraded areas** are not. Another important issue is the **safety of the equipment** used. Many studies have emphasized that neglected equipment in playgrounds causes mortality, which is the most important criterion that plays a role in parents' selection of parks. Also, one of the important indicators of environmental conditions in parks is that of "**visual cues" that make people perceive deprivation**. Here, deprivation is the indication that signals traditionally accepted norms and values have suffered from erosion. In this context, irregular physical environments (garbage, graffiti, etc.) and destructive social behaviors (alcohol use, wandering, etc.) can be given as examples [25].

Another important conceptual issue is **accessibility** which defines the ability to reach, and navigate through, the park. For enabling different social groups to benefit from the park, an area's accessibility can be examined under four headings such as "availability of the park", "equal accessibility", "individual accessibility" and "accessibility in the park" [25].

Availability of the park defines the amount of actual park space as measured park space per hectare or per capita in the city in question. Scientific research emphasizes that parks with an average area of 22.34 ha are the most preferred areas for PA but parks with an average area of 0.83 ha are not [26]. At the same time, it is necessary to provide the possibility of equal distribution of parks to all communities of different ethnic and income levels living in different neighbourhoods, i.e. **equitable access**. Research have shown that the types of public open green space and PA behaviors that different groups need according to their socio-demographic structure vary [28]. Population density is a key indicator of equality. As the number of people a region shelters increases, so does the need for public open spaces. Income levels and ethnicity are important decision points.

For example, low-income groups are populations with high levels of deprivation. Because they cannot benefit from open air recreation facilities with private businesses, be they sports clubs or otherwise, they need more public open green spaces such as valid park areas [23]. Some scientific studies have also emphasized that disadvantaged groups can access fewer small parks with much smaller surface areas, even though they use park areas more.

In the United Nations, for a research project in which an analysis of the equality distribution within park areas was made, it was emphasized that despite an allowance for PA, parks had a rich variety of facilities and even that they could widely meet user demand. Suburban Amenity Parks, which are placed in urban landscapes of neighbourhoods with low-density single family homes, are not used by the local community. It has been suggested that the reason for this is that public open spaces are not thought to be safe and the locals are more likely tend to use private open spaces, such as their own gardens. However, it has been emphasized that despite the fact they offer only a few facilities and amenities such as playgrounds (84%), football fields (73%), picnic areas (71%) and shade structures (62%), Green Mini Parks that are frequently found in neighbourhoods with dense populations have high demand potential [23].

Yet another form of accessibility, **individual accessibility / proximity** defines the distance that a person has to travel from his / her home to the nearest park area. Distance to the nearest park area is considered to be the most effective factor in increasing the level of PA [27,28].

By referring to a study of the correlation between PA and living in proximity to a park in a nationwide group of adolescents in the United States, it was emphasized that nearby people were more likely to "participate more in PA" than those who lived far away, and it was suggested that keeping roads connecting homes, workplaces and schools within walking distances, and planning such distances, using parks with walkways, for example, would provide a basic support for active lifestyles during daily work. Also, some of the studies suggest that the distance to the nearest park area is not directly related to the level of PA use within the parks, but the correlation between the number of parks within 1.6 km and the amount of park area, as well as the use of the park for PA is strong [24].

Some of the reasons for different outcomes are foreseeable. For example, people may not be able to walk more than 800 m to reach the park, which can be taken as a threshold measurement for supporting PA in determining the location of the park. The determination of such threshold measures is an important step in the design of walking friendly neighbourhoods. In their study to investigate the qualities of public open space and their effects on the level of PA, Kaczinski et al. (2008), while emphasizing that park spaces used for PA were located on average 955 m from the participants' homes, stated that park areas with an average distance of 990 m away were not preferred for PA [26].

Another reason why the results of the research differ is that the objective and perceptual measures of proximity to the park may vary. Of groups with different socio-demographic structures, the relationship between accessibility distance and level of use varies. For example, there is a negative relationship between the duration of the trip to the park for minority groups and the level of park use. That is, proximity to the park is an important criterion for the users in minority groups. Another reason is also that users who do not find the park interesting facility-wise prefer not to to use it whatever the walking distance to the park may be [24].

Byrne and Wolch (2009) also emphasize that the reason for the use of urban parks demonstrated by various models is not only the socio-demographic characteristics involved; but also that 1) the political, social, historical and economic context of park areas, 2) the park amenities and environmental characteristics (e.g. landscape, facilities, environment use) and 3) the differentiation of perceptions relating to park accessibility, security and other conveniences are influential. Understanding the impact of different dimensions which shape parks according to the needs of different sociodemographic structures, is important strategic information in assessing the growing need for humans to benefit from the ecological system, for the sustainability of public health [29].

Also, **access within the park** describes the ability of people to easily navigate within the park's boundaries. This type of access refers to the ease of transportation between support factors such as a facility that allows for PA and a car park, etc. For example, when it comes to decision making, ideally a person arriving for the match has a transportation system in which he or she can easily take along his or her equipment to and from the car park, or that disabled users there must be able to circulate accessible spaces, to ensure that they benefit from all of the facilities and opportunities. Even if not physically, a "Feeling of Inaccessibility" may also affect the PA behaviors in the area. In other words, "perceived access" is an important factor that also needs to be considered [25]. Also, the concept of the **aesthetic** is another important conceptual issue that influences the level of participation in PA in the area. "Aesthetics" describes the perceived attractiveness concerning various design elements of the space (size of the area, plan, landscaping, balance of shady and sunny areas, topography, accessibility to the area, number of main entrance doors, and availability of plastic objects such as pools or sculptures) and is a powerful motivational tool in the realization of PA in the field, like a high landscape [25]. As has been observed, even if the size of the area is an important indicator of the park quality, it is not the only criterion. In the context of the land, the theme and facilities of the park, landscaping/aesthetic properties and the character of the social-physical environment are important factors to be considered.

6. Discussion and Conclusions

Urban open green spaces are areas that contribute to the physical and mental health of individuals within a structured environment, that allow for the realization of various recreational activities, and that offer appropriate potential possibilities for use, especially for PA purposes. Thanks to their features, they have an impact on the level of habitability of the city. Meanwhile, the quality of service in open green areas, which reflects urban cultural accumulation and also plays a crucial role in the realization of social communication, designates the contribution of those areas to the quality of urban life. In our country, especially in recent years, increasing urbanization activities and correspondingly an increasing urban population have left urban green areas insufficient in terms of quality and quantity. In the plans, failure to reserve area sizes in accordance with the standards set by the legislation for various reasons, and failures that arise in practice can prevent existing urban green areas from being sustainable.

Socioecological models emphasize the influence of the built environment for public health and a livable environment, and expressly state that a built environment which can create healthy preferences in order to increase participation in PA, in the name of forming a healthy society, should primarily be established. Following this, society must be motivated and directed towards choosing these preferences [28].

Recently, factors that are effective in PA behaviors have begun to be investigated in detail through different disciplines. These factors include individual characteristics such as the amount of free time owned in general, financial possibilities, personal health, socio-demographic characteristics, and resource features such as the capacity of facilities available in the area, location, size and character of design, and the management policies being applied. Indeed, there are difficulties in comparing research results due to differences in definitions brought to urban open and green spaces which are proved to have an impact on PA, and following on from this, concepts, and approaches in perceptual/physical measurement methods related to the area make it difficult to precisely and clearly set forth the criteria that might be effective in encouraging PA. Nonetheless, the effects of the basic factors cannot be ignored. Parks are the most important type of Public Open Spaces with a proven effect on PA within the built environment. In this context, the acquisition of data in the process of planning and design, an evaluation of the qualities of available public open spaces and their effect on PA, the relation a park space has to other public open spaces and its immediate vicinity, its distribution within a city, its location and the design approaches within the park area must be handled separately, and the social/physical environment characteristics should be evaluated in terms of accessibility, security and landscape/aesthetic characteristics which are confirmed to be influential on PA.

7. References

1. World Health Organization QoL Group, 1995. The World Health Organization Quality of Life Assessment (WHOQOL), Position Paper from the World Health Organization, *Social Science And Medicine*, 41, 1403–1409.
2. Ceylan, A., 2007. Yaşam Kalitesinin Arttırılmasında Kentsel Yeşil Alanların Önemi ve Kentsel Dönüşüm İle İlişkilendirilmesi, İ.T.Ü., Fen Bilimleri Enstitüsü, Yüksek Lisans Tezi, İstanbul.
3. Kamp, I., Leidelmeijer, K., Marsman, G., Hollander, A., 2003. Urban environmental quality and human wellbeing. Towards a conceptual framework and demarcation of concepts: a literature study, *Landscape and Urban Planning*, 65, 5–18.
4. Çerçi, S., 1997. Konut Yakın Çevresinin Kullanıcı Bilişsel, Duygusal ve Davranışsal Parametrelere Bağlı Olarak Değerlendirilmesi, İ.T.Ü. Fen Bilimleri Enstitüsü, Doktora Tezi, İstanbul.
5. Oktay, D., 2007. Sürdürülebilirlik, Yaşanılabilirlik ve Kentsel Yaşam Kalitesi, *Mimarlık Dergisi*, 335, Mayıs-Haziran 2007,http://www.mimarlikdergisi.com/index.cfm?sayfa=mimarlik&DergiSayi=53&RecID=1329 (Accessed: 20.04.2017).
6. A. Gül, V. Küçük., 2001. Kentsel Açık-Yeşil Alanlar ve Isparta Kenti Örneğinde İrdelenmesi, *Süleyman Demirel Üniversitesi Orman Fakültesi Dergisi*, ISSN:1302–7085, Seri:A, Sayı:2, s.27–48, Isparta.
7. Karadayı, H., 2006. Yaşam Kalitesi ve Kullanıcı Memnuniyetinin Kentsel Tasarımdaki Etkisine Çok Boyutlu Yaklaşım, İ.T.Ü.Fen Bilimleri Enstitüsü, Yüksek Lisans Tezi, İstanbul.
8. Yuen, B., 1996, Creating The Garden City: The Singapore Experience, *Urban Studies*, Vol. 33, Issue 6.
9. Emür, S.H., Onsekiz, D, 2007. Kentsel Yaşam Kalitesi Bileşenleri Arasında Açık ve Yeşil Alanların Önemi – Kayseri/Kocasinan İlçesi Park Alanları Ana-

lizi, *Erciyes Üniversitesi Sosyal Bilimler Enstitüsü Dergisi,* Sayı : 22 Yıl : 2007/1 (367–396 s.)

10. Mazumdar, S., 2007. Kentsel Yaşam Kalitesi ve Yer Duygusu, *Mimarlık Dergisi,* 335, Mayıs-Haziran 2007. http://www.mimarlikdergisi.com/index.cfm?sayfa=mimarlik&DergiSayi=53 &RecID=1330 (Accessed: 20.04.2017).

11. Türkoğlu, H.D., 2007. İstanbul'da Konut Alanlarında Yaşam Kalitesinin Ölçülmesi, *Mimarlık Dergisi,* 335, Mayıs-Haziran 2007. http://www.mimarlikdergisi.com/index.cfm?sayfa=mimarlik&DergiSayi=53 &RecID=1328 (Accessed: 20.04.2017).

12. B. Etli, 2002. Edirne İli Merkez İlçe Yeşil Alan Sisteminin Peyzaj Mimarlığı İlkeleri Yönünden İrdelenmesi, *Trakya Üniversitesi Bilimsel Araştırmalar Dergisi,* B Serisi,Cilt 3, No 1, 47–59, ISSN 1302 647X.

13. Gedikli, R., 2004. Rekreasyonel Kullanımda Kent Ormanlarının Önemi ve Açık Yeşil Alan Standartları, I.Ulusal Kent Ormancılığı – 53, Ankara.

14. Yılmaz, H. ve ark., 2003. Kars Kent Halkının Rekreasyonel Talep ve Eğilimlerinin Belirlenmesi, *Atatürk Üniversitesi, Ziraat Fakültesi Dergisi* 34(4), 353–360.

15. Anonim, 1988. Bir Yörenin Turizm İmkanlarının Değerlendirilmesi için Rehber, Kültür ve Turizm Bakanlığı Turizm Planlama ve Yatırımlar Daire Başkanlığı Yayınları, Ankara.

16. Üstündağ, Ö. ve ark., 2011. Spor ve Rekreasyon alanlarının Şehir Planlamasındaki Yeri ve Önemi, 6th International Advanced Technologies Symposium (IATS'11), Elazığ.

17. Şahin, İ. ve Kocabulut, Ö., 2014. Sportif Rekreasyon aktivitelerine düzenli katılımı engelleyen faktörlerin İncelenmesi: Akdeniz Üniversitesi Turizm Fakültesi Öğrencileri Üzerine bir araştırma, *Journal of recreation and Tourism Research* (JTRT) 2014, 1 (2)

18. Chyi-Rong Chioy, Wei-Lun Tsai, Yu-Fai Leung, 2010. A GIS-dynamic segmentation approach to planning travel rotes on forest trail Networks in Central Taiwan, *Landscape and Urban Planning* 97, 221–228.

19. Sallis, J.F., Owen, N., 2015. Ecological Models of Health Behavior, HEALTH BEHAVIOR, Theory, Research and Practice, Fifth Edition Chapter 3, 43–65, ISBN 978-1-118-62900-0 (epub).

20. T.C. Sağlık Bakanlığı Türkiye Halk Sağlığı Kurumu, 2014. Sağlığın Korunması ve Geliştirilmesine Çok Paydaşlı Yaklaşım, Yönetici Özeti, Çok Paydaşlı Sağlık Sorumluluğunu Geliştirme Programı 2013–2023, Bakanlık Yayın No: 970,

ISBN: 978–975-590-510-5, Ankara. http://thsk.saglik.gov.tr/eDosya/04/cevre_saglik/Ynetici_zeti2.pdf (Accessed: 02.05.2017).

21. Aydemir, S.E., 2004. Kentsel Donatılar – Açık Alanlar ve Rekreasyon Kentsel Yeşil Alanlar, Kentsel Alanların Planlanması ve Tasarımı, İber Matbacılık, Trabzon.

22. Ergin, Ş., 1989. Kentsel Açık Yeşil Alanların Niteliksel Değerlendirilmesine Yönelik Matematiksel Bir Model Önerisi, DEÜ Müh.Mim.Fak.Yayınları, İzmir.

23. Ibes, D.C., 2015. A multi-dimensional classification and equity analysis of an urban park system: A novel methodology and case study application, *Landscape and Urban Planning* 137 (2015) 122–137.

24. Kaczynski et al., 2014. Are park proximity and park features related to park use and park-bades physical activity among adults? Variations by multiple socio-demographic characteristics, *International journal of Behavioral Nutrition and Physical Activity* 2014, 11:146.

25. Bedimo-Rung, A.L., Mowen, A.J., Cohen, D.A., 2005. The significance of parks to physical activity and public health: A Conceptual Model, *American Journal of Preventine Medicine* 2005; 28 (2S2): 159–168.

26. Kaczynski et al., 2008. Association of Park Size, Distance, and Features With Physical Activity in Neighborhood Parks, *American Journal of Public Health*, August 2008, Vol 98, No. 8.

27. Küçükali H., Küçükali İ., Taşdemir, M., 2016. Sağlığın yeşil belirleyicileri: Parklar, *Sağlık Düşüncesi ve Tıp Kültürü Dergisi*, Sayı: 38, p. 40–46, İstanbul. http://www.sdplatform.com/Dergi/935/Sagligin-yesil-belirleyicileri-Parklar.aspx (Accessed: 04.05.2017).

28. Koohsari, M.J., Mavoa, S., Villanueva K., Sugiyama, T., Badland, H., Kaczynski A.T., Owen N., Giles-Corti. B, 2015. Public open space, physical activity, urban design and public health: Concepts, methods and research agenda, *Health & Place*, 33 (2015) 75–82.

29. Byrne, J., Wolch, J., 2009. Nature, race, and parks: Past research and future directions for geographic research, *Progress in Human Geography*, 33(6), 743–765

B.Cemil Bilgili[1], Mustafa Ergen[2], İbrahim Aytaş[1]

Analysis of Urban Green Spaces; Case Study of Çankiri

[1]*Çankırı Karatekin University, Faculty of Forestery, Department of Landscape Architecture*
[2]*Siirt University, Faculty of Engineering and Architecture, Department of Architecture*

Introduction

Discussion about the green spaces of cities has grown in importance within the field of urban development [1, 2, 3, 4]. Current research has shown that rapid urbanization can cause serious problems in urban areas as well as in natural areas. A lack of urban planning can also result in insufficient social infrastructure, unsustainability in development, and most importantly, a lack of green space. These problems must be analyzed in order to determine the best approach to urban development, and to ensure sustainability in cities.

Planning for urban green space is one of the most important indicators in creating livable places in cities [5]. Green space is very important in making cities more livable. From a functional point of view, urban green space is identified as being a place where people spend their leisure time [2]. In addition, a lack of urban planning can result in insufficient green space; this can create barriers in the development of social infrastructures. To better understand this problem, the researchers used the Pearson correlation analysis in order to determine the relationship of green space to urban neighbourhoods.

Uncontrolled urban development brings the risk of insufficient green spaces. Urban planning is of great importance in controlling development. However, cities do not consider insufficient green space to be a major problem. The urbanization process is one that first considers the primary needs of the people in the cities. Meeting these primary needs is important because the growth of cities results in increasing population density, and this population needs housing, and accessibility to schools, hospitals, etc. All urban planners and researchers more or less accept the belief that green space is also very important in creating a balance for all neighborhoods. This concept places green space at the core of the analysis in this research. The roles of planning for green space in urban areas, including green space in development plans, are expected to increase [6, 7]. It is important to understand the current and future relationship between green space and neighbourhoods. This research focuses on analyzing green space in neighbourhoods, and better understanding the process of urban development in terms of green space conditions.

An urban development plan was used to analyse the state of green space in Çankırı city. The plan also examines the future of development in Çankırı city and how this relates to green space. In addition, the plan looks at how green space should be developed in neighbourhoods in order to improve the quality of life in the cities [2]. It is well known that land and natural resources cannot be produced in a short time [8]. A lack of planning in urban development can result in a lack of green space for the residents. Because of this, there are regulations in Turkey regarding green space in urban development, and means of controlling the urban sprawl. In addition, green space is very important in creating a quality environment for the residents of the city.

The problem that arises is that there is not enough planning for sufficient green space during urban development. This article discusses the distribution of green space within neighborhoods.

The main questions are constructed as follows for the hypothesis in our research: How is urban green space transformed before and after the implementation of urban development plans? Is green space equally distributed within all neighbourhoods? These questions will help prove that our research shows how urban green space is an important indicator regarding sustainability and the quality of life. We will also show how green space is of great importance for a healthy urban development and for creating desirable living places.

Area Description

Çankırı city was chosen as the area for study. Çankırı is located in the middle of Anatolia in Turkey. Çankırı shares a border with Çorum, Kastamonu, Karabük, Bolu, Ankara and Kırıkkale (Figure 1). Çankırı is located between 40° 30' and 41° on the northern latitude and 32° 30' and 34° on the eastern longitude [9]. Çankırı has the climate conditions of Middle Anatolia, and it borders the Black Sea Region [9].

Figure -1: Location of Study Area-Çankırı City

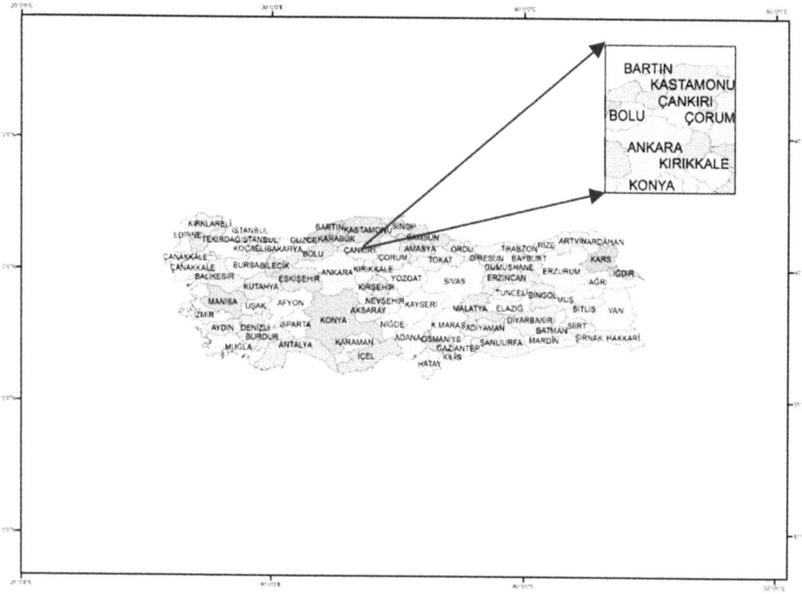

Regarding topography, most of the city is mountainous and has high hills [9]. The highest point is the Ilgaz Mountain, which is the natural border between the cities of Çankırı and Kastomunu [9].

Material And Method

We have chosen Çankırı city for analyzing green space conditions in urban development. The neighbourhoods and green spaces were determined for the analysis. The Geographical Information System was used to analyze the Çankırı development map. The analyses were conducted using the SPSS program. Spearman's correlation was used to determine the relationship between neighbourhoods and green space. Because there were less than 30 variables, one of the nonparametric methods, Spearman's correlation, was chosen for this research. It shows significant results when comparing the current development plan and future development plans regarding green space.

The study data were collected using ArcGIS software. The researcher described the neighborhoods and green space within the urban development plans of the Çankırı urban area with the assistance of this software. Accordingly, study data were collected. For the data analysis, the study used the urban development plans of the Çankırı province. Based on these plans, the study conducted a Pearson's correlation test to determine the distribution of the current green space in the neighbourhoods, using the SPSS program. Taking into consideration plans for the future, tests showed the amount of the green space within the neighbourhoods was the same. This operation showed the relationship between the neighborhoods and green space. The study also attempted to see the potential the city had to develop green space.

This study consists of three main sections. First, the neighborhoods and green spaces were described using the ArcGIS software, based on the urban development plans that were obtained from the municipality. In the second section, these data were analyzed using the SPSS software. Finally, the study evaluated the results that were derived from this analysis. The researcher intended to use the study results to demonstrate the amount and condition of the urban green space, and to understand the future of creating green space based on planning decisions for the future.

Results

This study is important for the comprehension of the planning status of urban green space in the province of Çankırı. It has been observed that there are continuing studies regarding creating a standard for the sufficient amount of green space within current urban planning. The areas highlighted in yellow on the map prepared by this study are the areas that are designated to be transformed into green space (Figure 2). This map shows how urban green space will be expanded through urban planning, particularly for the province of Çankırı. The map also shows that the goal is to include green space in all neighbourhoods, and to increase accessibility through the proper distribution of green space to all neighbourhoods in the city.

Table 1 shows that the current development plans do not indicate a correlation at a 95% confidence level between neighborhood areas and green space. This shows that the green space does not have a balanced distribution within the neighborhoods.

Figure 2. Green spaces

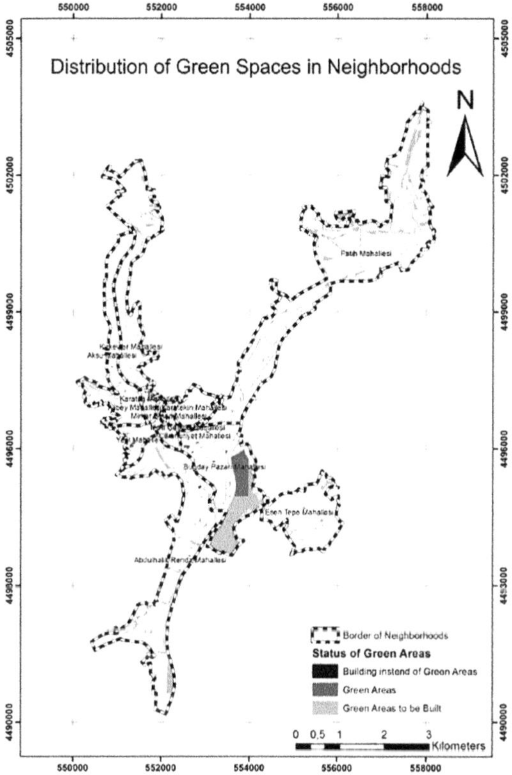

Table 1: The Analysis of the Neighborhood and Green spaces on the Current Development Plan

Correlations			
		Neighborhood areas	**Green space**
Neighborhood areas	Correlation Coefficient Sig. (2-tailed) N	1.000 . 14	0.474 0.087 14
Green space	Correlation Coefficient Sig. (2-tailed) N	0.474 0.087 14	1.000 . 14

An analysis made using the anticipated green space in Table 2 shows that there is a significant correlation at a 95% confidence level between neighborhood areas and green space. According to this result, there was an increase in the amount of urban green space, and a significant correlation appeared between these areas and the neighborhoods.

Table 2: The Analysis of the Anticipated Green Space for the Development Plan In Addition to the Neighborhood

Correlations			
		Neighborhood areas	**Green spaces to be built**
Neighborhood areas	Correlation Coefficient Sig. (2-tailed) N	1.000 – 14	0.893(**) 0.000 14
Green spaces to be built	Correlation Coefficient Sig. (2-tailed) N	0,893(**) 0.000 14	1.000 . 14

**. Correlation is significant at the 0.01 level (2-tailed).

Table 3 makes a collective analysis of the current status and the proposed green space, showing that there is a significant correlation at a 95% confidence level between neighborhood areas and green space. In conclusion, the study proves that the current and planned green space in the urban area of Çankırı are mainly included in the planning studies related to the urban area, and green space plays a major role in the urban development potential of Çankırı.

Table 3: The Collective Analysis of the Current and Proposed Green Space as Well As the Neighbourhoods

Correlations			
		Neighborhood areas	**Total green spaces**
Neighborhood areas	Correlation Coefficient Sig. (2-tailed) N	1.000 . 14	0.,920(**) 0.000 14
Total green spaces	Correlation Coefficient Sig. (2–tailed) N	0.920(**) 0.000 14	1.000 . 14

**. Correlation is significant at the 0.01 level (2-tailed).

The total surface area of the anticipated green space in the neighborhood development plan and the current green space shows that the correlation between the neighborhoods and green space is optimal. This situation indicates that the real distribution of the green space has increased systematically.

Discussion

This study has proved the importance of planning for green space. In this regard, this study has demonstrated the distribution of green space among the neighbourhoods, and supported this distribution on statistical grounds. A review of the previous studies examining green space showed that an increase was discovered in the residents' satisfaction regarding the proximity and distance of the green spaces [10]. In consistency with the present study, they measured satisfaction with the distribution of green spaces based on users. Another study drew attention to the point that expanding a livable and sustainable urban environment depends on the multifunctional green background [11]. Obviously, increasing urban livability and creating sustainable urban areas depends on the creation of urban green space, and on making this green space accessible and sufficient for everyone. In this regard, this study draws attention to the fact that the prospect of developing adequate green space is important. This concept was tested in this study based on development plans, particularly those for the province of Çankırı. Expanding green space needs to be part of new development plans. Another relevant study has emphasized that the characteristics of urban green space are important for users, and these should include both cultural and historical characteristics [12]. In the implementation to be made in the province of Çankırı, the distribution of urban green space should not be tackled considering only parks. It is necessary to plan for these areas from a broader perspective, in order to make them beneficial for a variety of users. This consideration is important for user satisfaction and the quality of urban life. A comparison with other studies showed the importance of the urban green space analysis of Çankırı. Accordingly, the changes in the development plan for the urban area, as well as expanding the green space, are both very important for increasing urban livability and ensuring the satisfaction of the public. The contribution of these areas to urban ecology and their pleasing visual effect are the other factors that show the importance of the proper distribution of urban green space.

Conclusion

This study reveals the significance of urban green space and emphasizes the importance of a balanced distribution of green space in cities. The goals of urban development include creating more livable places in cities. It is inevitable that the balanced distribution of green space within neighborhoods is crucial. This study chose Çankırı as an example in the analysis that demonstrated the significance of green space development. Planning a balance of green space is one of the ways to ensure a more sustainable approach in urban planning. The best planning approach is to determine where the green spaces are within the urban environment, and make plans to add green space where it is needed. At this point, the most important factor in the analysis of the urban area in Çankırı is that the urban green space created due to planning decisions plays a very important role for the neighborhoods of the city.

References

[1] E. BESUSSI, N. CHIN, M. BATTY and P. LONGLEY: *The Structure and Form of Urban Settlements. In Remote Sensing of Urban and Suburban Areas. 13 edited by T. Rashed and C. Jürgens,* 13–31, Remote Sensing and Digital Image Processing Series. Vol. 10, Springer Science+Business Media, DOI: 10.1007/978–1-4020-4385-7_2, (2010).

[2] F. J. GISBERT, I. C. MARTI, E. GIELEN: *Clustering Cities through Urban Metrics Analysis. Journal of Urban Design,* DOI: 10.1080/13574809.2017.1305882, (2017).

[3]J. LOWRY and M. LOWRY: *Comparing Spatial Metrics that Quantify Urban Form. Computers, Environment and Urban Systems,* 44:56–97, (2013).

[4]N. SCHWARZ: *Urban Form Revisited-Selecting Indicators for Characterising European Cities. Landscape and Urban Planning,* 96 (1): 29–47, (2010).

[5] Y. B. ERGEN and M. ERGEN: *Application of Geographical Information System to the Analysis of Urban Green Areas in Urban Development Plans – A Case Study of Zonguldak, Turkey.* Advances in Landscape Architecture, Dr. Murat Özyavuz (Ed.), Intech, DOI: 10.5772/55752, (2013).

[6] K. TSUCHIYA, Y. HARA, D. THAITAKOO: *Linking Food and Land Systems for Sustainable Periurban Agriculture in Bangkok Metropolitan Region. Landscape and Urban Planning.* Vol 143, 192–204, (2015).

[7] K. C. SETO, M. FRAGKIAS, B. GÜNERALP, M. K. REILLY: *Meta-Analysis of Global Urban Land Expansion. PloS One.* (6):e2377, (2011).

[8] Y. B. ERGEN: *Deprivation of Ecological Balance As a Result of Opening of Agricultural Lands Due to the Pressure of Urbanisation. Urban Ecology.* Edited by Münir A. Öztürk, Ü. Erdem, G. Görk, Ege University Press, 197–208, (1991).
[9] TURKSTAT, *Seçilmiş Göstergelerle Çankırı 2013, Türkiye İstatistik Kurumu Matbaası-Ankara.* ISBN: 978–975-19-6120-4, (2014).
[10] R. A. BIJKER and F. J. SIJTSMA: *A Portfolio of Natural Places: Using a Participatory GIS Tool to Compare the Appreciation and Use of Green Spaces inside and outside Urban Areas by Urban Residents. Landscape and Urban Planning,* 158: 155–165, (2017).
[11] M. MESIMAKI, K. HAURU, D. J. KOTZE, S. LEHVAVIRTA: *Neo-spaces for Urban Livability? Urbanites' Versatile Mental Images of Green Roofs in the Helsinki Metropolitan Area, Finland, Land Use Policy,* 61, 587–600, (2017).
[12] A. P. KASZYNSKA, M. CZEPKIEWICZ, J. KRONENBERG: *Eliciting non-monetary values of formal and informal urban green spaces using public participation GIS. Landscape and Urban Planning,* 160, 85–95, (2017).

Kemal Demir[1]

Designing of a Summer House in the Vineyards of Kayseri

[1]*Department of Architecture, Faculty of Architecture, Erciyes University, Kayseri, Turkey*

1. Introduction

Viticulture has been one of the most important areas of agricultural activity in Anatolia from the time of the Hittites to the present day. In Roman times, wide vineyards were planted around the villas of summer resorts. The fact that wine is considered as a holy drink in Christianity also promoted the advancement of viticulture in Anatolia [1]. As the conquest of Anatolia took place after the majority of Turks had accepted Islam, there was then a more distant approach to wine from the religious perspective. On the other hand, as there was a need for grapes to make molasses and vinegar and as they enjoyed consuming dry or fresh grapes, just like the Greek and Armenian communities, Turks also gave importance to viticulture.

When viticulture in Anatolia is talked about there are two types of situations which are considered. The first is concerned with large vineyards with an agricultural purpose, as in Manisa Alaşehir, and the second refers to vineyards that have the quality of a summer resort with a house within them inhabited only by the owners in summer time, as in Kayseri, where vines are grown to meet the requirement for winter grapes, molasses, and vinegar. Until the time of the Republic, in various Anatolian cities such as Ankara, Kayseri, Niğde, Konya, Diyarbakır, Antep Urfa, Tokat, and Tosya, "vineyard" residences were established to which owners went in the summer time for the purpose of a holiday.

In Kayseri, for many centuries it has been traditional practice for the inhabitants of the city, which is situated on a plain, to move to settlement areas known as vineyards on the surrounding mountain slopes, particularly near Erciyes in the summer months, and to return to the city for the winter [2].

The main reason for this act of migration is that it is difficult to live on the plain as the summer season is hot and dry, while the vineyard areas on the slopes have natural, cool and healthy climatic conditions. In this study, the design of a new summer house realized by the author of this study in the city of Kayseri, in the district of Melikgazi and the quarter of Erenköy, is reviewed.

2. Location of the Land

The land is situated in the city of Kayseri, in the district of Melikgazi, and the quarter of Erenköy is in an area of summer residences, located very close to the city center. On the deed registry, it is specified as being a "vineyard with a house". The size of the land is 3940 m^2. The land is in a neighborhood with four vineyard houses.

The smallest plot grade (0.00 m) is in the north and the highest plot grade (+20.00 m.) is in the south. As per these particulars, the structure has a remarkable landscape, being situated on a slope with a 30% gradient. It looks onto the city center in the northern direction, Mt Erciyes in the south, Mt Ali in the east and Mt Yılanlı in the west. (Figure 1 and Figure 2).

Figure 1. Land on the zoning plan (Municipality of Melikgazi, 2017)

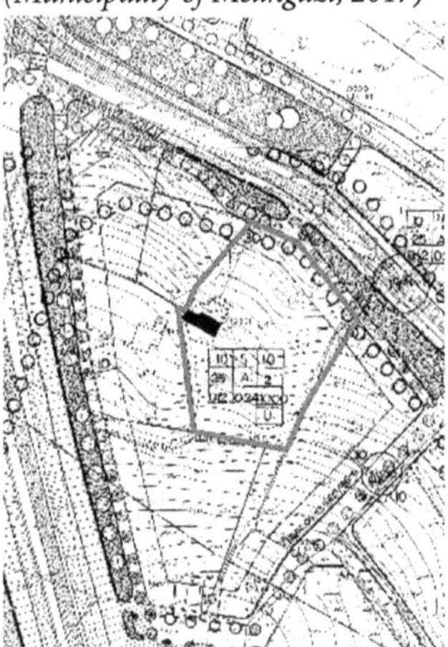

Figure 2. Satellite image of the land (Google Maps, 2017)

While previously traditional vineyard houses were rare, with vineyards having no order and being larger than nearly 5000 m^2, with the implementation of zoning plans after 1990, the parcel sizes were reduced to 1000 m^2. On the zoning plans, the building coverage ratio (TAKS) was determined as 0.12 and the floor areas ratio (KAKS) was determined as 0.24, and a doubling of the number of structures was foreseen.

In the vineyard areas in Kayseri at present; in the design and implementation for the building of summer houses, the approach whereby the zoning right is totally assumed is widely accepted. The summer vineyard house which was designed and implemented, built as a summer house and a weekend house for the author's own family in the year 2007, is different from this approach. In the remaining part of this study, this different approach will be evaluated.

3. Stages of Development

3.1. First stage in the year 2007 (Reuse the existing vineyard house)

The land containing a ruined vineyard house was bought in the year 2007. As the existing structure on the land was repaired, it was turned into a summer house for use at the weekends. For this purpose, the structure of the existing vineyard house was firstly investigated and a building survey was conducted (Figure 3). The ruined vineyard house (Figure 4) consisted of a structure with a size of 77 m^2, with its main walls standing, and with two rooms, one kitchen and two open terraces at the front (95 m^2). Inside the terrace a snow well was established and in the lower parts of the vineyard, a storehouse was built (Figure 5). The grape vines, almond and apricot trees in the vineyard, which had not been taken care of for many years, were in very poor condition.

Within the framework of the original structure reliefs, an alteration project was prepared (Figure 6). However, when the repair works commenced, it could be seen that the existing structure was severely damaged and that it was not earthquake-proof. Therefore, it was decided that it should be demolished and rebuilt.

Figure 3. Reliefs of old vineyard house

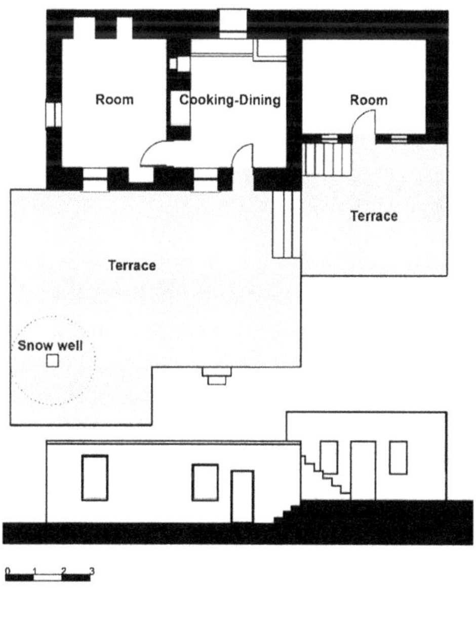

Figure 4. Old vineyard house

Figure 5. Old storehouse

Figure 6. Alteration Project for the old vineyard house

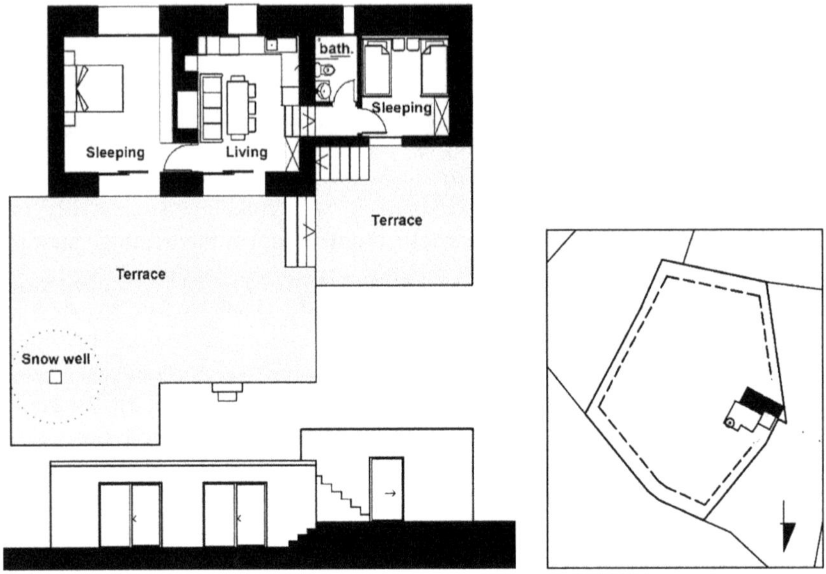

Within the framework of the existing zoning rules in Kayseri, since the old structure was built before 1957, the opportunity was given for it to be repaired and used or for it to be rebuilt on condition that it would keep its original shape. In the light of these data, it was decided to design a new summer house on the land where there was the ruined structure and snow well by considering the structure approximation distances as per the zoning regulation.

3.2. Second stage in the year 2007 (Designing of the new summer house and its implementation)

At this stage, as traditional Kayseri vineyard houses were investigated, it was observed that most of them had an anteroom in the middle and that they had an open plan scheme. While the "Open House", designed with an open plan idea met the requests and needs of users at the outmost level, it also enabled the harmonization of the building, which was stable as regards the dynamic changing environmental conditions [3].

In this respect, the open house plan is the housing type that meets user requirements in the most suitable way and enables the participation of the user in the most effective way. The open house is a strategy, supporting the dynamic decision-making process [4]. Expressing it in a different way, it is the way to find spatial equivalence in line with the future requirements of the users.

The concepts making up the open plan approach are flexibility, applicability, and changeability [4]. Depending on the requirements changing during the period when the structure is used, in order to avoid interventions realized without any supervision and to provide functional quality (functionality), a variety of different usages (flexibility), and adaptability to the requirements that may change in time (adaptability) are criteria that must be considered during the design stage [5].

The new design consisted of a single flat as the main space for living and dining functions with an open plan scheme and a total area of 55 m^2. It also had a sleeping area and a wet area on both sides of this space.

The main living space was enriched with two open terraces. The existing old large terrace, with a snow well, and an area of 30 m^2, had a view of the city to the north. The other terrace, with an area of 15 m^2, faced the south. As with the outer space, arrangements for the usage of the vineyard and garden were left for the future (Figure 7).

Figure 7. Second Stage of the Design

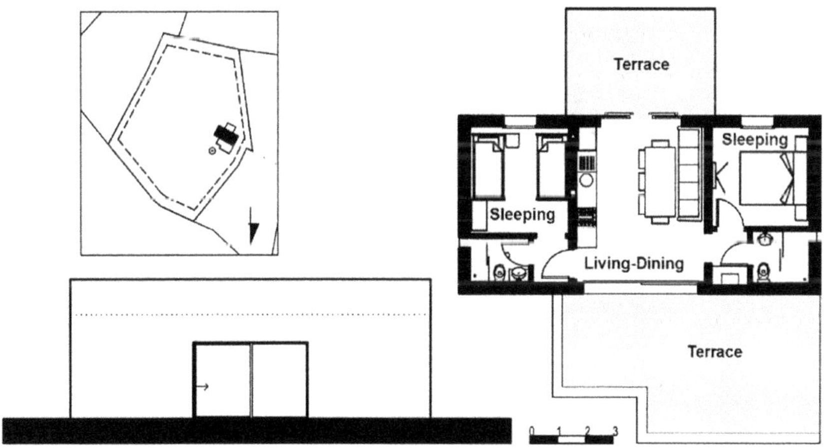

3.3. Third stage in the year 2007 (Addition of a second house for other family relatives)

In the implementation stage of the design, a vegetable garden and grass area were arranged around the structure and the remaining part was left in its natural form. In the third stage, it was necessary to establish a second housing unit for other family members. The functional shaping of the structure was realized in line with the program specified below.

In the first house, considering that the family and children would grow and as it was thought that the middle anteroom would only be suitable for an eating and drinking function, it was foreseen that a southern terrace could be added to the middle anteroom. Thus, the living space was enriched and the closed area was enlarged to 70 m^2. In addition, it was decided that the city facing terrace in the northern part should be extended to a size of 70 m^2 and to be semi-open, considering the possibility of enlargement and development.

In the second structure which was planned for other family members and had a size of 30 m^2, a living/sleeping/eating and drinking space, wet space, and a semi-open terrace with a size of 20 m^2 were established. Within the second structure, a wet space unit suitable for the use of guests, a cellar for the storage of food both in winter and summer time, and a warehouse measuring 10 m^2 were built (Figure 8).

Figure 8. Third stage of design and its implementation.

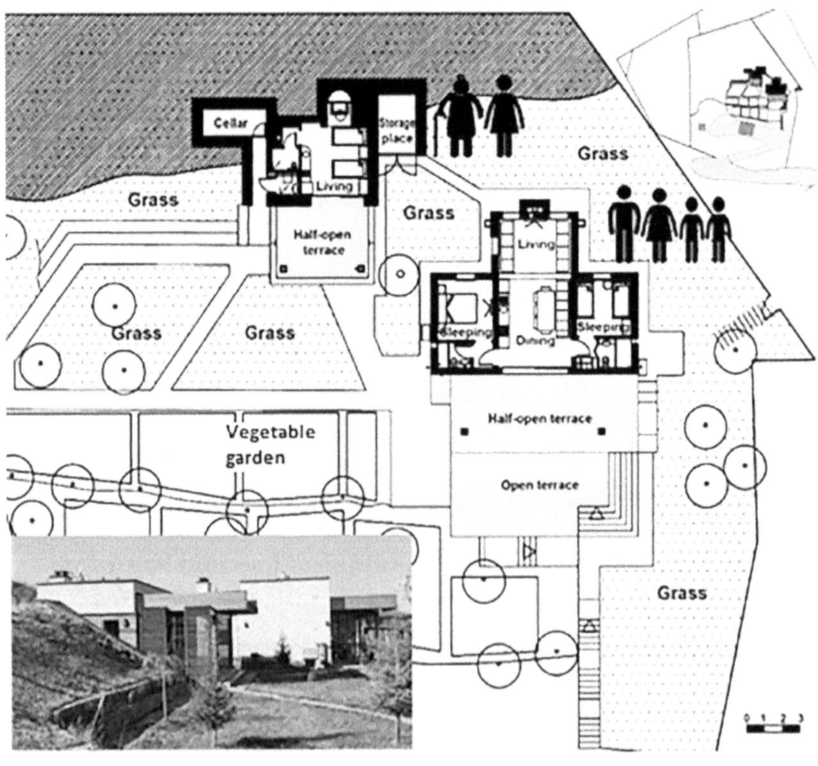

3.4. Fourth stage in the year 2014 (Enlargement of the structure as it develops)

According to the new domestic requirements which arose during the period from 2007 to 2014, the following alterations were made to the structure.

Changes and additions to the first family house

In the first unit, as a third child joined the family in the year 2009, and as a sleeping area was required for him during his period of growth, it was necessary to provite a third sleeping unit. For this reason, the sitting space with chimney which was situated in the southern part was separated from the cooking and eating space with the help of a horizontal sliding door.

The semi-open terrace which was covered with round timbers was provided with heat insulation both internally and externally. It had wood sheathing, and the side surfaces were closed by being fixed. In this way, the total closed area of this unit was increased to 110 m^2.

Foldable wooden doors were placed on the surface looking towards the open terrace. Thus it became suitable for seasonal usage as the doors could be opened, and in the summer months, the open terrace and the main space were integrated. In the other seasons, as the doors could be closed, the living space was protected from the cold.

The open front terrace was turned into a semi-open terrace with a cover that was carried by studs. The side terrace which was connected to the garden and the housing area for family relatives was turned into a common open terrace with a furnace and barbecue, by being enlarged to 70 m^2.

Changes and additions to the second structure

The size of the second structure which was built for other family members and guests reached 50 m^2 as alterations were made due to insufficiency of space, because living/sleeping/cooking functions were realized in the same place. Accordingly, just like in the family house of the author, the sides of semi-open front terrace were covered with fixed walls with wooden constructions and the front façade was closed with two doors. In addition, a semi-open terrace of 20 m^2 was also added to the structure (Figure 9).

Figure 9. Fourth stage of the design and its implementation.

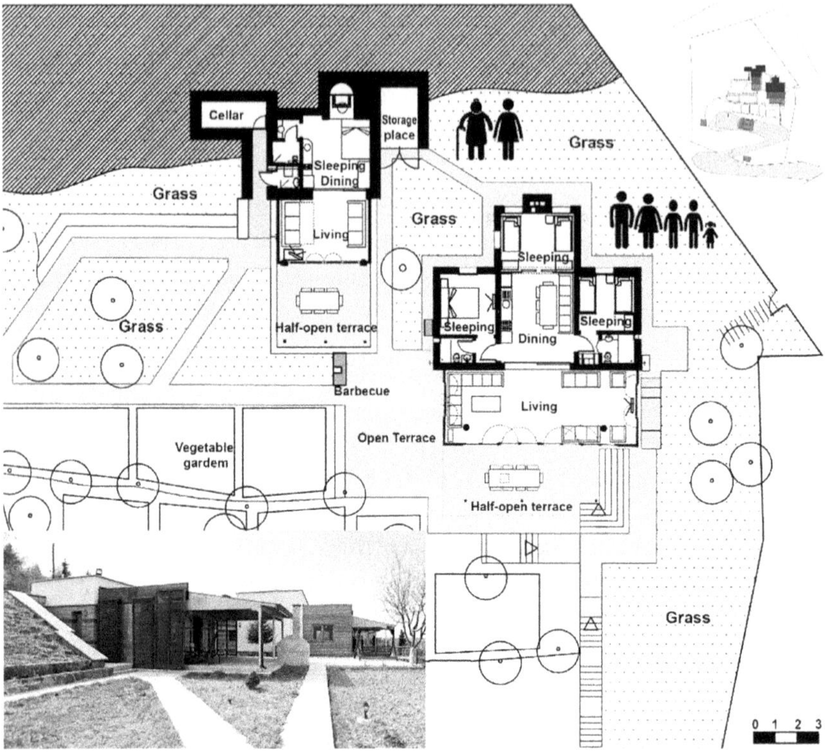

4. Construction and Structure

The two structures were established on a reinforced concrete base using a masonry construction technique. A roof cover of wooden pine logs was made on top of the peripheral ties on the masonry walls. The pine logs, the shells of which were stripped, had diameter of 30–35 cm and a length of 12 m. Wooden logs with a length of 12 m were also laid on top of the front terraces of the two houses by using consoles with a size of 4 and 4.5 meters. In this way, semi-open upper cover for the terraces was formed. A horizontal wooden suspended ceiling was formed by covering the top of the logs with wooden floorboards and a horizontal rafter. On the roof, a free-standing roof structure with a slope was built to coincide with the wooden logs, and the top was covered with wooden panels. Water insulation was applied on the parapet walls starting from the roof slope in a vertical direction and water disposal was provided by gargoyles (Figure 10).

Figure 10. Roof systems of the first and second houses.

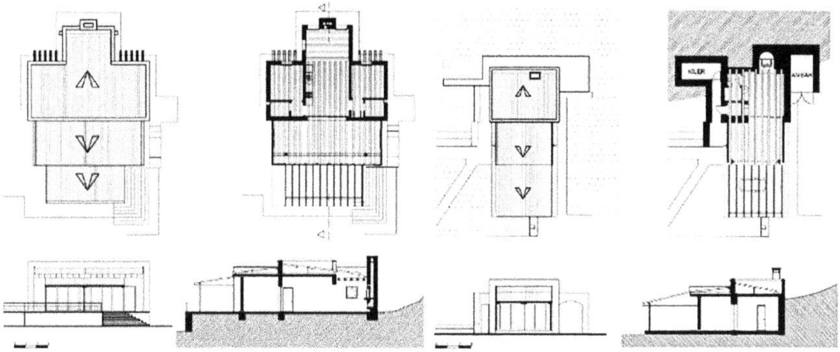

In the first family house, the height of the internal space was 3.40 meters up to the coating on the wooden logs. However, as an addition was made to the living area with a chimney later on, the height was 3.00 meters. The ceiling height of the sleeping units was set as 3.40 m and the height of the wet spaces was set as 2.30 m.

In the second house for other family members, the height of the internal space was 3.00 meters to the coating on the wooden logs and the food cellar, which was entered from the outside, was covered with a reinforced plaque and embedded in the sloped topography. The garden storage place and the shelter unit were made so that they leaned against the sloped topography from the southern front.

5. Conclusion

The desire to come together with nature, by getting away from the dense, dynamic, and stressful environment of city life, has become a need instead of a luxury [7]. This lifestyle, which has traditional roots, has become increasingly popular and is particularly widespread in Kayseri.

Even though the summer houses which are used in the nearby region are densely situated in a natural environment, it is observed that they are approached as city dwellings in their schemes and structuring.

In this study, the design stage and implementation experience regarding a new summer house which was established in place of an old and unrepairable vineyard house situated in the vineyards of Kayseri has been shared.

In this design, the traditional vineyard lifestyle was interpreted with a modern approach. A design which was suitable for the socio-cultural structure of the family and their lifestyle, as well as being suitable for development and for use in a flexible way, was foreseen. Accordingly, as the aim was to integrate the house with nature, the structuring permit given with the zoning plan was not completely used. In the implementation stage of the structure, sustainable construction materials were used.

On the other hand, it is expected that this vineyard house, being situated very close to the city, can to be used all year long without leaving city life in the near future. In this study, a different experience which is user focused is presented to designers as regards the houses they will design in vineyard areas.

References

[1] E. Doğer: *Antik Çağ'da Bağ ve Şarap*. İletişim Yayınları, İstanbul, (2004).

[2] V. İMAMOĞLU: *Kayseri Bağ Evleri*. Türkiye İş Bankası Kültür Yayınları, İstanbul, (2001).

[3] Z. ANDIÇ: *Türkiye'de Açık Ev Uygulaması Olanaklarının İncelenmesi*. (Yayınlanmamış Yüksek Lisans Tezi), İTÜ Fen Bilimleri Enstitüsü, İstanbul, (1999).

[4] K. DEMİR, B. ERGEN, Z. ERGEN, S. ÇABUK: *Collaborative 3D design with BDMUD method: the effects of building on cityscape reflections on urban planning*. Architecture, City and Environment, 11 (32): 61–80 (2016).

[5]. F. D. V. WERF: *Practice On Open Building*. Open building Foundation, Rotterdam, (1990).

[6] S. SAYIN, S. Z. KORKMAZ: *Konutlarda Kullanıcı Gereksinimlerine Bağlı Olarak Yapılan Cephe Müdahalelerinin Fiziksel Çevre Kalitesine Etkisi*. Selçuk Üni. Müh.-Mim. Fak. Dergisi, 21, 3–4, 177–189 (2006).

[7] E. GÖKYER, M. ÖZTÜRK, Y. DÖNMEZ, S. ÇABUK: *Bartın İli Dağlık Alanlarında Coğrafi Bilgi Sistemleri Kullanılarak Ekoturizm Faaliyetlerinin Değerlendirilmesi*. İnönü Üni. Sanat ve Tasarım Dergisi 5, 12, 25–35 (2015).

Yasin Dönmez[1]

Participatory Approach Model in Protected Areas: Biosphere Reserves[1]

[1]*The University of Karabuk, Faculty of Forestry, Department of Landscape Architecture, Karabük/Turkey*

1. Protected Areas

Protected areas are among the important subjects to be investigated for people working in different scientific disciplines. Many definitions have been made for these areas based on their various properties. The International Union for Conservation of Nature (IUCN) developed a global definition to eliminate confusion in this field. According to the definition of the International Union for Conservation of Nature made in 2008, Protected Areas are "the areas managed with laws or other efficient methods involving clearly defined geographical borders and devotion, and were well-known areas managed with laws or other efficient methods for the purpose of preserving nature and relevant ecosystem services/cultural values" [1].

Recently, increased urbanization has led to the tendency of dwellers to escape from urban problems. The fast-growing population, and depending on this, the pressures on natural resources and the variety of the demands, unplanned and unhealthy growth and the environmental problems brought by these factors have made mankind start searching for management methods that might cover the needs of future generations without damaging them. Today, the efforts on this field are continuing with an increase in developing protection methods and systems for natural and cultural resources [2,15].

Protected areas constitute the core of preservation efforts all over the world. The preservation area categories are given in Chart 1. In the past four decades, the number of the preservation areas included in the UN list increased ten-fold, and it was reported that there were over 104 000 areas in the last analyses. The number of preservation areas also increased at a similar rate, and reached 2.4 million km in 1962, and 20 million km^2 in 2004. Roughly 12% of the land areas of the world are defined as protected areas as of today [3].

1 This study was produced from the Doctoral Thesis with the title "Planning of Seyfe Lake Nature Preservation Area as Biosphere Reserve" submitted to Ankara University, Science Institute in 2013.

Chart 1. IUCN Protected Area Categories

Category I (a)	Absolute Preservation Area: Protected areas managed for scientific purposes.
Category I (b)	Wildlife areas: Protected Areas managed for the purpose of preserving wildlife.
Category II	National Parks: Protected areas for the purpose of preserving the ecosystem and recreation.
Category III	Nature Monument: Protected areas for the purpose of preserving certain nature characteristics.
Category IV	Habitat/Species management Areas: Protected Areas for the purpose of preserving the nature through Management Intervention Method.
Category V	Land/Sea Landscape Preservation Areas: Protected areas for the purpose of nature preservation and recreation in land/sea landscape areas.
Category VI	Resource Management Areas for the Purpose of Preservation: Areas Protected for the purpose of providing sustainable use of natural ecosystems.

2. Management Styles of Protected Areas

There are many kinds of management styles for protected areas in Turkey and in other countries. In terms of management targets, protected areas have a wide spectrum. The IUCN has divided the protected areas into six categories based on the management targets [4,5]. These categories may have different names according to the laws and regulations of each country (Chart 2).

In defining the protected areas, many different terms are used at the national level. For example, there are 50 categories in Australia and twelve categories in England. Meanwhile, there are also international networks like Biosphere Reserves and Nature 2000 and some regional ones like World Heritage Areas and Ramsar Areas [3,6].

Chart 2. IUCN Categories and Corresponding Statues in Turkey [7].

IUCN	**TURKEY**
Ia. Absolute Nature Reserve	Nature Preservation Area
Ib. Wildlife Area	1st Degree Natural Sit Areas
II. National Park	National park
III. Nature Monument	Nature Monument

IUCN	TURKEY
IV. Habitat/Species Management Area	Wildlife Preservation Areas
V. Land/Sea Landscape Preservation Area	Special Environment Preservation Areas/Nature parks
VI. resource Preservation Area with Management	Drinking water basins

In recent years, there have been discussions regarding possibilities of using new preservation and planning tools instead of classic ones in nature preservation work (Chart 3)., because it has been understood that it is not possible to manage preservation areas with an approach in which local people are pushed aside with mere concepts of preservation [4].

In order to perform efficient preservation in preservation areas, activities must be planned in the context of aims and targets. However, these actions must have three basic principles: being scientific, participative, and sustainable. Efficient management of preservation areas requires that the groups that are interested in the areas, known as interest groups/stakeholders, are included in the management process. People who are informed, included and authorized make better choices for themselves, and help managers and decision-makers make better choices. Today, the in-situ preservation categories are widely accepted at an international level and are applied to the preservation areas of IUCN and UNESCO's biosphere reserves and world heritage areas. [2,4,8].

When the present preservation status is analyzed it can be observed that in general local people are excluded from the planning processes and the desired preservation is not realized. It is extremely important to include local people in such preservation activities. Based on this, the Biosphere Reserves approach has been considered in UNESCO's Man and Biosphere programme.

Chart 3. New Approach in Protected Areas [4,9].

Areas Protected in the Past	New Approach in Protected Areas
Planning and management against local people (living in and around the Protected area; foreseeing an exclusion of local people from planning and management)	Participative planning and management
Central Management	Mutual, local management

Areas Protected in the Past	New Approach in Protected Areas
Mere protection concept	Management by considering social and economic targets
Developing on its own	Planning as part of national, regional and international systems
Management like "a separate island"	Development within a network (absolute preservation areas connected to each other with green corridors and surrounded with buffer zones)
Preserving mostly due to visual and landscape properties	Preserving mostly for scientific, economic and cultural reasons
Management for visitors and tourists	Management mostly considering local people
Management that creates reactions due to short-term approaches	Management consistent with long-term approaches
Preservation works	Restoration and rehabilitation works as well as preservation
Prior national values	as a social value
Areas that receive interest only at national level	Including areas that receive interest at international level

3. Biosphere Reserves

Biosphere Reserves are the areas included in the Man and Biosphere Programme (MaB),which has international importance, and have continental and/or coastal ecosystems. *Biosphere reserves are intended basically for the resolution of conflicts between the preservation of biological variety and continuing economic growth and cultural values in a sustainable way.* In other words, biosphere reserve areas are the areas where practices for the preservation of biological variety, economic development and cultural values are tested, chosen, presented and developed [3,10,11].

Biosphere Reserves include ecosystems which we may define as specific or typical formed by land, coastal areas and the sea, separately or together. Biosphere Reserves ensure that the defined ecosystems are protected in an efficient way and local people are encouraged to use natural resources in a sustainable way to obtain regional development. In order for a region to be selected as a Biosphere Reserve, the specific area must represent one of the bio-geographical characteristics of the world and must contain natural ecosystems that are damaged at the lowest level. The area must be natural: it must represent the ecosystem it is located in; include rare species; and be rich in biological variety to be included as Biosphere Reserve area. These areas must have the following criteria:

- Ecosystems which have natural or nearly-natural properties and are under threat
- Ecosystems with natural properties needed for scientific aims, genetic sources and endemism centers
- Areas that may have the quality of providing sustainable development
- Areas that have traditional samples of area use
- Areas that have the size that is adequate to perform the three basic aims of biosphere reserves, which are preservation, development and logistics [3,11,12].

In the context of the establishment aims of MaB, three aims have been defined in designing Biosphere Reserves [3,10].

Preservation: a Biosphere Reserve area preserves the natural resources and specific characteristics of an area. There is at least one legally protected area in each Biosphere reserve area. These areas may be used for activities such as walking, diving, bird-watching, educational trips, scientific research, observing plant and animal life. One of the aims of a biosphere reserve area is to support the local people in preserving the plant and animal variety that is specific to the area.

Development: an ideal Biosphere Reserve area is an area that is managed at local scale for economic development as well as its preservation aims. These areas may be used in an audited manner for tourism, agriculture, fishing, forestry and recreation. The mechanism (the managers) that control the Preservation Area encourage the development of a cooperation area near the Preservation Area. This area includes villages, businesses, agricultural areas, fishery areas and other human activities. Economic and cultural development may also help the preservation of the natural area. Another aim of a Biosphere Reserve is to improve the economic and cultural development for future generations.

Logistics: A Biosphere Reserve area is an area where changes in the physical, biological and human environment are observed in the long run. The scientists and administrative staff working ata Biosphere Reserve share the data they obtain from their research and monitoring programs with one other to better understand the effects of the area on the environment and humans. A Biosphere Reserve area is an area for education and experiments. The area management methods are opened for practice. Local people, land owners and organizations cooperate to influence the region in terms of preservation and development. The interest and devotion of local people in cooperation are important and necessary factors for the success of the Biosphere Reserve area. One of the aims of BiosphereReserve areas is to encourage and improve cooperation between local people and land owners for research, development, and training activity plans [3,6,10,11,13].

In order to realize all these functions, a zoning system is appliedin biosphere reserves, which is also the case in other nature preservation areas. The most important difference of biosphere reserves from other nature preservation areas is the fact that local people and residential areas are also included in the zoning system. In this way, the aim is to establish a systematic relation between the protected area and local people [3,10,14]. In order to receive the expected benefits of biosphere reserves and to reach the targeted aims, planning must be performed in accordance with the principles set by UNESCO-MaB. The most important part of this planning consists of the separation of the reserve area into zones that have preservation status at different intensity levels. With this planning, which is also called zoning, the reserve area is separated into 3 areas (Figure 1): the Core Zone, the Buffer Zone and the Transition Zone [3,6,11].

The ***Core Zone*** covers the landscape areas and ecosystems that are included in the preservation system. This zone must have a size that is adequate for the preservation of nature in the area. In the Core Zone, human activities are out of the question, except for scientific studies and monitoring activities and – when necessary – traditional use of local people.

The ***Buffer Zone*** is the area that surrounds the Core Zone and its borders are clearly defined. The activities in its zone must not contradict with the preservation aims and must support the preservation of nature. Scientific studies in this zone may be intended to reveal the natural vegetation structure of the zone, and increase the quality of the agricultural areas, forestry activities, and fishery. A Buffer Zone enables the use of the zone for several purposes such as education, tourism and recreation.

The ***Transition Zone*** is the extension of Core and Buffer Zone and includes agricultural activities, residential areas and other usage areas. Local people in this zone have to work in accordance with nature preservation organizations, scientists, NGOs, private investors etc., on using the zone in a more sustainable manner. Scientific studies that are intended to reclaim the disrupted ecosystems may be allowed in Transition Zones. Biosphere reserves must play important roles in supporting these zones in a sustainable manner for the development of the area [3,6,11].

Figure 1. Zoning System in Biosphere Reserve Areas [11].

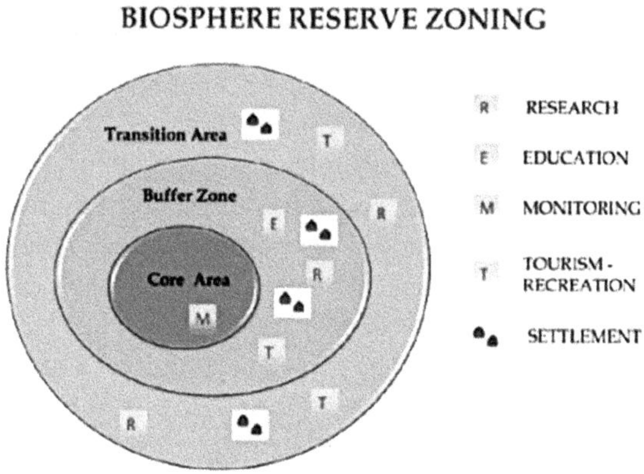

The selection and declaration of biosphere reserves is slightly different from other preservation practices. Biosphere reserves involve areas that represent different ecosystems from various biogeographical areas of the world (mountains, areas disrupted by the effects of human intervention, coastal areas, islands, thick forest areas, deserts, tropical zones and polar areas) and must have several basic properties. The following criteria are considered in defining Biosphere Reserve areas:

- ✓ The area must represent one of the bio-geographical areas of the world.
- ✓ It must include landscape areas, ecosystems, and plant and fauna species that need preservation.
- ✓ It must be a model for sustainable area use in the area in which it is located.
- ✓ It must have an adequate size for performing the three basic functions of biosphere reserves.
- ✓ There must be three main areas in a Biosphere Reserve plan. These are the Core Zone that is protected in legal terms, the Buffer Zone whose borders are defined clearly and the Transition Zone that surrounds these.

4. Discussion and Results

For the purpose of preserving the flora and fauna of existing ecosystems, many nature preservation areas have been established with various statuses so far. In many of these areas, it is necessary to empower the management systems to reduce the conflicts between preservation and development, and to obtain positive results in practice. Biosphere reserves are needed because they have the properties to preserve natural potential in wider areas, and to develop the areas in social, economic and cultural terms.

The pressures on protected areas will continue as the population of the world increases. It is necessary to reduce the pressures on protected areas to the lowest possible level to ensure the balance between preservation and usage. Management models that were formed for these areas have not achieved the desired level because of legal gaps and several other reasons. When recent studies are analyzed, it is observed that participative models have been more successful.

Biosphere reserves are an important model in which local people are included. They involve a system in which local people are involved more in the processes in a system that may be established by public institutions as a special committee. A participative approach is needed to achieve success and the desired level in protected areas.

References

[1] URL-1: http:// www. milliparklar. gov.tr / korunanalanlar / kavramlar . htm, Erişim Tarihi:19.04.2017

[2] O. KURDOĞLU: *Dünyada Doğayi Koruma Hareketinin Tarihsel Gelişimi ve Güncel Boyutu*. Artvin Çoruh Üniversitesi Orman Fakültesi Dergisi, 8 (1), 59–76 (2007).

[3] Y.DÖNMEZ: *Seyfe Gölü Tabiatı Koruma Alanının Biyosfer Rezervi Olarak Planlanması*. Ankara Üniversitesi, Fen Bilimleri Enstitüsü, Doktora Tezi (Basılmamış), 173s, Ankara, (2013).

[4] G. GÜNEŞ: *Korunan Alanların Yönetiminde Yeni Bir Yaklaşım: Katılımcı Yönetim Planları*. Ekonomi Bilimleri Dergisi Cilt 3, No 1, 2011 Issn: 1309–8020 (Online), (2011).

[5] N. DUDLEY, KJ. MULONGOY, S. COHEN,S. STOLTON, CV. BARBER, ve SB. GİDDA. *Etkin Korunan Alan Sistemlerine Doğru. Biyolojik Çeşitlilik Sözleşmesi Korunan Alanlar İş Programı Uygulama Kılavuzu*. Türkçeye Çeviri: Dr. S. Kalem, Montreal, Teknik Seriler No. 18,108 sayfa. (2005).

[6] Y.S.SARIÇAM: *İzmir İl Bütününde Biyosfer Rezerv Alanları ve Saptanması Üzerine Araştırmalar "Karaburun Yarımadası Örneği"*. Ege Üniversitesi, Fen Bilimleri Enstitüsü, Doktora Tezi, İzmir (2007).

[7] P.YEŞİL: *Kelkit Havzası Tokat Kesiminin Biyosfer Rezervi Olarak Planlanması*, Atatürk Üniversitesi Fen Bilimleri Enstitüsü, Doktora Tezi, Erzurum (2009).

[8] S. ÖZBAY: *Doğa Koruma Alanlarında Planlama Çalışmaları ve Ayvalık Adaları Tabiat Parkı Yönetim Planı Önerisi*. İTÜ, Fen Bilimleri Enstitüsü, İstanbul, 147 s. (2008)

[9] A. PHİLLİPS:. *Management Guidelines for IUCN Category V Protected Areas:* Protected Landscapes/Seascapes. IUCN Gland, Switzerland and Cambridge, UK. xv + 122pp, (2002).
[10] M. HADLEY: *Biosphere Reserves Special Places for People and Nature.* United Nations Educational, Scientific and Cultural Organization (UNESCO), Paris, (2001).
[11]. M. OZYAVUZ: *Biosphere Reserves.* The Biosphere, Dr. Natarajan Ishwaran (Ed.), ISBN: 978-953-51-0292-2, InTech, (2012).
[12]. F.F.ALBAYRAK: *Korunan Alanların Ekoturizm Gelişimine Etkileri: Camili Biyosfer Rezervi Örneği,* Yüksek Lisans Tezi, Çoruh Üniversitesi, Fen bilimleri Enstitüsü, Artvin, (2010).
[13]. M. OZYAVUZ and M. E. YAZGAN: *Planning of İğneada Longos (Flooded) Forests as a Biosphere Reserve.* Journal of Coastal Research, 26(6):1104–1111. (2010).
[14]. M. OZYAVUZ Y. ATASEVEN and E. E. SİSMAN: *Biosphere Reserve Approach: A Case Study Igneada Longoz (Flooded) Forests National Park (Kirklareli).* Journal of Environmental Protection and Ecology 12, No 3A, 1383–1385 (2011).
[15] E. GÖKYER, M. ÖZTÜRK, Y. DÖNMEZ, S. ÇABUK: *Evaluation of Ecotourism Activities in the Mountain Areas of Bartın city, by using Geographical Information Systems.* İnönü University, Journal of Art and Design, 5, 12, 25–35 (2015).

Asist. Prof. Dr. Mahire Özçalik[1]
Assoc. Prof. Dr. Dr.Murat Özyavuz[2]

Water Garden Design in Landscape Architecture

[1]*E_mail: mahira.me@gmail.com*
[2]*E_mail: mzyavuz@gmail.com*

1. The History of Landscape Architecture

Peyzaj (the Turkish word meaning landscape) is imported from the French language to Turkish. Its original version in the French language is *paysage. Paysage* literally is "countryside", and it is derived from *pays*, (*pays* is "country"). According to Öztan (1992), landscape architecture is a science in which the uses and plans of the land principles are defined, as well as being the art of creating a useful, functional and aesthetic environment.

The American Society of Landscape Architects (ASLA) offers the following definition and application of the profession of landscape architecture: "Landscape architecture is the profession which applies artistic and scientific principles to the research, planning, design and management of both natural and built environments. Practitioners of this profession apply creative and technical skills and scientific, cultural and political knowledge in the planned arrangement of natural and constructed elements on the land with a concern for the stewardship and conservation of natural, constructed and human resources. The resulting environments shall serve useful, aesthetic, safe and enjoyable purposes. In facing the growing urgency of environmental issues confronting human societies, we must do more than sustain the earth; we must heal, enhance and manage the life-sustaining processes of the planet and ensure the integrity and strength of the global landscape which connects them."

The emergence of the current landscape architecture profession is based on exterior design characteristics of the ancient Persian, Egyptian and Roman cultures.. With the emergence of the Renaissance, our involvement in exterior spaces has increased. This profession started to evolve with French Landscape architectural design taking a new form in the 17th Century.

Since ancient times, humans have changed and used the environment according to their requirements in the preparation of living areas. This change in activities regarding the environment was on a low scale at the beginning, then evolved with the passing of some special processes in subsequent years, to finally take on the wide-scale form it has today, as landscape architecture. Landscape architecture started its journey as garden design in the beginning, and then continued its journey as garden art to the 19th century. In the Industrial Revolution, the emergence of social and physical changes in the cities required that the environment exterior to the cities should be designed for providing benefit to people. In view of these emerging requirements, the design of outer-city green areas came to be termed landscape architecture. Picturesque English Landscape Garden design theory was effective in this period. This design approach expanded with the American landscape architect Frederic Law Olmsted and moved from the countryside to city areas. Naturalistic design, providing natural scenery to the city areas was applied in public areas and city parks. Landscape architecture continued its journey with rivalry from other environmental science disciplines (architecture, engineering, horticulture, etc.) after 1863. Landscape architecture as a term has seen a positive and speedy development in its content and scope, with the influence of social and environmental changes, hence causing it to develop its current wide-scale form.

2. The Importance of Water Use in Garden Arrangement, in History

Water is the basic element for the formation of the natural habitat. The existence and continuity of the world is dependent on water. Water increases the quality of the living area by providing impressive and comforting features. Water has the feature of being relaxing to the soul in mythology. The use of water in history has been compiled in the dissertation of Akdoğan (1974) and Korkut (1995), which we shall summarize.

The oldest representative water gardens are seen in ancient Egypt. There was at least one pool in the Ancient Egypt Garden and it had a square form, a staggered shape and ground water feeding. It is seen in frescoes from the tombs of the Pharaohs that the pool was designed in formal style for water storage, with lotus growing inside. Dead bodies were moved to the pool before burial at funerals. Hence, we can conclude that water carried a religious and mystical meaning in ancient Egypt. Water was used in a lot of different manners in the Mesopotamia civilizations, Persian, Aegean, and Greek, as well as in the gardens of the château, the monastery and the Far East in history. However, it had its most spectacular form in Italy with fountains, statues and cascades during the Renaissance.

The most spectacular gardens of Mesopotamia were seen during Babylon. These were a kind of roof garden built on the terrace of the buildings at the Royal Palace. These gardens generally had formal design features. Cool areas for entertainment, pools with fountains and jetting water, trees providing shade and ornamental flowers were seen in these gardens. The Assyrians built spectacular gardens and also developed irrigation systems in Mesopotamia. As we can see, the main canal with lateral perpendicular canals on its two sides was built in this age.

Figure 1: The Hanging Gardens of Babylon, well-known and characterized as one of the Seven Wonders of the Ancient World, were built to cheer up Nebuchadnezzar II's homesick wife of Persian origin, Amyitis, in the Iraqi city of Babylon.

At the beginning, water gardens started to be seen as part of religious belief. Towards the end, they were seen as pleasure gardens enriched with greenery and water features. Hellenistic Aristocrats held up oriental gardens as an example, and they decorated these with grottoes and pools. They gave priority to summer-houses, statues, seating features and pools in their gardens.

The Romans attached great importance to fountains in the form of monuments, and regarded them as symbols of the city's power. Their view and enjoyment began in the antique age and continued into the medieval age, with fountains in the shape of monuments built in the medieval age. The water resources of Roman cities were distributed to public fountains, big homes and public bath-houses via cisterns. The importance Romans gave to water can be seen clearly in the successful engineering of the aqueducts and monumental thermal bath-houses (Cendere 1998). Gardens of monasteries and châteaux from the Middle Ages were primarily designed for functional usage in lieu of aesthetic beauty. Every monastery generally had one garden in its yard; this garden was divided into four parts with two lanes perpendicular to each other. These châteaux were surrounded by high walls. These walls in addition to the water channels surrounding these walls caused a reduction of the garden area. Also, these walls were used as a safety feature.

Generally, a well, pool or fountain and reservoir for irrigating the plants existed in the center of the garden. Because the monasteries were generally built on high hills, they faced difficulties in terms of water supply. The aesthetic use of water was seen in little pools and water basins.

The Renaissance was a period in European history, from the 14th to the 17th century, which came to be regarded as the cultural bridge between the Middle Ages and modern history. This new thinking became manifest in art, architecture, politics, science and literature. "Renaissance" comes from the French word for "rebirth." Italy was the birthplace of the Renaissance. Magnificence, splendor and flashiness was at the forefront in the Renaissance. Water was used abundantly and in dynamic forms for cascades, pools and canals.

Figure 2: Water use at Villa d'Este, Italian Renaissance Garden.

The excessive scale and glamorization we see in details of the French Baroque garden were the most typical indication of its divergence from Renaissance design. The characteristic features of baroque gardens were those of a centrally positioned building, elaborate parterres, fountains, basins and canals. Orderliness, plainness and balance had the priority for baroque garden design. Water was generally used for stationary water channels. On top of this, artificial pools in different geometric shapes were generally preferred.

Gardens were seen as the symbol of the people's warfare in the Far East. "Shan-shui" is used for landscape in Chinese; it literally means mountain and water. However it does not mean only mountain. `Mountain' comprises everything in its surroundings; forests, water, clouds, mist and any other of a billion details that are an integral part of it without any divisions. "Mountains without clouds are bald and if there is not water, where is its attractiveness?" (Eberhard, Guo xi.2000). Water and mountain then, which are the basic features of the garden, become merged in the garden.

Figure 3: A view from Summer Palace (Yiheyuan) of Chine.

Gardens are a mysterious art in Japan. Gardens have not been used as a symbol of wealth in Japan. Water gardens are a style of Japanese gardens, and these gardens attract attention with their water features. We can see that water is flowing through them, dividing the garden into two parts, and also that a circulating system is used. Bridges are not only used for the crossing of pools; they are also used for viewing the scenery. Dry Rock Gardens are a style of Japanese gardens. We can see a strong expression of water created with stone, sand and gravel in these gardens.

3. Using Types of Water in Landscape Architecture

Water has always supported the landscape in every civilization. Water has been used for swimming pools, decorative pools, small fountains, waterfalls, streams, fountains, cascades and ponds in garden design. We can see that water, which is the most important element in nature, has been used as a static and dynamic form in landscape architecture design.

Stationary water structures (artificial pools and ponds) make an effect of the water mirror in its situation. Plants, clouds, and light from the moon take on a great many meanings in the surfaces of these water structures (Öztan 1970). Stationary water surfaces provide tranquility and a restful mood. They have capturing and gathering features and spread their moisture and coolness to the surroundings (Cendere 1998).

Dynamic water structures consist of cascades, streams, canals, water jets, water curtains and fountains. The visual aspect of water is reflected by the light. Through the drops leaving the dynamic water, they present bright and colorful displays by reflecting the light. In addition to the visual features, the voice of the splashing water presents a pleasant musical sound and gives vitality to the area.

People are drawn to water, which is the primary element of life and provides continuity to life. Water has a strong magnetic force. Because of this reason, we live in the vicinity of water or we carry water with pipes to our living areas. Other means of using water in the landscape are swimming, using it as an ice resource, hunting and surfing. Also, very small water-based structures can provide dynamism and vividness in any place. It is seen that water which is the source of the life is used as a feature in art in a stationary and dynamic form. Unlimited design types are created by using different isolation materials.

4. Design of Water Gardens

Water is the most ideal and preferable element in formal and informal landscape planning work. Each of the elements contained in landscape planning, such as ponds, streams, pools (ornamental, cascades, biological), water jets, fountains, islands, water features, water walls and rivers, together constitute a water garden. If applied to the right area, water adds richness to the garden, and constitutes a tranquil and natural place and acharming feature.

As a result of today's excessive population increase and an increase in concrete structures sourced from unplanned urbanization and industrialization, people are in the situation of searching for suitable living areas which have social, aesthetic and functional features. In this perspective, water is very important for maintaining the life of flora and fauna. Using water in different forms gives tranquility to humans in respect to physiology and physics. Also, water is a very effective design item in terms of creating a healthy environment for humans. Water comes in different shapes and forms, and when it is in the field of vision, water attracts humans towards it. Furthermore, the sound of running water arouses the people's interest. Water interacts simultaneously with all five senses. In brief, if this water is in the composition of plants, we see the emergence of water gardens. N. Şentürk defined the water garden as: "the water garden is a water element, in which it is aesthetically designed with acoustic plants in harmony with surrounding landscape features according to planning principles and application techniques, and in which it constitutes a complete ecological balance with its flora, fauna and microorganisms."

To create a successful water garden, the primary principle is that time should be allocated for planning and the right water items should be chosen. We should think very comprehensively even if we create a design for a very small area. For reaching the targeted people and achieving the desired effects, we should pay attention to the size of the water features, their arrangement and position, the harmony of the water features with plants and other items, accessibility for maintenance, light and shade effects on the water's surface, and choices for compatible electrical and mechanical systems during the planning phase. There is a very common belief that water gardens create problems related to pollution. Actually, if the design is made according to water garden design principles, self-cleaning systems can be obtained by choosing the right plants. To achieve this result, in addition to designing decorative features for the water garden, plants for the waterfront, plants of the marsh and swamp, plants preferring humid and saturated soils and plants living in water are very important. Children's safety is very important in the design of a water garden. In short, meticulous planning, preliminary work, and the right equipment and techniques for the design of the water garden are the key to achieving successful results. If the water garden is designed with attention to every detail, it will yield the desired services: providing tranquility, giving excitement and submitting to easy maintenance.

References

1. Akdoğan, G. 1974. Book of Garden and Landscape Art History. Publication of Ankara University Faculty of Agriculture, Ankara. ss: 290
2. American Society of Landscape Architects Declaration on Environment and Development – January 1993
3. Cendere, A. 1998. Using the water items in city area and green area. Thesis of Master of Science, (not published). İstanbul Teknik Üniversitesi, İstanbul. ss:203
4. Erdal, Z. 2003. Using of the water items in city area – Example of İstanbul". Thesis of Master of Science, (not published. İstanbul Teknik Üniversitesi,İstanbul, ss: 2217
5. Eberhard,W. 2000., Images of Chinese, Kabalcı Yayınevi, İstanbul, ss:428

6. Hans,S. 2003.,History of Gardens and Parks., Publication of the Dost Bookstore., Ankara. ss: 275
7. Korkut, A.B. 1995., Landscape Architecture with plant examples., Hasad Yayıncılık Ltd.Şti. İstanbul. ss:271
8. Öztan, Y.1992. The responsibilities and duties of the landscape architecture for the next century. Journal of Landscape Architecture, Sayı 30:13–18, Ankara.
9. Öztan, Y. 1970. Water in Landscape Architecture. Journal of Landscape Architecture, Sayı:6–8, Ankara.
10. Şentürk, N. 1990., Water Gardens, Journal of Architecture, Interior Architect, Visual Art, sayın:9, s: 100–105
11. İnternet: https://www.asla.org/ Access date: 20.04.2017
12. http://www.kansas.net/~tjhittle/ladef.htm Access date: 28.04.2017
13. İnternet: https://en.oxforddictionaries.com/definition/paysage Access date: 05.05.2017

Assist. Prof. Dr. Özgür Burhan Timur,
Buse Yazici (Landscape Architect)

Use of School Gardens for Sustainable Urban Agriculture

1. Introduction

Urban agriculture activities started with planting vegetables in the gardens of urban public spaces such as school gardens, playgrounds for children, parks and church gardens, in order to meet people's food needs during the Second World War. Today, urban agriculture activities have come back for different reasons and have attracted attention in cities in recent times.

In particular, the urban man, who is a part of nature and has moved away from nature due to intensive human settlement in today's cities, has turned towards urban agriculture with the aim of getting closer to nature and also producing safe, organic and cheap food with his own hands.

Urban agriculture, which corresponds to cheap and safe food production in home gardens, public spaces and within urban walls, also reinforces community relations. Urban agriculture develops the sense of belonging to something and it comes with the use of an alternative space that allows people to get rid of the daily stress brought by life in the city.

This study explains the concept of urban agriculture, through an examination of the history of urban agriculture typologies and urban agriculture. Examples are given from all over the world which emphasize the importance of urban agriculture for sustainable cities, and school gardens for urban agriculture.

2. Concept and History of Urban Agriculture

Urban agriculture is an agricultural activity in the city within the increasing urbanization phenomenon. It is possible to see agricultural activities in the Hanging Gardens of Babylon, the domus gardens in the Roman period and many civilizations throughout history. However, public urban agricultural activities were first carried out within the walls of Medieval Europe for the purpose of protection during the Derebeylik period [1].

The event that caused the emergence of urban agriculture activities in today's modern world was that European people who were hungry grew vegetables in common areas such as public gardens, parks and church gardens during and after World War II [2]. Alongside the beginnings of urban agriculture in Europe, America saw a great migration of villagers from cities due to the air pollution caused by industrial facilities, and these villagers emerged from unemployment and poverty after the economic crisis in 1893. Work was initiated to prevent population declines in urban areas due to immigration, and for this purpose, vacant spaces and abandoned industrial facilities in urban areas were given support with the aim of being used for urban agriculture activities. At this point, urban agriculture was proposed as a solution to social problems caused by migration and to abandoned areas which had resulted from migrations [3,4,5]. Urban agriculture activities were forgotten after the World War due to the increase in wealth for people living in cities and also the crowdedness of these cities. Nowadays, it concerns ensuring safe food production and the results of food transport costs [6].

3. Typologies of Urban Agriculture

There are a wide range of urban agriculture varieties and scales according to urban conditions (Table 1). Urban agriculture activities, which are favored by groups of people who want to meet their food needs in a healthy and reliable manner, are on the agenda and increasing day by day with different uses and different names. These activities range from food production by families with low incomes for their own consumption to community, hobby and guerrilla gardens, and to agricultural organizations that are larger than this [7].

Table 1. Urban agriculture typologies – Adapted from American Planning Association(APA) [7].

	Noncommercial	**Commercial**
1	Private garden	Market farm
2	Community garden	Urban farm
3	Institutional garden	Peri-urban farm
4	Demonstration garden	Bee-keeping
5	Edible landscape	
6	Guerilla gardening	
7	Hobby-bee keeping	
8	Hobby-chicken keeping	

School gardens (Institutional gardens) are among these urban agricultural typologies, one of the most important in terms of transferring urban agriculture activities to future generations.

4. The Importance of Urban Agriculture for Sustainable Cities

Following the concern for the environment expressed at the Stockholm Conference in 1972, the phrase `sustainable development' which emerged in the United Nations Working Paper of 1987 triggered enormous environmental movements in many countries around the World. This was how these demands found an answer in the field of urbanism. New environmentalist research was spreading in the field of urbanism, as it was in every area [8,9].

Sustainability according to UNESCO's MOST (United Nations Educational, Scientific, and Cooperation Organization–Social Transformation Program) (1996); "The relationship between society and nature that is shaped socially can be maintained in the long run." "Sustainable development; has been expressed as a development that can provide sustainability" [10].

The "Sustainable Cities Development Report", prepared by the United Nations in 2009, addressed a very important issue at this point. The report specifies that "more ambitious policies are needed to limit development and energy consumption for sustainable development", emphasizing that "first of all it is necessary to reduce pollution, protect natural areas and arable land" [11].

Towards the end of the 20th century, it came to be understood that environmental problems were important in terms of the economy, and so they became a focus orf many people and sciences [12]. A recognition of the negative developments threatening human life was the basis of the concept of 'sustainability' [10].

Geenhuisan & Nijkamp (1994) write of sustainable cities, that in order to provide change in continuity these are "cities where socio-economic interests are harmonized with environmental and energy concerns " [13].

According to Huang (1998), the sustainable city is not only a city where the environment is protected; it also requires economic and social changes to reduce the need for protection of the environment [14].

By ensuring the sustainability of cities, it will be possible not only to solve existing environmental problems, but also to increase the quality of life of the existing population and to create habitable places where future generations can easily live their lives [15].

Urban areas that have been inhabited by the majority of the world population are the most appealing settlement areas in terms of offering various opportunities for city people [16]. According to estimates made by the United Nations and many other sources, it has been clearly stated that this tendency will continue throughout this century and by 2050 70% of the world's population will be living in cities [17]. For this reason, sustainability will become much more important for cities. Given the fact that today about 50% of the population on earth live in cities and the effects that cities have on other areas, it is important to emphasize that the most important areas for sustainability are cities [7.15].

The structuring of cities increases the value of urban areas. The limited income available from agriculture prevents agriculture in city centers and in peri-urban areas, and these areas are converted into industrial or mass housing areas with a higher profit rate. In cities that have been transformed within this conception large amounts of fertile agricultural land are disappearing. Not only is it not possible to replace the declining agricultural land but there is also not enough space in the cities to form new areas. At this point, the most appropriate scenario for urban agriculture is the sustainability of urban agriculture activities by taking into account health conditions in the remaining areas (urban space, public building gardens, house gardens, etc.) in industrial facilities and public housing areas.

Despite the large number of areas for urban agricultural activities, even in the most intense areas of metropolitans, the supply of necessary food from outside the city leads to a decrease in the reliability of food as well as its freshness [18]. In addition, shipping costs and food prices are also rising.

Related to this, FAO (2014) notes that urban agriculture spreading in towns and cities is beneficial in terms of its contribution to the recycling of urban waste, mitigation of the urban climate, production of employment and provision of fresh food. Urban agriculture provides cheap access to safe and fresh food for urban people as well as bringing great potential for economic and environmental advantages [19].

As Smit (1996) points out, urban agriculture activities make cities greener and cooler. At the same time, the biological transformation of urban waste and the transformation of uncultivated areas into productive landscapes, contribute to the formation of cleaner and healthierenvironments and contribute significantly to the sustainability of cities [7,20].

5. The Importance of School Gardens for Urban Agriculture

School gardens are a local, community-based and potentially useful interface for the promotion of healthy living and environmental sustainability.

The interest of school gardens in urban agriculture is accepted for their sociological, psychological, and physical dimensions and emerging from community gardens are observations that they provide a number of health benefits [21].

According to Drescher (2000), there are also differences in the approaches of developed countries and developing countries to urban agriculture activities. In developing countries the main objectives of urban agriculture activities are crisis management and food security, whereas in developed countries recreational and educational aspects as well as environmental issues related to urban open spaces and greenbelt management are gaining importance [20].

At this point, school gardens and children's farms in urban areas in developed countries are very important areas in terms of urban agriculture education, building awareness of the ecological and economic benefits of urban agriculture for the city and transferring this consciousness to future generations (Fig. 1).

Fig. 1. Edible School Garden [22].

In the meantime, students in school gardens receive significant training in environmental issues such as permaculture, waste minimization, organic food production and sustainable living [23]. In today's cities, people can easily access fruit and vegetables thanks to widespread supermarkets, but beyond community-based urban farming, activities in school gardens provide education for future generations about food hygiene, community awareness and the fight against obesity (Figure 1). At the same time, urban agriculture leads students to healthy vegetable production and consumption. With these features, school gardening programs have become popular as action-oriented learning environments in many countries, often driven by environmental sustainability and the essential components of healthy nutrition.

Fig. 2. Conceptual model of the impact of school garden programmes[24]

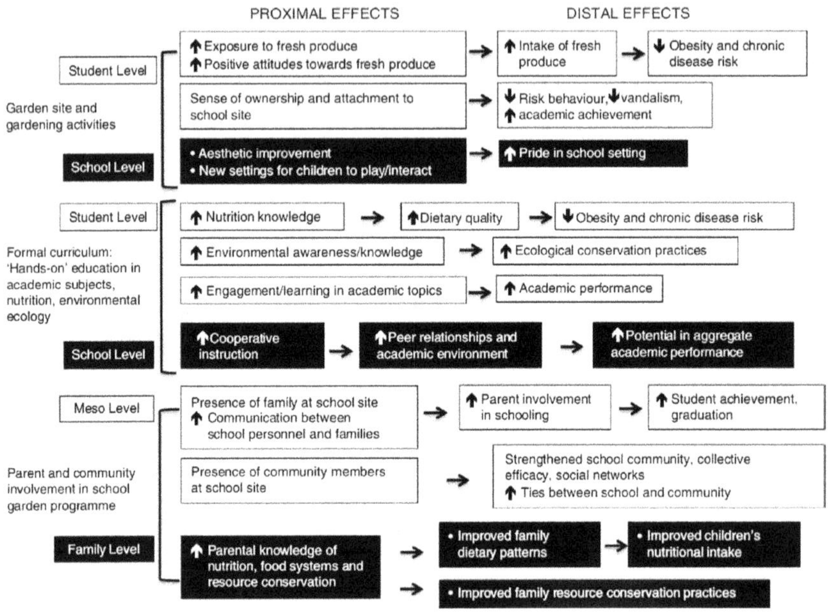

6. Application Examples

In many countries school garden programs have been implemented with different design approaches, school levels of different age groups and various curriculum programs. In these applications the results are different because of the different designs and processes [24]. The overall aim of school gardening programs is to enable students to adopt healthier eating behaviors, to learn fruit and vegetable cultivation, and to taste the foods they produce. In addition, the program aims to transform students into active participants rather than being passive learners in programs such as art, mathematics, literature and chemistry. The increase in student achievement in school garden programs has been statistically proven [24].

Initial work on school-based gardening in the United States has shown some positive effects on nutritional knowledge and behavior. As a result of the experiments, it was observed that the nutritional habits of the children who were educated in nutrition and gardening changed. Their tendencies to choose vegetable-rich food increased, they became able to define fruits and vegetables better, and they had more information about nutritional values [25].

Children spend most of their time on school campuses, so the school environment is seen as a place where children can engage in programs to increase their fruit and vegetable consumption. The fact that many schools have gardens ensures that one of the best environments in which the programs can be implemented is schools [26].

A study was conducted in 10 different schools across California to create, use and sustain educational school gardens. In the study conducted, it was determined that the best practice model in school gardens should be related to four main factors (people, funds, materials and teaching) [26].

A California-based pilot study, the "Farming Food and Fitness Program" and the "Children's Health Improvement" program have linked together school, family and community activities with the inclusion of agriculture in the school environment [27]. The program aims to establish cafeteria salad bars, link curricula to activities appropriate to subjects and classes, improve the composting system to reduce lunch waste and create a garden laboratory that would help to combine nutritional education with food choices. The multi-functional nature of the program ensures a rise in awareness, as well as the development of nutritional knowledge and nutritional options for students, teachers and staff [28].

Later, the Berkeley Unified School District in California was established and became a model for school lunch programs nationwide. With the help of the Chez Panisse Foundation, they serve locally grown fruits and vegetables (as organic as possible) and local organic hormone-free milk [29].

At the same time, the Davis Joint Unified School District (DJUSD) implemented a composting program as a part of the "Farm with School Connection", linking together farmer's salad bars, class education and local farms with students, through farm tours and waste management [30].

In the 2000–2001 school year, a food waste composting system was tested in three schools in DJUSD to reduce lunch waste and include students in the compost and recycling process. In the pilot year, the program reduced waste disposal costs by $ 6 230 and contributed to sustainability. At present this work is being implemented in every elementary school in the district.

The Edible Schoolyard is a 1 acre (4.047 m^2) garden in the middle school area in Berkeley, California. It was founded in 1995 with the support of the Chez Panisse Foundation [30].

The Edible School Garden is a pioneer among school-based gardens and has many members throughout the country. Gardening is part of the basic curriculum for each student and practical experiences are combined with mathematics and science lessons. Lesson plans have been developed for the garden. The program is a host to the Edible Schoolyard Academy where educators can participate in workshops and gather ideas for the implementation of new programs.

7. Conclusion

Urban agriculture, defined as agricultural activities carried out in the city together with increasing urbanization phenomena, enables healthy and organic foods in cities to be produced more cheaply than products on the markets. At the same time urban agriculture provides both employment and cheap access to food for low-income families living in the city. Urban organic waste is used as compost, and urban waste costs are reduced thanks to urban agriculture activities which have positive effects on the urban climate and nature.

As a result of the research and experiences gained; urban agriculture activities have a very important place in the creation of sustainable cities. These activities support sustainability through environmental, social and economic contributions. School garden programs, on the other hand, are the primary priority areas for the implementation of urban agriculture in terms of raising the awareness of future generations and for directing urban agriculture.

Urban agricultural activities at the school provide children with action-oriented learning while allowing children to become active participants rather than passive students. With these activities, the motivation levels in the school are being increased while the students gain knowledge and experience. School garden programs, at the same time, give healthy nutritional habits to future generations by giving knowledge about what nutrients should be consumed for what purpose and what nutrients the human body needs for nutrition, while giving nutritional production and care information to the students.

School garden programs provide organic food production. These programs provide students with academic knowledge on waste management and composting, and these programs are important training areas for minimizing waste and contributing to the sustainability of cities. These programs should be supported and expanded by turning them into state policy.

References

1. O.B. TİMUR: Peyzaj Sanat Tarihi Ders Notları. Çankırı Karatekin Üniversitesi. ÇANKIRI, (2015).
2. D. VİTİELLO AND D. MORLEY: History of Urban Agriculture in Urban Agriculture PAS Report, 563, p. 10–13, (2011).
3. L.J. LAWSON: City Bountiful: A Century of Community Gardening in America, Berkeley, University of California Press, (2005).
4. D. VİTİELLO AND M. NAİRN: "Community Gardening in Philadelphia: 2008 Harvest Report." Penn Planning and Urban Studies, University of Pennsylvania, (2009).
5. K. HODGSON, M.C. CAMPBELL AND M. BAİLKEY: Urban Agriculture: Growing Healthy, Sustainable Places, American Planning Association, Planning Advisory Service Report, 563, Chicago, IL, (2011).
6. J. KLOPPENBURG, J. HENDRİCHSON AND G.W STEVENSON: Coming in to the foodshed, Agriculture and Human values,13, 33 42, (1996).
7. A.E. CENGİZ, U. PEKİN TİMUR, A.A. KARADAG, D. DEMİROGLU: A Contribution to Sustainable Urban Development: Urban Agricultural Lands. Urban and Urbanization, ISBN: 978-954-07-3772-0, p. 85–104, ST. Kliment Ohridski University Press, Sofia. (2014).
8. F. AKLANOĞLU: Geleneksel Yerleşmelerin Sürdürülebilirliği ve Ekolojik Tasarım: Konya-Sille Örneği, Ankara Üniversitesi Fen Bilimleri Enstitüsü Peyzaj Mimarlığı Anabilim Dalı, Doktora Tezi, Ankara. (2009).
9. B. Ağca: Dünya Sürdürülebilir Kalkınma Zirvesi. Uluslararası Ekonomik Sorunlar Dergisi. Sayı 7. Dışişleri Bakanlığı Yayınları, (2002).
10. A. ÇAHANTİMUR, T.H. YILDIZ: Sürdürülebilir Kentsel Gelişmeye Sosyokültürel Bir Yaklaşım: Bursa Örneği. İTÜ Dergisi/a 7 (2), 3–13. (2008).
11. P. NAESS: "Urban Planning and Sustainable Development", European Planning Studies, 9 (4):503–524, ISSN 0965–4313, print/ISSN 1469–5944 online/01/040503–22 Ó 2001, DOI: 10.1080/0965431012004987, (2001).
12. D. Toprak: Sürdürülebilir Kalkınma Çerçevesinde Çevre Politikaları ve Mali Araçlar. Süleyman Demirel Üniversitesi Sosyal Bilimler Endtitüsü Dergisi Volume 2 Issue:4 Isparta, (2006).

13. A. ALTUNTAŞ: Sürdürülebilir Toplumlar ve Metropollerin Baskılarından Kurtulmak İçin Alternatif Bir Yol: Sürdürülebilir Kentler. Mustafa Kemal Üniversitesi Sosyal Bilimler Enstitüsü Dergisi 9 (17), 135–148, (2012).
14. Z.A. GÜREL ÜÇER: Kentsel Politikaların Belirlenmesinde Bir Araç: Sürdürülebilirlik Göstergeleri. Çağdaş Yerel Yönetimler, Cilt 26 Sayı 1, 103–124, (2017).
15. A. ATIL, B. GÜLGÜN, İ. YÖRÜK: Sürdürülebilir Kentler ve Peyzaj Mimarlığı. Ege Üniversitesi, Ziraat Fakültesi Dergisi 42 (2), 215–226, (2005).
16. S. VASCONCELOS: Web Site: http://www.ece.aucland.ac.nz/~sinnen/Vasconcelos Silva2006.pdf, (2007).
17. J. AHERN: From fail-safe to safe-to-fail: Sustainability and resilience in the new urban World. Landscape and Urban Planning Volume 100, Issue 4, (341–343), (2011)
18. S. RASOULİ: Kent Topraklarının Tarımsal Amaçlı Kullanımı: Kentsel Tarım. Web Site: www.skb.org.tr/kent-topraklarinin-tarimsal-amacli-kullanimi-kentsel-tarim, (2015).
19. E. BRYLD: Potentials, problems, and policy implications for urban agriculture in developing countries. Agriculture and Human Values, 20: 79–86, (2003).
20. H. KOÇ: Daha Yaşanabilir Yerleşmeler Arayışında Kentsel Tarım. Planlama Dergisi. 2003/1: 34–40, (2003).
21. D. ARMSTRONG: A Survey of Community Gardens in Upstate New York: Implications for Health Promotion and Community development. Health Place 6, 319–327, (2000).
22. Anonymous: Edible School Garden. Web Site: http://growinggardeners.net/wp-content/uploads/2010/07/master-plan-poster_finallowres.jpg. (2017).
23. S. SOMERSET and A. BOSSARD: Variations in Prevalence and Conduct of School Food Gardens in Tropical and Subtropical Regions of North-Eastern Australia. Public Health Nutr 12, 1485–1493, (2009).
24. J. N. DAVİS, M.R. SPANİOL, S. SOMERSET: Sustenance and Sustainability: Maximizing the Impact of School Gardens on Health Outcomes. Public Health Nutrition, 18(13), 2358–2367, (2015).
25. S. M. PARMER, J. SALİSBURY-GLENNON, D. SHANNON, B. STRUEMPLER: School Gardens: An Experiential Learning Approach for A Nutrition Education Program to Increase Fruit and Vegetable Knowledge, Preference, and Consumption Among Second-grade Students. Journal of Nutrition Education and Behavior, 41(3), 212–217, (2009).
26. E. L. HAZZARD, E. MORENO, D.L. BEALL, S. ZİDENBERG-CHERR: Best Practices Models for Implementing, Sustaining, and Using Instructional

School Gardens in California. Journal of Nutrition Education and Behavior, 43(5), 409–413, (2011).

27. K. HENEMAN, S. K. JUNGE, C. SCHNEİDER, S. ZİDENBERG-CHERR: Pilot Implementation of the Improving Children's Health Through Farming, Food, and Fitness Program in Select California Schools. Journal of Child Nutrition and Management, 32(1), (2008).
28. E. OXENHAM, A. D. KİNG: School Gardens as a Strategy for Increasing Fruit and Vegetable Consumption. J Child Nutr Manag, 34(1), (2010).
29. K. SEVERSON: Food Joins Academic Menu in Berkeley School District. San Francisco Chronicle. Retrieved June 15, 2009, from http://www.sfgate.com/cgi-bin/article.cgi?file=/chronicle/archive/2004/08/29/MNG8T8FL1C1.DTL%20, (2004).
30. H. GRAHAM, G. FEENSTRA, A. EVANS, S. ZİDENBERG-CHERR: Davis School Program Supports Life-long Healthy Eating Habits in Children. California Agriculture, 58(4), 200–205, (2004). http://www.edibleschoolyard.org/

Ömer Lütfü Çorbaci*

A Review of the Karabuk University Iron and Steel Campus According to Planting Design Principles

*Recep Tayyip Erdogan University, Faculty of Art, Design and Architecture, Department of Landscape Architecture, 53100, Rize, Turkey, omerlutfu.corbaci@erdogan.edu.tr

1. Introduction

A University Campus is a multifunctional educational area which includes university buildings, student dormitories, housing for faculty members, social facilities, and apart from these, main roads connecting these units to each other, walking paths, green areas, inner courts, plazas, and squares. The term "campus" was first used for Princeton University, which is located in a large park outside the city [1]. Campuses have gone beyond being educational institutions where students and academic staff do scientific work and managers and administrative staff provide services, and have become living spaces for socialization and cultural and sporting activities. Today, however, "campus" is used as an umbrella term defining the whole ensemble found within the boundaries of a university, including all buildings and the spaces between these buildings [2]. A city is a place with a wide variety of functions and living spaces. University campuses are also one of these areas that affect the quality of the citizens' lives positively, make the city attractive, and have positive effects on the identity, character, and image of the city.

University campuses are designed areas within the boundaries of cities. They mainly consist of soft surfaces such as soil, ground cover, bushes, shrubs, and trees. Campuses are important for human health and quality of life as well as for urban ecology. One of the most important elements of these areas is the plants. The views created by taking advantage of the size, form, texture, line, and color characteristics of the plants on the campuses is precious and attractive for urban people.

Plants make places lively. Thanks to the work of planting design, people are able to meet nature by getting rid of the pressures of daily life [3].

The fact that plants are living beings and they constantly undergo change and development is ignored in most planting design studies. The plants are usually planted without considering their future dimensions and characteristics but only considering their dimensions at the time they are taken from the nursery. Although this does not constitute a problem in terms of planting design during the first application phase, it is observed that it leads to problems in terms of the healthy life and habitus formation of the plant species used.

Plants are living organisms that continuously develop throughout their lifetimes and periodically present various habits and forms. Therefore, it is necessary to follow design principles when making a planting design. The functional and aesthetic characteristics of plants should be taken into account in planting designs in accordance with the characteristics of the project area. The plants' suitability for climatic conditions (natural plants are preferred), and their temperature, humidity, light, soil, and shade requirements etc. should be considered in terms of functional characteristics. In terms of aesthetics, however, facts such as habitat, form, texture, line, and color characteristics should be considered.

The subject of this research is Karabuk University Iron and Steel Campus, which is located between the Karabuk city center and Safranbolu County, covering an area of about 2500da. The research started by identifying the natural or exotic plant species in the campus area. In the study, once the concept of planting design and design elements were explained, they were evaluated according to planting design criteria, and finally, solutions were proposed regarding planting design studies, considering spatial and design principles. The planting design areas in the research area were photographed and solutions were proposed for the erroneous compositions.

Vegetation is the most important factor that reveals the natural characteristics of landscaping. It creates spaces with nice views in terms of the physical and the aesthetic, and adds richness to the environment. Designing these places by using various plant species creates the concept of planting design. Planting design is expressed as an art in which science and art are evaluated jointly. Planting design requires the creation of an abstract concept. A proper planting design requires the environmental requirements of the site to be combined with this concept of abstraction. A good planting design is complemented by proper application and maintenance. In other words, this process requires artistry, vision, and patience [4]. Planting design is the process of bringing plants of different species together into a whole. Once the concept is determined and the garden elements are introduced, the next step is to examine the individual characteristics of the plants in order to determine the effects they make throughout the design [5].

Plants are design elements that have both aesthetic and functional roles in landscape architecture. In planting design, there are design elements and principles like in every branch of art. The designer creates designs based on these elements and principles. These are: dimensions (ratio/proportion), form, shape/ground, continuity, diversity, repetitions, emphasis, suitability and contrast, koram (hierarchy), and balance. As the materials used in planting designs are alive, many changes occur in plants over time. These changes may be related to the dimensions over the years or to the shape and color depending on the season.

Size is one of the most important design elements for the plants and should be evaluated initially during plant selection for a design. Plant size directly influences the size of gaps, the level of interest, and the entire structure of the design. Plant size is one of the most remarkable and visible characteristics of all. It is very important, especially when viewed from a certain distance. Plants give diversity or uniformity to the design through their heights and widths [6].

One of the most important principles in plant designs is determining the correct planting distance in accordance with the size of the plants. Considering the planting distance, classification characteristics of plants (tall tree, medium-sized tree, shrub, tall bush, medium-sized bush, short bush, and ground cover), and taxa characteristics, differences may be seen in the use of the plant such as solitary or group use. Solitary or group uses of plants have different psychological effects. In particular, group use brings some problems with it. When using plants of the same or different taxa, it is necessary to consider morphological characteristics as well as design characteristics.

Habits are of great importance in terms of the measurement of planting distance. When determining planting distances in planting designs, the main criterion is the height of plant canopies. This varies according to the characteristics of plant taxa. Therefore, it is very important to know the general characteristics of plant species in order to use them as well as possible. In particular, one must be very aware of the maximum height they can reach, the widest top diameter they can have, their interaction, and their inclination for pruning etc.

The visual quality of solitary or group plant compositions is discussed in terms of the lines, textures, forms and colors which make up the visual characteristics of the plants. Although these elements are more abstract than the particular effects of fruits and flowers, they are essential for understanding the entire composition. Considering these concepts is the main condition for creating a visually impressive and understandable composition [7].

The lines are the edges that determine the shape and the form. The lines are seen everywhere. All materials, both natural and man-made, form the source of the line. Branches, the edge of a flower bed, make a line formed by plants, and the contour of a group of plants can be a linear unit [8].

The texture depends on the distance one is to the plant being observed. If the distance is far enough to be able to see the vegetation between all plants or between the tree communities and bushes, the texture will depend on the area and the volume of individual bushes, trees, or tree communities. A plant group of different textures may look alive and more interesting to the observer, but too much diversity creates complexity and excessive richness [7].

Form can be defined as the general shape, architectural structure, or silhouette of plants/plant groups. The form of the plants has various functions such as forming the structure of the plant composition and becoming an access point, forming the background, and providing integrity between other design elements and vegetation. The basic forms of plants are classified as columns, fastigiata, pyramidal/conical, round/spherical, spreader/horizontal, sagger and landscape picturesque [6]. In other words, the form is defined as the contour or silhouette of plants. The body, branches, and leaves come together to create the form. The forms of plants are perceived more strongly when they are leafy. Their influence and the extent of their perception level go down when they shed leaves [9].

Color is a design element that has a visual effect related to the absorption and reflection of light waves by objects. Color and vision are achieved by the light from the objects. Color is a visual feature that depends on the wavelength of the light. Different colors are perceived according to the wavelength of the light [10]. The color differences in nature created by seasonal changes give us enough information about different colors. It is necessary to have sufficient knowledge on color theory and harmony in order to be able to use a color successfully in landscaping. The designer must correctly perceive the shade and intensity of the color and the shadows it creates on plants [4].

3. Materials and Methods

The Karabuk University Iron and Steel Campus is located in the Western Black Sea Region. It is located on the way to Safranbolu and approximately three km away from the city center. The university campus area is about 2500 da. The buildings of the campus area in use are the: Rectorate Building, Faculty of Technical Education, School of Foreign Languages, Faculty of Science and Letters, Faculty of Economics and Administrative Sciences, Faculty of Engineering, Faculty of Theology, Indoor Sports Hall, Burhanettin Uysal Stadium, Canteen, Kindergarten, R&D, Social Facility, and the Cafeteria.

In this study, photographs taken using digital camera have been preselected in the digital environment to decide which plants to use.

This study, which aims to determine planting design mistakes and offer solutions, consists of five consecutive stages. In the first stage, the concept of planting design and its components are defined. In the second stage, the work area is introduced. In the third stage, the plant species in the campus are identified and the general characteristics of these plants such as leaf colors, and time in leaf, flower colors, flowering time, mature fruit colors, fruit ripening time, heights, canopies, ecological requirements, and whether plants are natural or exotic are determined. In the fourth stage, mistakes are determined according to planting design criteria, and then photographed. The current situation is evaluated and then the illustrations of the present and future dimensions of the taxa are drawn by using Autocad 2017 program and the illustrations of the final stage of the areas and some additional illustrations are drawn by using SketchUp 2016 and Lumion 6. Finally, in the fifth stage, the current placing of the species and their future dimensions are discussed and various suggestions are introduced.

4. Findings

There are 118 species of 36 families in the campus (Table 1,2,3,4). 42 of these are natural and 76 are exotic species. The families and the numbers of species are as follows: Aceraceae (6), Agavaceae (1), Anacardiceae (1), Apocynaceae (1), Araliaceae (1), Berberidaceae (4), Betulaceae (2), Bignoniaceae (3), Buxaceae (2), Caprifoliaceae (5), Celastraceae (3), Cornaceae (2), Cupressaceae (11), Fagaceae (3), Ginkgonaceae (1), Hamamelidaceae (1), Hippocastanaceae (1), Hydrangeaceae (1), Juglandacea (1), Labiatae (2), Leguminosea (6), Lythracea (1), Magnoliaceae (2), Malvaceaea (1), Moraceae (1), Oleaceae (7), Pinaceae (10), Plataneceae (1), Rosaceae (24), Saliceceae (5), Scrophulariaceae (2), Simoribaceae (1), Tamaricaceae (1), Taxaceae (1), Tiliaceae (2), Vitaceae (1).

Table 1: The lists and properties of trees (Angiospermae) [11,12,13,14].

	TREES (ANGIOSPERMAE)								
	Plant Name	Family	General feature of the plant	Time/Color				Plant height /diamete(m)	Natural /Exotic
				Leaf	Flower	Fruit	Trunk		
1	*Acacia dealbata* Link	Leguminosae	E	1-12/Green	2-3/Yellow Greenish	9-10/Brown	Gray	25-30/6-12	E
2	*Acer buergerianum* Miq.	Aceraceae	D	3-10/Green/Red	2-3/Yellow Greenish	9-10/Brown	Gray	8-10/4-6	E
3	*Acer campestre* L.	Aceraceae	D	3-10/Green/Yellow	4-5/Yellow Greenish	9-10/Brown	Gray	15-20/8-12	N
4	*Acer negundo* L.	Aceraceae	D	3-10/Green/Yellow	4-5/Yellow Greenish	9-10/Brown	Gray	20-25/8-16	E
5	*Acer platanoides* L.	Aceraceae	D	3-10/Green/Yellowish Orange	4-5/Yellow Greenish	9-10/Brown	Gray	25-30/8-16	N
6	*Acer pseudoplatanus* L.	Aceraceae	D	3-10/Green/Yellow	4-5/Yellow Greenish	9-10/Brown	Gray	15-20/8-12	N
7	*Acer saccharum* Marsh.	Aceraceae	D	3-10/Green/Orangish Red	4-5/Yellow Greenish	9-10/Brown	Gray	25-35/8-16	E
8	*Aesculus hippocastanum* L.	Hippocastanaceae	D	4-10/Green/Yellowish Orange	5-6/White	9-10/Brown	Gray	20-30/8-16	E
9	*Ailanthus altissima* (Mill.)	Simoribaceae	D	4-10/Green/Orangish Red	5-6/Greenish Yellow	9-10/Brown	Gray	20-30/8-12	E
10	*Betula alba* L.	Betulaceae	D	4-10/Green/Yellow	3-4/Green	6-7/Brown	Gray	15-20/4-6	N
11	*Carpinus betulus* L.	Betulaceae	D	4-10/Green/Yellow	7-8/Greenish Yellow	11-12/Brown	Gray	20-25/6-8	N
12	*Buxus sempervirens* L.	Buxaceae	E	1-12/Dark Green	3-4/Yellow	8-9/Brown	Gray	6-8/3-4	N
13	*Castanea sativa* Mill.	Fagaceae	D	4-10/Green/Yellow	5-6/Yellow	8-9/Brown	Gray	20-25/6-8	N
14	*Catalpa bignonioides* Walt.	Bignoniaceae	D	4-10/Green/Yellow	5-6/White	9-10/Brown	Gray	10-15/6-10	E
15	*Cercis siliquastrum* L.	Leguminosae	D	3-10/Green/Yellow	3-4/Purple	9-10/Brown	Gray	6-8/4-6	E
16	*Cornus mas* L.	Cornaceae	D	3-11/Green/Yellow	3/Yellow	8-9/Red	Gray	6-8/3-4	N
17	*Cotinus coggygria* Scop.	Bignoniaceae	D	3-10/Yellow	4-5/Pink	9-10/Black	Gray	4-6/3-4	N
18	*Crataegus monogyna* Jacq.	Rosaceae	D	3-11/Green/Yellow	4-5/Yellow	7-8/Red	Gray	8-10/4-6	E
19	*Cydonia oblonga* Mill.	Rosaceae	D	4-11/Green/Yellow	5-6/Pink-White	10-11/Yellow	Gray	6-8/4-5	N
20	*Elaeagnus angustifolia* L.	Oleaceae	D	4-11/Green/Yellow	5-6/ Yellow	9-10/Orange	Gray	6-8/4-6	N
21	*Eriobotrya japonica* (Thunb.) Lindl.	Rosaecae	E	1-12/Green/Yellow	5-6 /White	8-9/Yellow	Gray	6-8/4-5	E
22	*Fraxinus excelsior* L.	Oleaceae	D	4-11/Green/Yellow	3-4-5/ Purple	7-8/ Brown	Gray	12-15/6-8	N
23	*Gleditsia triacanthos* L.	Leguminosae	D	4-10/Green/Yellow	4-5/Yellow	9-10/Brown	Gray	20-25/16-20	E
24	*Hibiscus syriacus* L.	Malvaceae	D	4-10/Green/Yellow	7-9/ Purple	11-12/Brown	Gray	4-5/2-4	E
25	*Juglans regia* L.	Juglandacea	D	4-10/Green/Yellow	4-5/Light Green	8-9/Purple/Brown	Gray	20-25/12-16	N
26	*Lagerstroemia indica* L.	Lythracea	D	4-10/Green/Orangish Red	7-8/Pink	10-11/Brown	Gray	6-8/6-8	E
27	*Ligustrum japonicum* Thunb.	Oleaceae	E	1-12/Green	4-5/White	8-9/Purple/Black	Gray	3-6/3-5	E
28	*Liquidambar styraciflua* L.	Hamamelidaceae	D	4-10/Green/Yellow	5-6/Yellow	9-10/Brown	Gray	30-40/8-12	E
29	*Magnolia grandiflora* L.	Magnoliaceae	E	1-12/Dark Green	5-6/White	9-10/Red	Gray	20-25/12-16	E
30	*Magnolia x soulangiana* Soul. Bod.	Magnoliaceae	D	3-10/Green/Yellow	3-4/Pinky White	9-10/Red	Gray	6-8/4-6	E
31	*Morus alba* L.	Moraceae	D	4-10/Green/Yellow	4-5/Greenish Yellow	6-7/White	Gray	15-20/8-12	N
32	*Malus floribunda* Siebold ex Van Houtte	Rosaecae	D	3-10/Green/Yellow	3-4/Pink	8-9/Red	Gray	8-10/4-6	E
33	*Paulownia tomentosa* (Thunb) Steud.	Scrophulariaceae	D	4-10/Green/Yellow	3-4/Purple	9-10/Brown	Gray	12-16/6-8	E
34	*Platanus orientalis* L.	Platanaceae	D	4-12/Green/Yellowish Orange	4-5/Greenish Yellow	9-10/Brown	Gray	40-50/12-30	E
35	*Populus alba* L.	Salicaceae	D	4-10/Green/Yellow	4-5/ Light Green	7-8/Brown	White	20-25/10-12	N
36	*Populus nigra* L.	Salicaceae	D	4-10/Green/Yellow	4-5/ Light Green	7-8/Brown	Gray	30-35/12-14	N

	TREES (ANGIOSPERMAE)								
	Plant Name	Family	General feature of the plant	Time/Color				Plant height /diameter (m)	Natural /Exotic
				Leaf	Flower	Fruit	Trunk		
37	*Populus tremula* L.	Salicaceae	D	4-10/Green/Yellow	4-5/Light Green	7-8/Brown	Gray	20-25/10-12	N
38	*Prunus amygdalus* (L.) Batsch	Rosaceae	D	3-10/Green/Yellow	3-4/White	7-8/Brown	Gray	8-10/4-6	N
39	*Prunus armenica* L.	Rosaceae	D	3-10/Green/Yellow	3-4/White	7-8/Light Yellow	Gray	6-8/4-6	N
40	*Prunus avium* L.	Rosaceae	D	3-10/Green/Yellow	3-4/White	7-8/Light Red	Gray	6-8/4-6	N
41	*Prunus cerasifera* JF Ehrh. cv. "Atropurpurea"	Rosaceae	D	3-10/Red	3-4/White	7-8/Light Red	Gray	8-10/6-8	E
42	*Prunus cerasus* L.	Rosaceae	D	3-10/Green/Yellow	3-4/White	7-8/Light Red	Gray	6-8/4-6	N
43	*Prunus laurocerasus* L.	Rosaceae	E	1-12/Green	4-5/White	8-9/Red-Black	Gray	8-10/6-8	N
44	*Prunus persica* (L.) Batsch	Rosaceae	D	3-10/Green/Yellow	3-4/Pink	7-8/Red	Gray	6-8/4-6	N
45	*Prunus serrulata* Lindl. "Kanzan"	Rosaceae	D	3-10/Green/Yellow	3-4/Purple	7-8/Red	Gray	8-10/4-6	E
46	*Quercus robur* L.	Fagaceae	D	4-12/Green/Yellowish Brown	4-5/Green	9-10/Brown	Gray	25-35/12-30	N
47	*Quercus rubra* L.	Fagaceae	D	4-11/Green/Red	4-5/Green	9-10/Brown	Gray	20-30/12-20	E
48	*Rhus typhina* L. "Laciniata"	Anacardiaceae	D	4-11/Green/Yellow	4-5/Red	9-10/Red	Gray	4-6/3-4	E
49	*Robinia pseudoacacia* L.	Leguminosae	D	4-11/Green/Yellow	5-6/White	9-10/Brown	Gray	15-20/6-8	N
50	*Salix alba* L.	Salicaeae	D	4-11/Green/Yellow	3-4/Yellow	7-8/Brown	Gray	25-30/8-16	N
51	*Salix babylonica* L.	Salicaeae	D	4-11/Green/Yellow	3-4/Yellow	7-8/Brown	Gray	10-15/6-8	E
52	*Sophora japonica* L.	Leguminosae	D	4-10/Green/Yellow	6-7/White	10-11/Brown	Gray	10-12/6-8	E
53	*Tamarix tetrandra* Pallas	Tamaricaceae	D	4-10/Green/Yellow	4-5/Pink	7-8/Brown	Gray	3-4/2-4	N
54	*Tilia rubra* DC.	Tiliaceae	D	4-11/Green/Yellow	6-7/Yellow	9-10/Brown	Gray	25-30/16-20	N
55	*Tilia tomentosa* Moench.	Tiliaceae	D	4-11/Green/Yellow	6-7/Yellow	9-10/Brown	Gray	25-30/16-20	N

Table 2: The lists and properties of climbing (Angiospermae) [11,12,13,14].

	CLIMBING (ANGIOSPERMAE)								
	Plant Name	Family	General feature of the plant	Time/Color				Plant height /diameter (m)	Natural /Exotic
				Leaf	Flower	Fruit	Trunk		
1	*Campsis radicans* (L.) Seem.	Bignoniaceae	D	4-10/Green/Yellow	7-8/Orange	9-10/Brown	Brown	8-10/0,5-1	E
2	*Hedera helix* L.	Araliaceae	E	1-12/Green	10-11/Greenish Yellow	1-2/Black	Gray	25-30/0,5-1	N
3	*Lonicera caprifolium* L.	Caprifoliaceae	E	1-12/Green	5-6/Whitish Pinky	8-9/Orange	Brown	6-8/1-2	N
4	*Parthenocissus quinquefolia* (L.) Planch	Vitaceae	D	4-10/Green/Red	5-6/Green	9-10/Black	Brown	6-8/0,5-1	E
5	*Rosa rampicante* L.	Rosaceae	D	4-10/Green/Yellow	5-6/ Pink	9-10/ Brown	Brown	6-8/1-2	E
6	*Wisteria sinensis* (Sims) DC.	Leguminosae	D	4-10/Green/Yellow	4-5;9-10/ Purple	9-10/ Brown	Brown	15-20/1-2	E
7	*Vinca major* L.	Apocynaceae	E	1-12/Green	4-5/ Purple	9-10/Brown	Green	0,3-0,6/0,1-0,3	E

Table 3: The lists and properties of trees and shrubs (Gymnospermae) [11,12,13,14].

	TREES and SHRUBS (GYMNOSPERMAE)								
	Plant Name	Family	General feature of the plant	Time/Color				Plant height/diameter (m)	Natural /Exotic
				Leaf	Flower	Fruit	Trunk		
1	*Abies bornmulleriana* Mattf.	Pinaceae	E	1-12/Dark Green	5-6/Yellow	9-10/Brown	Brown	15-20/4-6	N
2	*Abies nordmanniana* (Stev.) Spach	Pinaceae	E	1-12/Dark Green	5-6/Yellow	9-10/Brown	Brown	40-50/6-8	N
3	*Calocedrus decurrens* Pritchard & Baker	Cupressaceae	E	1-12/Dark Green	10-11/Yellow	2-3/Brown	Brown	50-60/8-12	E
4	*Cedrus atlantica* (Endl.)	Pinaceae	E	1-12/Blue	5-6/Grayish Brown	9-10/Brown	Brown	15-20/4-6	E
5	*Cedrus deodara* Royle ex D. Don.	Pinaceae	E	1-12/Light Green	5-6/Grayish Brown	9-10/Brown	Brown	30-35/6-8	E
6	*Cedrus libani* A. Rich.	Pinaceae	E	1-12/Dark Green	5-6/Grayish Brown	9-10/Brown	Brown	30-40/6-8	N
7	*Chamaecyparis lawsoniana* (A. Murr.) Parl.	Cupressaceae	E	1-12/Green	5-6/Bluish Green	9-10/Brown	Brown	50-60/8-10	E
8	*Cupressus arizonica* Greene	Cupressaceae	E	1-12/Blue	5-6/Grayish Brown	9-10/Brown	Gray	15-20/4-6	E
9	*Cupressus sempervirens* L.	Cupressaceae	E	1-12/Dark Green	5-6/Brown	9-10/Brown	Gray	30-35/2-4	N
10	*Juniperus chinensis* L. "Pfitzeriana Glauca"	Cupressaceae	E	1-12/Bluish Green	3-4/ Bluish Black	9-10/Blue	Gray	3-4/1-2	E
11	*Juniperus communis* L.	Cupressaceae	E	1-12/Green	3-4/ Bluish Black	9-10/Blue	Gray	15-20/3-4	E
12	*Juniperus horizontalis* Moench.	Cupressaceae	E	1-12/Green	3-4/ Bluish Black	9-10/Grayish Blue	Gray	0,5-1,5/0,5-1	E
13	*Juniperus sabina* L.	Cupressaceae	E	1-12/Green	3-4/ Bluish Black	9-10/Grayish Blue	Gray	3-5/0,5-2	N
14	*Juniperus oxycedrus* L.	Cupressaceae	E	1-12/Green	3-4/ Bluish Black	9-10/Orange	Gray	15-20/3-4	N
15	*Ginkgo biloba* L.	Ginkgonaceae	D	3-10/Yellow	4-5/Green	10-11/Dark Yellow	Gray	25-30/8-16	E
16	*Picea abies* (L.) Karst.	Pinaceae	E	1-12/Green	5-6/Green	9-10/Brown	Brown	40-50/6-8	E
17	*Picea pungens* Engelm.	Pinaceae	E	1-12/Blue	5-6/Dark Red	9-10/Brown	Brown	25-30/4-6	E
18	*Picea pungens* Engelm. "Glauca"	Pinaceae	E	1-12/Blue	5-6/Dark Red	9-10/Brown	Brown	20-25/4-6	E
19	*Picea orientalis* (L.) Link	Pinaceae	E	1-12/Green	5-6/Green	9-10/Brown	Brown	50-60/6-8	E
20	*Pinus sylvestris* L.	Pinaceae	E	1-12/Green	5-6/Yellowish Brown	9-10/Brown	Brown	35-40/6-8	N
21	*Taxus baccata* L.	Taxaceae	E	1-12/Green	5-6/Yellow	10-11/Brown	Brown	15-20/6-8	N
22	*Thuja occidentalis* L.	Cupressaceae	E	1-12/Green	5-6/Brown	9-10/Brown	Brown	15-20/4-6	E
23	*Thuja orientalis* L.	Cupressaceae	E	1-12/Green	5-6/Brown	9-10/Brown	Brown	10-15/2-4	E

Table 4: The lists and properties of shrubs (Angiospermae) [11,12,13,14].

	SHRUBS (ANGIOSPERMAE)								
	Plant Name	Family	General feature of the plant	Time/Color				Plant height /diameter (m)	Natural /Exotic
				Leaf	Flower	Fruit	Trunk		
1	*Abelia grandiflora* (Andre) Rehd.	Caprifoliaceae	SE	1-11/Garish Green /Claret Red	6-9/White	10-11/Brown	Brown	1-2/0,5-1,5	E
2	*Berberis julianae* Schneid.	Berberidaceae	E	1-12/Dark Green	3-4/Light Yellow	8-9/Red-Black	Gray	2-3/1-2	E
3	*Berberis thunbergii* DC "Atropurpurea"	Berberidaceae	D	4-10/Red/Red	3-4/Light Yellow	8-9/Red-Black	Gray	2-3/1-2	E
4	*Buddleja davidii* Franch.	Scrophulariaceae	SE	1-11/Green/Yellow	5-7/Purple	9-10/Brown	Gray	2-3/1-2	E
5	*Buxus sempervirens* L. "Rotundifolia"	Buxaceae	E	1-12/Dark Green	3-4/Yellow	8-9/Brown	Gray	0,4-1/0,3-1	E
6	*Chaenomeles japonica* (Thunb.) Lindl.	Rosaceae	D	4-10/Green/Yellow	3/Pink	9-10/Yellow	Gray	2-3/1-2	E
7	*Cornus alba* L.	Corneceae	D	3-10/Green/Red	4-5/White	7-8/White	Red	2-3/1-2	E
8	*Cotoneaster horizontalis* L.	Rosaceae	D	3-10/Green/Red	4-5/White	10-11/Red	Gray	0,5-1,5/0,5-1,5	E
9	*Cotoneaster bullatus Bois.*	Rosaceae	E	1-12/Green	4-5/White	10-11/Red	Gray	1-2-1,5/1-2	E
10	*Euonymus alatus* (Thunb.) Sieb.	Celastraceae	D	4-10/Green/Red	4-5/Yellow	9-10/Orange	Gray	1-3/1-2	E
11	*Euonymus fortunei* (Turcz.) Hand.-Maz.	Celastraceae	E	1-12/Green	4-5/Greenish White	9-10/Pinky Red	Green	0,3-0,6/0,5-1	E
12	*Euonymus japonica* Thunb.	Celastraceae	E	1-12/Green	4-5/Green	9-10/Red	Green	4-6-/2-4	E
13	*Forsythia intermedia* Zabel.	Oleaceae	D	4-10/Green /Yellow	2-3/Yellow	7-8/Brown	Yellow	1-2,5/1-2	E
14	*Lavandula angustifolia* Mill.	Labiatae	E	1-12/Green	5-7/Blue	8-9/Black	Green	0,4-0,8/0,4-0,6	E
15	*Ligustrum ovalifolium* Hassk.	Oleaceae	D	4-10/Green /Yellow	6-7/White	11-12/Black	Gray	1-3/1-2	E
16	*Ligustrum vulgare* L.	Oleaceae	D	4-10/Green /Yellow	6-7/White	11-12/Black	Gray	1-3/1-2	E
17	*Mahonia aquifolium* (Pursh) Nutt.	Berberidaceae	E	1-12/Green	3-4/Yellow	8-9/Purple-Black	Gray	1-2/0,3-1	E
18	*Philadelphus coronarius* L.	Hydrangeaceae	D	4-10/Green /Yellow	4-5/White	9-10/Brown	Gray	3-4/0,5-2	E
19	*Photinia fraseri* "Red Robin" [J]	Rosaceae	E	1-12/Green	4-5/White	10-11/Red	Gray	3-6/2-4	E
20	*Pyracantha coccinea* M. J. Roemer	Rosaceae	E	1-12/Green	4-5/White	10-11/Red	Gray	3-5/2-4	E
21	*Rosa canina* L.	Rosaceae	D	4-10/Green/Yellow	5-7/White	10-11/Red	Brown	2-4/2-4	N
22	*Rosmarinus officinalis* L.	Labiatae	E	1-12/Green	4-5/Purple	8-9/Black	Green	1-2/0,5-1,5	E
23	*Sambucus nigra* L.	Caprifoliaceae	D	4-10/Green/Yellow	4-5/Dark Pink	8-9/Black	Gray	3-4/1-2	N
24	*Spirea bumalda* Burv.	Rosaceae	D	4-10/Green/Yellow	4-5/Pink	8-9/Brown	Brown	2-3/1-2	E
25	*Spirea vanhouttei* Zabel.	Rosaceae	D	4-10/Green/Yellow	4-5/White	8-9/Brown	Brown	0,5-1,5/0,5-1,5	E
26	*Syringa vulgaris* L.	Oleaceae	D	4-10/Green/Yellow	4-5/Purple	8-9/Brown	Gray	4-6/2-3	N
27	*Viburnum opulus* L.	Caprifoliaceae	D	4-10/Green/Red	4-5/White	8-9/Red	Gray	3-4/1-2	E
28	*Viburnum tinus* L.	Caprifoliaceae	E	1-12/Dark Green	4-5/White	8-9/Black	Gray	4-5/1-3	E
29	*Yucca filamentosa* L.	Agavaceae	E	1-12/Dark Green	5-6-11/Cream	8-9/Green	Green	4-5/1-3	E

Research area no.1 is the planting design work made using alley trees in the green areas around the Burhanettin Uysal Stadium on theIron and Steel Campus. In this area, neither the planting distances of the species (*Prunus cerasifera* JF Ehrh cv. "Atropurpurea" and *Acer pseudoplatanus* L.) nor their transition in terms of form were observed. It is apparent that these species are planted at around 1.20–2m intervals. This will lead to functional and aesthetic problems as the plant species grow and interlace over time. Besides preventing the healthy growth of the plant, it will also cause unappealing views to emerge. In addition, it will block the light through the rooms of Burhanettin Uysal Stadium due to the rough texture it will have when it comes into leaf. The plans of the current and last stages of research area no.1, and illustrations of the final stage of the area are as follows:

*Figure 1. Research area no.1 (**Left**). The plan of the present and final state of research area no.1 (**Right**).*

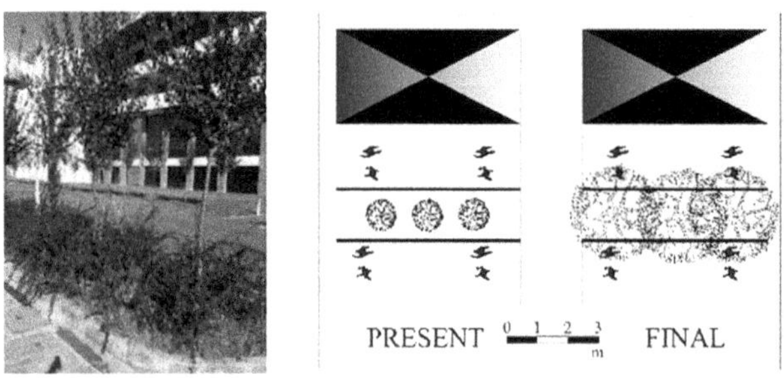

*Figure 2. The illustration of the present state of research area no.1 (**Left**). The illustration of the final state of research area no.1 (**Right**).*

Research area no.2 includes the planting design studies made using a group of bushes in the green areas around the Faculty of Economics and Administrative Sciences. The use of the bushes in groups leads to the formation of massive areas. However, the sociological characteristics of the plants had not been taken into account. Accordingly, they were planted at very frequent intervals to make the field look intense, due to aesthetic concerns. This caused the plants to interlace and affected their forms and lives according to sociological aspects. The planting area for plants used here (*Berberis thunbergii* DC "Atropurpurea", *Photinia fraseri* "Red Robin" [J], *Viburnum tinus* L.) had not been observed. It is apparent that these species were planted at around 0.2–0.3m intervals. This will lead to functional and aesthetic problems as the plant species grow and interlace over time. It will also prevent the healthy growth of the plant.

*Figure 3. Research area no.2 (**Left**). The plan of the present and final state of research area no.2 (**Right**).*

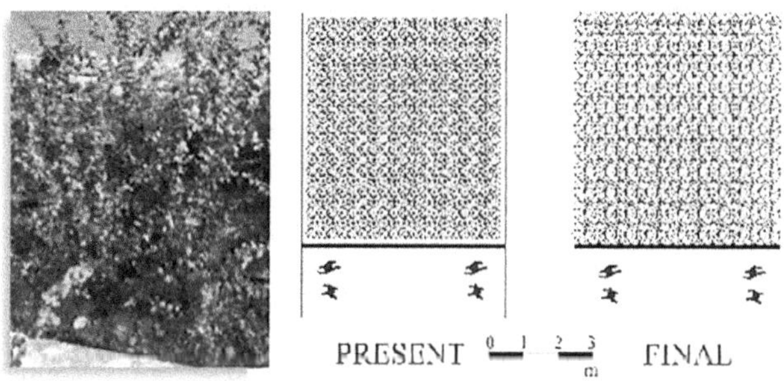

*Figure 4. The illustration of the present state of research area no.2 (**Left**). The illustration of the final state of research area no.2 (**Right**).*

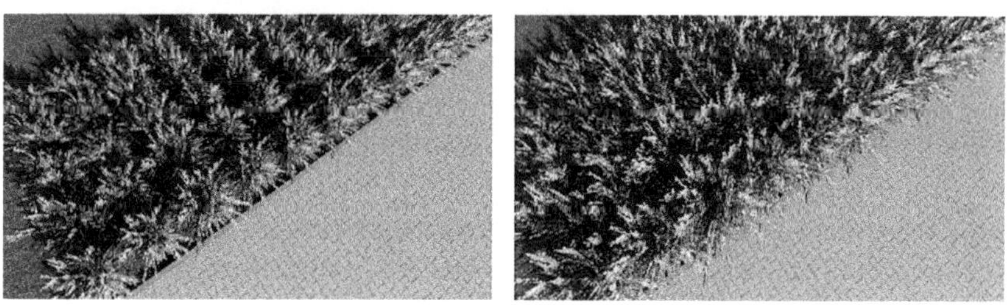

Research area no.3 is the road verge planting design work carried out on the two-way on-campus byroads around the Rectorate Building. *Cedrus libani* A. Rich. are often used in planting designs as they are aesthetic, functional, and natural. In this area, neither the planting distance of *Cedrus libani* A. Rich, nor its suitability for the road verge – in terms of form – have been observed. It is determined that too many *Cedrus libani* A. Rich were used on the road verge, the width of which is only 1.40 m. This will cause the plant species to grow over time and create problems in passing in vehicular and pedestrian areas. In this context, the *Cedrus libani* A. taxon, which reveals a unique beauty with its habit branching out from the ground, is pruned to allow for the passing of vehicles and pedestrians, which in turn disturbs the health and aesthetics of the plant. The plans for the current and latest stages of these research areas and the illustrations of the final stages are as follows:

*Figure 5. Research area no.3 (**Left**). The plan of the present state of research area no.3 (**Middle**). The plan of the final state of research area no.3 (**Right**).*

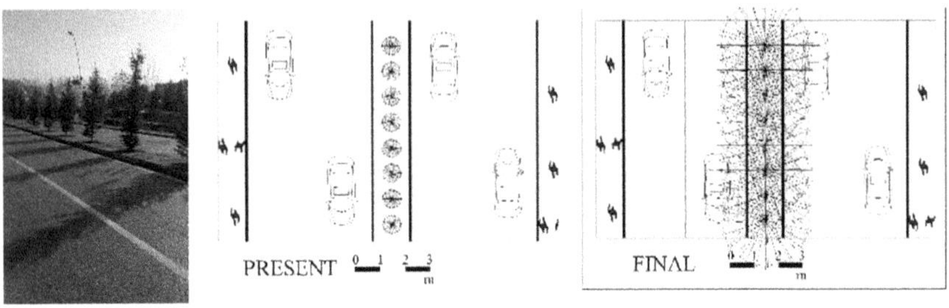

*Figure 6. The illustration of the present state of research area no.3 (**Left**). The illustration of the final state of research area no.3 (**Right**).*

Research area no.4 includes the planting design work between the pedestrian and vehicle pathways around the Faculty of Engineering building. As the plants used in this area (1: *Pinus griffithii* McClel., 2: *Pinus nigra* Arnold) are on the edge of both the pedestrian and the vehicle pathway, in the future it is very likely that they will cause problems for passage due to their prospective size and form. According to planting design criteria, *Pinus griffithii* McClel. and *Pinus nigra* Arnold are not suitable for pruning from the bottom since these plant species are attractive due to branching outwards from the ground.

*Figure 7. Research area no.4 (**Left**). The plan of the present state of research area no.4 (**Middle**). The plan of the final state of research area no.4 (**Right**).*

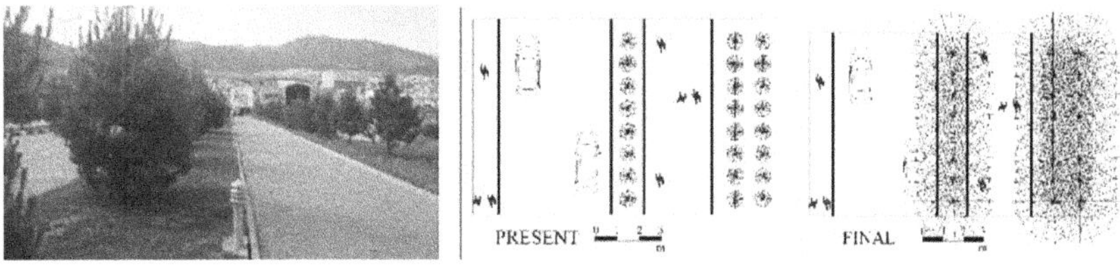

*Figure 8. The illustration of the present state of research area no.4 (**Left**). The illustration of the final state of research area no.4 (**Right**).*

Research areas no.5 and no.6 include the planting design work carried out between the pedestrian and vehicle pathways around the Burhanettin Uysal Stadium. In the future, it is very likely that the plants used in this area (*Acer pseudoplatanus* L. and *Prunus cerasifera* JF Ehrh cv. "Atropurpurea") will cause problems and prevent the lighting elements (Research area no.5) and guide signs (Research area no.6) from functioning properly as they are will be placed very close to these elements.

*Figure 9. Research area no.5 (**Left**). The plan of the present and final state of research area no.5 (**Right**).*

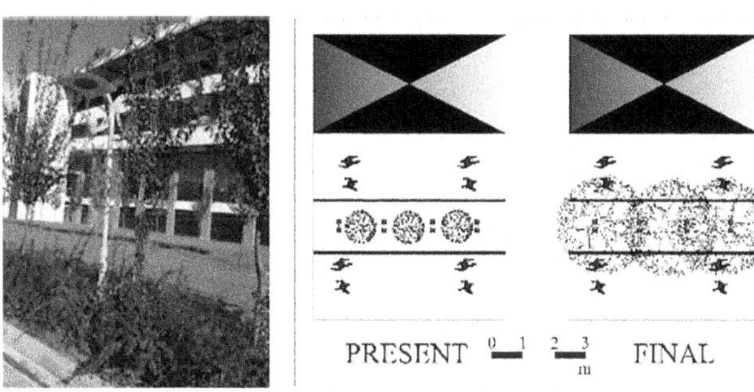

*Figure 10. The illustration of the present state of research area no.5 (**Left**). The illustration of the final state of research area no.5 (**Right**).*

*Figure 11. Research area no.6 (**Left**). The plan of the present and final state of research area no.6 (**Right**).*

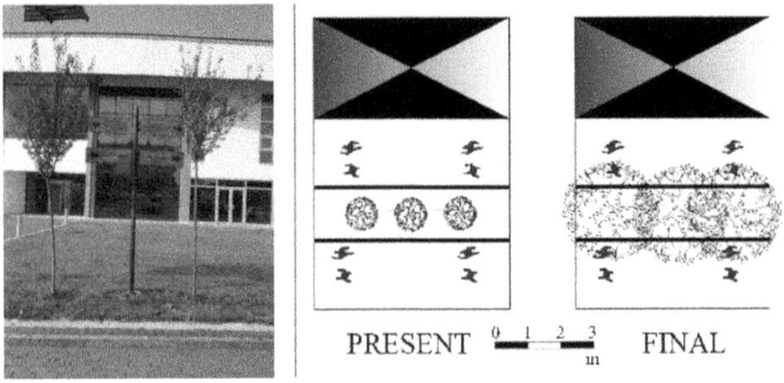

*Figure 12. The illustration of the present state of research area no.6 (**Left**). The illustration of the final state of research area no.6 (**Right**).*

Research area no.7 includes the planting design work of the retaining walls around the Burhanettin Uysal Stadium. *Acer pseudoplatanus* L. was used here but the planting distance and planting locations were not observed. It is apparent that these species were planted at around 2–2.5m intervals and about 0.4m away from the retaining wall. This will cause the pressure that root applies on the wall to increase, and accordingly, lead to destruction of the wall.

*Figure 13. Research area no.7 (**Left**). The plan of the present state of research area no.7 (**Middle**). The plan of the final state of research area no.7 (**Right**).*

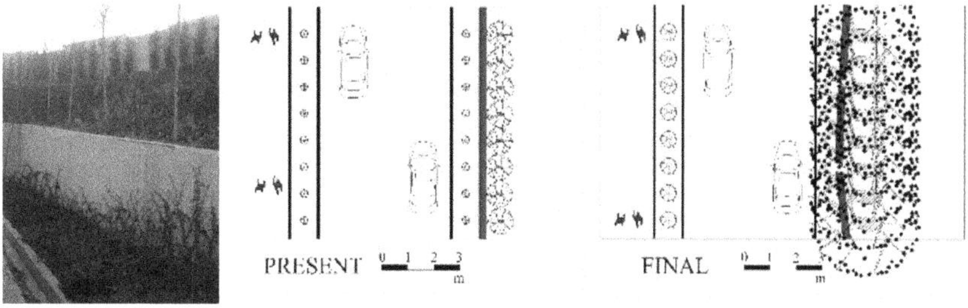

*Figure 14. The illustration of the present state of research area no.7 (**Left**). The illustration of the final state of research area no.7 (**Right**).*

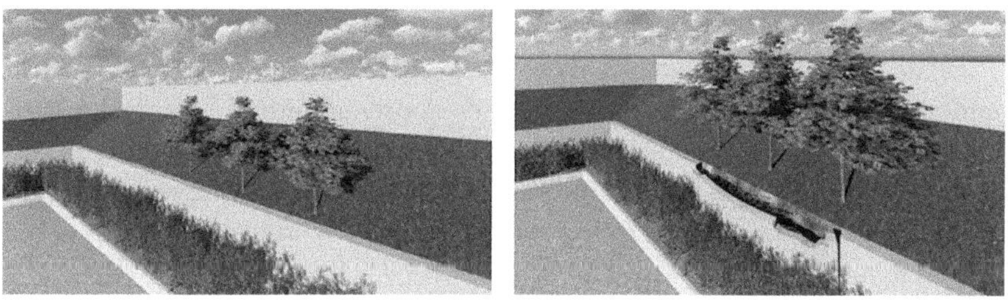

Research area no.8 includes the planting design work carried out to construct the crescent section of the main concept of the Burhanettin Uysal Stadium. The Burhanettin Uysal Stadium is designed to represent the Turkish flag, and it appears as a crescent and star when viewed from the top. In order to support the concept and create the look of the flag's colors, plants that show red and white color characteristics were used. In the design, the *Acer pseudoplatanus* L. plant taxon was used for the red area near the crescent part of the flag, however, the plant does not have a red color.

The research area no.9 is the planting design carried out to construct the star part to form the main design for the Burhanettin Uysal Stadium. The *Forsythia intermedia* Zabel plant species was used in the white part that forms the star of the flag. However, the plant does not have a white color.

Research area no.10 involves the grass planting design work carried out throughout the Iron and Steel Campus. In the campus area, part of the grass application was made by the seeding method, and the rest was made by using roll-out grass. In the areas where the seeding method was used, irregularities are observed due to watering problems and irregular seeding. On the other hand, a vast area of dry land is observed due to mistakes made in the application of the roll- out grass, as well as watering problems. The fact that the team that applied the roll-out grass did not consist of experts brought with it many problems. Failure to perform the lapping process properly caused a partial pitting of the surface. As a result of the team's failure to pay attention to details of the grass rolls joining one another, the rolls overlapped, preventing air intake and causing drying. These problems in the grass areas, which make up a large part of the campus area, disturb the integrity in terms of design.

Strengthening the influence of planting designs is possible by performing the necessary maintenance procedures after application. This can be achieved by the coordinated work of a specialized team. However, in our country, one of the biggest problems faced by the campuses is that the landscaping services do not exist, or function inadequately under the Department of Construction and Technical Affairs.

At the Iron and Steel Campus, there is only one landscape technician and two gardeners working under the Department of Construction and Technical Affairs. It is not impossible to maintain 2500 da of green areas on the campus with this team. Additionally, it is understood that the staff is not sufficiently knowledgeable in identifying and maintaining plants. This causes problems in terms of the continuity of the applied planting design.

5. Findings and Discussion

It has been determined that the plants used in the research area no.1 (1: *Prunus cerasifera* JF Ehrh. cv. "Atropurpurea": 6–8m /8–10m, 2: *Acer pseudoplatanus* L.:8–12m/15–20m) were planted without considering their forms and the final canopy diameters and heights. For this reason, the plants used in the research areas will not only block the view of the building and prevent the building from getting light but will also cause the plants to be damaged.

It would be correct to use these types of alleys in planting designs considering the planting distances. Using *Prunus cerasifera* JF Ehrh. Cv. "Atropurpurea" and *Acer pseudoplatnus* L. species with minimum 4m and 6m planting distances respectively would be better in terms of design, and ensuring the healthy growth of the plant. Moreover, the prospective size, form, and texture characteristics of plant species must be taken into consideration.

It is determined that the plants used (1:*Berberis thunbergii* DC "Atropurpurea":1–2 m /2–3 m, 2:*Photinia fraseri* "Red Robin" [J]: 2–3 m /3–6 m, 3: *Viburnum tinus* L.:1–3 m /4–5 m) in research area no.2 were planted without considering the final canopy diameters and heights. In such a case, the plants used in the research areas will not only block the view of the building and prevent the building from getting light but also cause the plants to be damaged. It would be correct to use these types of bush groups in planting design work with a consideration of their planting distances. Although the planting distance of these shrub species, which can be shaped by pruning and can be kept within the desired dimensions, varies according to the design purposes, using *Berberis thunbergii* DC "Atropurpurea" with a minimum of 0.4 m, *Photinia fraseri* "Red Robin" [J] with minimum 0.5 m, and *Viburnum tinus* L. with minimum 0.5 m planting distances would be better in terms of design and the healthy growth of the plant.

It is determined that the plants used (*Cedrus libani* A. Rich. 6–8 m/ 30–40 m) in research area no.3 were planted without considering the final canopy diameters and heights. In such a case, the plants used in the research areas will not only cover a large part of the road and block vehicular and pedestrian passage but will also cause the vehicles and the plants to be damaged. Therefore, it would be appropriate to remove the *Cedrus libani* A. species from the area before the pile root growth increases further, and move it to more suitable areas. It would be appropriate to use plant species with a conical/pyramidal and round/spherical bush species suitable for pruning in planting design studies of such road verge areas. If trees and shrub species should be used, it would be better to use plant species that are stemmed and that branch from the top.

It is determined that the plants used (1:*Pinus griffithii* McClel. 6–8/20–25 m, 2:*Pinus nigra* Arnold 6–8/35–40 m) in the research area no. 4 were planted without considering the final canopy diameters, heights, forms and textures. The plants used in the research area will block vehicular and pedestrian passage. In such areas, where the vehicular and pedestrian traffic is busy, it would be more appropriate to use bush species that do not block passage. Besides, in summer in order to block the sunlight, it would be more appropriate to choose a plant with coarse-textured leaves, and in winter one in order to transmit the light, with fine-textured large leaves.

It is determined that the plants used (1: *Prunus cerasifera* JF Ehrh. cv. "Atropurpurea": 6–8 m/8–10 m, 2: *Acer pseudoplatanus* L.: 8–12 m/15–20 m) in the research area no.5 and area no.6 were planted without considering the final canopy diameters, heights, and forms. The plants were planted without considering the size of lighting elements and guide signs and it was overlooked that in the future they could block the lighting function or the visibility of the guide signs. In such areas, it would be more appropriate to choose bush species that are compatible with tall lighting elements and that will not interfere with the guiding function. In this context, determining the location of the guide signs to consider the future growth of the plants is important for the healthy growth of the plants as well as the visibilityof the signs.

It is determined that the plants used (*Acer pseudoplatanus* L.: 8–12 m/15–20 m) in the research area no.7 were planted without considering the final canopy diameters, heights, closeness to the wall, or the root structure. In the research area, it would be correct to use bush species that will not create loads so that the root structure of which will not affect the retaining wall. For this reason, transplantation of *Acer pseudoplatanus* L. taxa located on the upper side of the retaining walls is necessary to prevent loss of life.

The leaves of the plant (*Acer pseudoplatanus* L.), which is used in research area no. 8, are green at the first emergence during the vegetation period and are spilled with a yellowish orange color in September-October. However, this situation is incongruous with the concept of "red". For this reason, trees or bush species which have extensive amounts of longer-lasting red color, leaves, flowers, or fruit, should be preferred in this area.

The plant (*Forsythia intermedia* Zabel.) used in the area no.9 is a species that emerges with golden yellow flowers around February-March and turns pale in the autumn. It is determined that this characteristic of the plant does not comply with the white element of the design. For this reason, it is necessary to use plant taxa which show white color characteristics. As it is a covered area, it requires the use of bush species rather than trees. Therefore, bush species with an extensive amount of longer-lasting red leaves, flowers, or fruit should be preferred in this area.

The tree taxa such as *Prunus cerasifera* JF Ehrh. cv. "Atropurpurea", *Acer pseudoplatanus* L., and *Cedrus libani* A. Rich., which are misused in the design management and have high market value, along with bush taxa such as *Berberis thunbergii* DC "Atropurpurea", *Photinia fraseri* "Red Robin" [J], and *Viburnum tinus* L., which are used extensively, should be transplanted to ensure the healthy growth of the plants, aesthetically better forms, and more accurate use. In case these valuable plant taxa are used in new landscaping studies or in revisions of existing areas, they will reduce the number of plants taken from outside the university and contribute to the university budget. In this context, it would be more appropriate to get these plants ready for transplantation and move them to appropriate areas in the appropriate period using the campus transplantation truck.

As a result, it has been pointed out that the plant species used in these 10 research areas are used incorrectly in terms of planting design criteria, and that they were planted without considering their size, form, texture, color, and conformity with the areas. This leads to major problems. In order to ensure that these problems do not occur in future planting designs, expert landscape architects should be assigned for the design, application, and maintenance stages of the sites. In this context, it is absolutely necessary to establish Landscaping Services under the Construction and Technical Affairs Department at the universities, and the staff should have adequate levels of education to be able to perform these duties. The number of staff in these landscaping services should be planned according to the size of the campus area and the staff should be encouraged to undergo training when necessary.

References

1. TURNER, P., "Campus: An American Planing Tradition", The Architectural History Foundation/MIT Pres, Series No. 7., 1990.
2. ARKUN, A. K. VE ÖZKAL, Ç., Bilkent Üniversitesi Kampüsü Peyzaj Analizi, Yüksek Lisans Dersi Sunum Ödevi, Ankara, 2005.
3. KARAŞAH, B., VAR, M., Trabzon ve bazı ilçelerinde kent dokusundaki bitkilendirme tasarımlarının ölçü-form açısından irdelenmesi. Bartın Orman Fakültesi Dergisi, 2014:1–11., 2012.
4. LESZCZYNSKI, N., Planting the Landscape – A Professional Approach to Garden Design, John Wiley&Sons, Inc., New York, 1999.
5. SCARFONE, S., Professional planting design, Hoboken, N.J. : John Wilcy & Sons, 2007.
6. BOOTH, N., Basic Elements of Landscape Architectural Design, Waveland Pres, Inc.Illinois, USA, 1990.
7. ROBINSON, N., The Planting Design Handbook, Ashgate Publishing Company, Hampshire, England, 2004.
8. AYAŞLIGIL, Y., Bitkisel Tasarım Ders Notu, İstanbul, 1998.
9. UZUN, G., Temel Tasarım, Çukurova Üniversitesi Ziraat Fakültesi Genel Yayın No:196 Ders Kitapları Yayın No: A-62, Adana, 1999.
10. WALKER, T. D., Planting Design, Van Nostrand Reinhold, New York, 1991.
11. ANSIN, R. AND OZKAN, Z. C., Spermatophyte, woody taxa. Karadeniz Technical Univ. Faculty of Forestry Publication No: 167–19, Trabzon, Turkey, 1997.
12. ARSLAN M., BARIŞ M.E., Ankara Park ve Bahçeleri Egzotik Ağaç ve Çalıları Kitabı, Ankara-Kasım 2012, ISBN: 978–605-4067-43-5., 2012.
13. YALTIRIK, F., Dendrology I, Gymnospermae 2nd Edition, Istanbul Universty Faculty of Forestry Publication, No: 3443–386, Istanbul, Turkey, 1993.
14. KORKUT A., ŞİŞMAN E. E., ÖZYAVUZ M., Peyzaj Mimarlığı, Yayın Yeri: Verda Yayıncılık & Danışmanlık, 2010.

Türker Oğuztürk[1], Ercan Gökyer[2], Ömer Lütfü Çorbacı[3]

Evaluating Landscape Changes in a Coastal City: Case Of Amasra City, Turkey

[1]*Recep Tayyip Erdogan University, Faculty of Art, Desing and Architecture, Department of Landscape Architecture, 74100, Rize, Turkey, turker.oguzturk@erdogan.edu.tr*

[2]*Bartın University, Faculty of Forestry, Department of Landscape Architecture, 74100, Bartın, Turkey*

[3]*Recep Tayyip Erdogan University, Faculty of Art, Desing and Architecture, Department of Landscape Architecture, 74100, Rize, Turkey*

1. Introduction

Over the ages, people have built thousands of cities and settlements in the world where they live. Important settlements, tourism and trade areas are located on coasts throughout the world. Coastal areas are affected positively or negatively by these activities depending on whether human beings are using the coastal areas for protection, nutrition, energy, trade or tourism activities. Due to intensive usage pressure, there is constant change and development on the coast.

There are various changes to coasts related to human activities and natural factors (wind, climate, earthquakes, etc.). Population density and human activities in coastal areas are among the main factors influencing landscape change [1,2,3].

The changes to coastal areas can be revealed as a result of the evaluation made through old and new photographs of a selected area dating from two different time periods [4,5].

In this study, Amasra, which is located in northwestern Turkey and is an important coastal settlement with historical, touristic and natural features, along with its immediate surroundings, was selected in order to determine and analyze changes in the landscape. We can categorize the data used in the study under three main headings below:

1- Printed and visual data about the study area
2- Satellite images
3- Old and new photos (1955, 1970, 2016).

Amasra's town center and its immediate surroundings were determined as the boundaries of the research area. Satellite images from Google Earth (2003–2016) obtained on different dates were used in order to be able to reveal landscape changes in and around the town center of Amasra.

At the same time, in order to evaluate the changes in the field of research in a visual way, new photos were taken from the places where the old pictures had been taken. the photographs were compared and the changes were analyzed.

2. General Features of Research Area

Amasra, which is the research area, is the district of the Bartın province that has a coast on the Black Sea in the north of Turkey and is known for its historical and touristic features. The length of the shoreline in the study area is approximately 7.41 km.

The population of the Amasra district center was determined as being 6666 people in 2015 [6]. In the summer months and during longer holidays, hotels and hostels are fully booked and there are times when the population increases by 2–3 times with day-trip visitors.

Amasra attracts tourists and day-trip visitors with its historical monuments, museums and other historical and cultural features, the areas that allow various recreational activities (The jetty, Küçük Harbor), the Çekiciler bazaar where ornamental items made of wood can be bought, and its natural beauty [7].

Road transportation is provided from three different routes from Bartın center to Amasra. The Bartın to Amasra route is 20 km long when the old road is used, 26 km when the new road is used and 17 km when the Tunel road is used. The map of three different routes providing access to Amasra is shown in Figure 1.

Figure 1: Location and transportation map of research area.

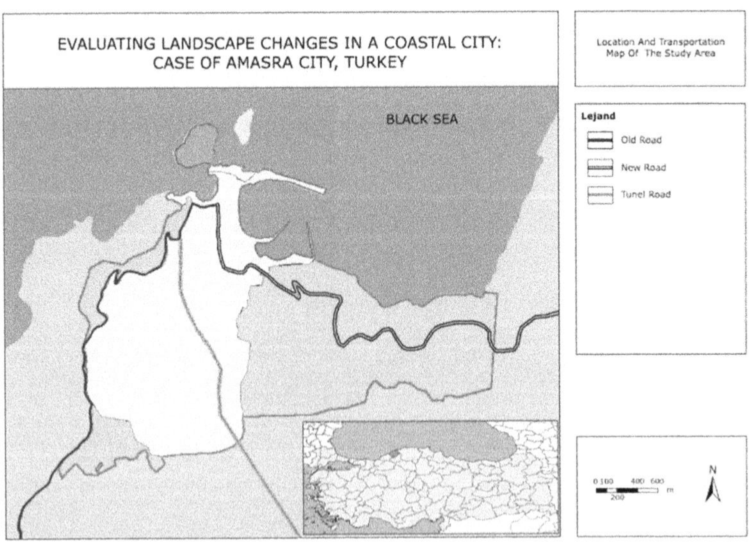

3. Landscape Change

In the scope of the study, satellite images and photographs of different dates were used in order to determine landscape change.

3.1 Landscape Change Analysis Using Satellite Images

Google Earth satellite images were used to detect changes over time in the research area. Google Earth satellite imagery for 2003 and 2016 was reviewed and changing areas were flagged up and shown on the images. When the 2003 and 2016 data on the study area were analyzed, it was determined that the changes in the urban area mostly originated from structuring for settlement purposes and tourist facilities. The areas where a change was detected are numbered from 1 to 22 and shown in Figure 2 and Figure 3.

Figure 2: Satellite image of Amasra dated 04.10.2003 [8].

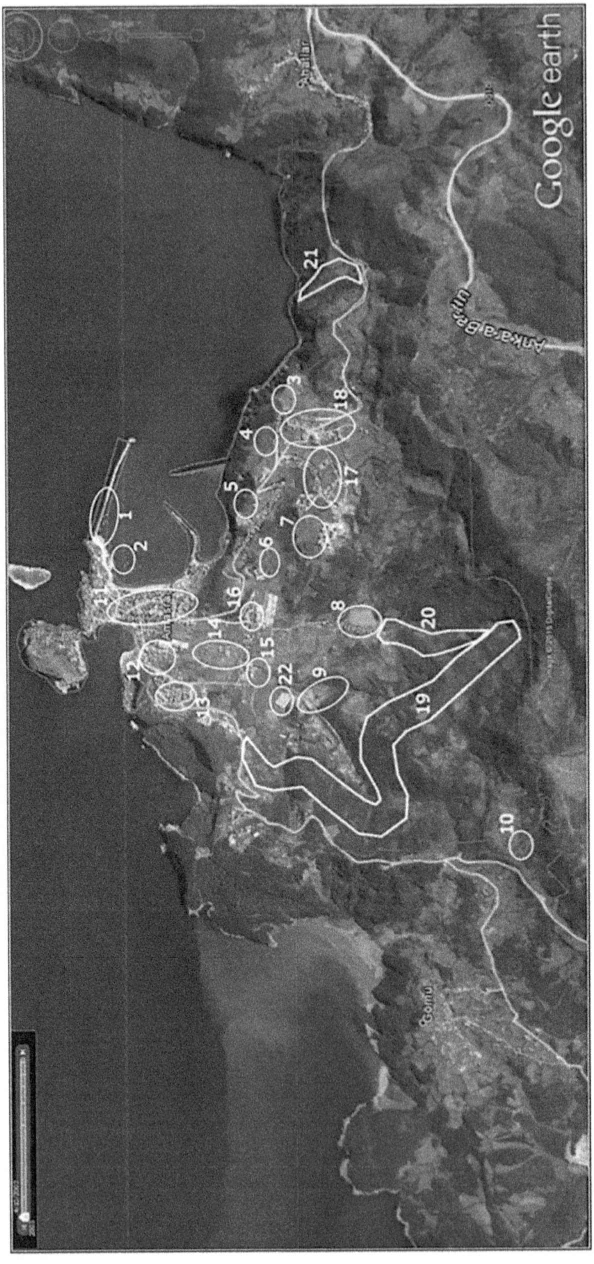

Figure 3: Satellite image of Amasra dated 12.10.2016 [8].

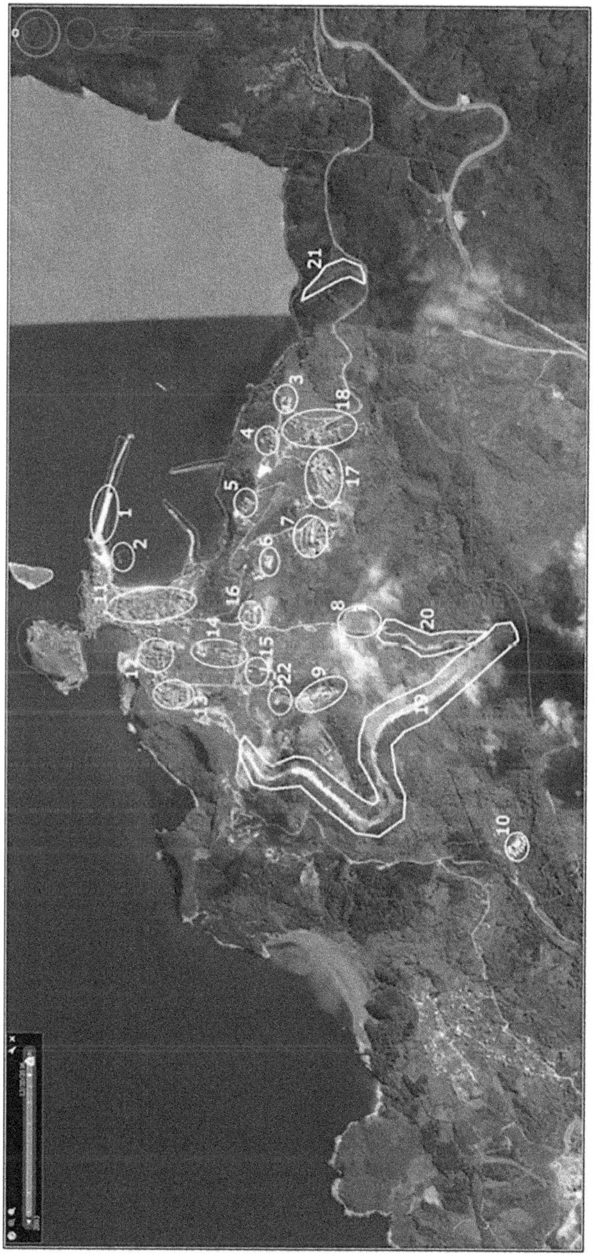

3.2 Landscape Change Analysis with Differently Dated Photographs:

Changes were determined through photographs taken on different dates from past to present. Important changes in the area took place around the Büyük Harbor area, the Küçük Harbor area and the Kaleşah district.

The Büyük Harbor area: In Figure 4, we can see that the highway along the Büyük Harbor was unpaved. Today, the Büyük Harbor, as it was called in the past, is used as the public beach and the area where boats are moored. We can see that the beach area in the Büyük Harbour has decreased over time. When we examine the differently dated pictures of the Büyük Harbour (Figure 4), we can see that while there were a few low buildings with gardens in the old days, high buildings close to each other have now been built. We can determine that the building density has increased in the city center of Amasra and the city silhouette has changed within the last 61 years.

Figure 4: Visual change analysis in Büyük Harbour area [9].

Kale District, Kum District Location:

It was observed that flora in the city center of Amasra has disappeared over time and that it has been replaced by high buildings and public institutions (Figure 5). In addition, we see that a marina was built at the beginning of 2016 and the breakwater was expanded (Figure 5).

Figure 5: Visual change analysis of the Kale District and Kum District [10].

Kaleşah District, Küçük Harbour, Boztepe District Location:

It is apparent that the number of houses in the sloping areas above the sea level is nowadays increasing in an irregular and unplanned way (Figure 6). We can see that settlement areas have increased towards hills and some land has turned into woods or residential areas over time. We can see that second houses have been built in scenic hilly areas.

Figure 6: Visual change analysis of Kaleşah District and Küçük Harbor Area [11].

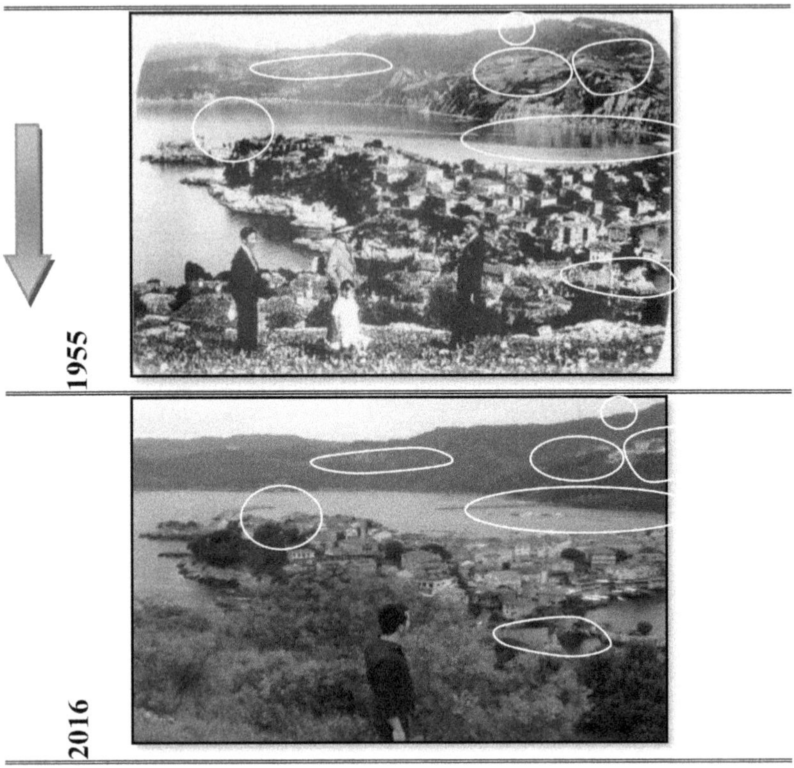

4. Results

Amasra is extensively visited by people living in big cities, especially in summer because of the fact that it is an important tourist area. Amasra is a charming coastal settlement preferred for touristic activities and settlement purposes due to its closeness to big cities such as Ankara and Kocaeli and its natural, historical and cultural features.

The fact that urban settlements in Amasra have accumulated in a narrow area has resulted in irregular settlement, incompatible with the historical texture in the vicinity of the protected and historic areas Therefore, the aesthetic appearance of the city is distorted and this situation has brought Amasra to the point of losing its silhouette as a town. Besides, the sites and historical city images have been damaged. In Amasra, the urban area is constantly developing and growing in an unplanned manner due to tourist facilities and second houses. In Amasra, new residential areas, hotels and secondary residential areas have been built in the scenic areas above the sea. The constriction of the city due to the restriction of the topographic conditions affects not only structures but also green areas, transportation axes and recreation areas. The roads in the town, which are trapped in narrow areas due to the topographical nature of the land, cannot be left wide enough and nowadays have proven almost inadequate by falling on one side. Thus, circulation in the town itself is difficult and in many places the traffic has to flow in one direction. With the concentration of the town, the size of green areas in the city center has also shrunk. The green areas that underwent changes over time were made available to the people of Amasra as well as visitors. Festivals held in Amasra also increase the value of Amasra in terms of tourism and increase the amount of day-trip tourists by attracting people in the surrounding districts. Following the recent tunnel road in Amasra, steep slopes and retaining walls have formed. These slopes and retaining walls create a rather poor appearance at the entrance to Amasra, which destroys the forest areas and nature. The unplanned construction in Amasra has led to the destruction of the coastal area, causing the identity of the city to be adversely affected and its silhouette to deteriorate.

5.4 Suggestions on Urban Development and Planning in Amasra

Amasra is better known for its historical features and coastal tourism. For this reason, these coastal and historic areas must be protected and kept clean. Tourism should be developed by identifying alternative tourism activities that can be performed besides coastal tourism. The participation of people is important for the development of tourism [12] in Amasra. Demonstrations and projects should be carried out to emphasize the significance of Amasra's historical values dating over 3000 years, to ensure the development of tourism.

Second houses built without planning affect the identity of a town in a negative way. Landscape repair work must be performed in these areas by destroying buildings that are harmful and not suitable to the town's identity. Second residential constructions should be made within the scope of plans that are appropriate to Amasra's silhouette. Negative effects can be reduced by avoiding natural damages caused by landscape repair work, the new tunnel road and unplanned construction. Based on the examination made within the scope of this study, we can say that Amasra City of Bartın Province should be re-planned and developed using a historical and tourist town model, with its natural and cultural features, coast, and historical areas.

References

1. GÖKYER, E., Evaluating landscape changes and environmental threats in a coastal landscape case study of Bartın coastal area, Turkey. Fresenius Environmental Bulletin, 23(7), 1683–1688, 2014.
2. GÖKYER, E., Bartın Kenti ve Arıt Havzası'nda Peyzaj Değerlendirme. Doktora Tezi, Ankara Üniversitesi Fen Bilimleri Enstitüsü, Peyzaj Mimarlığı Anabilim Dalı, Ankara, 143 s, 2009.
3. ANTROP, M., Handling landscape change. Landscape Change Conference Proceedings. ECLAS 2005. Ankara. 3–15, 2005.
4. SESLİ, F. A., KARSLI, F., Monitoring Coastal Land Use Changes on the Turkish Black Sea Coast with Remote Sensing: an Example from Trabzon/Turkey, *2nd Fıg Regional Conference and 10th Anniversary of Onıght*, Marrakech, 2003.
5. ÇÖLKESEN, İ., SESLİ, F. A., Kıyı çizgisinde meydana gelen zamansal değişimlerin bilgi teknolojileri ile belirlenmesi: Trabzon örneği. TMMOB Harita ve 169 Kadastro Mühendisleri Odası, *Ulusal Coğrafi Bilgi Sistemleri Kongresi*, 30 Ekim-2 Kasım 2007, KTÜ, Trabzon, 2007.
6. TÜİK, Türkiye İstatistik Kurumu, http://www.tuik.gov.tr/Start.do, 11.07.2016.
7. ÇORBACI, Ö.L. Bartın Doğal ve Kültürel Değerleriyle Amasra'nın Peyzaj Planlaması. Doktora Tezi, Ankara Üniversitesi Fen Bilimleri Enstitüsü, Peyzaj Mimarlığı Anabilim Dalı, Ankara, 2014.
8. URL-1 https://www.google.com/earth/ Google earth, 22.05.2016.
9. URL-2 http://www.amasra.com.tr/eski-amasra-fotograflari/, 2017.
10. URL-3 http://cesmicihan.tr.gg/Galerim/pic-1000154.htm, 2017.
11. URL-4 http://cesmicihan.tr.gg/Galerim/pic-1000122.htm, 2017.
12. TÜRKMEN, F., DÖNMEZ, Y., Korunan Alanların Turizme Açılmasına İlişkin Yerel Halkın Görüşleri (Yenice Örneği) Karabük Üniversitesi Sosyal Bilimler Enstitüsü Dergisi, Sayı 2, Cilt 5, 2015.

Zuhal Dilaver[1], Emel Baylan[2]

Conservation of Steppe Vegetation and its use in The Processes of Adaptation to Climate Change

[1]*Ankara University, Faculty of Agriculture, Department of Landscape Architecture, Turkey.*

[2]*Yüzüncü Yıl University, Faculty of Agriculture, Department of Landscape Architecture, Turkey.*

1. Introduction

Activities carried out by societies aiming at an improved environment and a comfortable life have caused environmental pollution at different levels and the degradation of limited resources. Increasing pressures in parallel with the growth of the world's population have led to these problems reaching tangible dimensions on a global scale. Predicting that resources could be exhausted and the environment could irreversibly deteriorate has demanded an environmentally conscious approach to human life and revealed the concept of sustainability.

Sustainability can briefly be defined as ensuring the continuity of productivity for many years in optimal conditions [1].

The concept of sustainability was used together with that of development in the Bruntland Commission's 1987 report named Our Common Future, where sustainable development was defined as "the development that meets the needs of present without compromising the ability of future generations to meet their own needs" [2]. This definition points to the importance of equality between generations and emphasizes the need to protect resources for future generations.

The overall goal of sustainable development is the long-term stability of the economy and environment and this is only achievable through the integration and acknowledgement of economic, environmental, and social concerns throughout the decision-making process [3].

Today, climate change is regarded as one of the most important environmental problems facing the world [4] and appears as an important threat to sustainable development. Nevertheless, there are plenty of opportunities to be considered in terms of climate change mitigation and adaptation [5].

In this study, climate change is discussed in relation to sustainability, the characteristics of steppe vegetation are addressed, and recommendations are made for the conservation of steppe vegetation and its use in adaptating to climate change.

2. Climate Change

Climate change as used by the Intergovernmental Panel on Climate Change (IPCC) refers to a change in the state of the climate that can be identified (e.g. using statistical tests) by changes in the mean and/or the variability of its properties and that persists for an extended period, typically decades or longer. It refers to any change in climate over time, whether due to natural variability or because of human activity. A different definition came from the United Nations Framework Convention on Climate Change (UNFCCC), where climate change refers to "a change of climate that is attributed directly or indirectly to human activity that alters the composition of the global atmosphere and that is in addition to natural climate variability observed over comparable time periods" [6].

The global climate has tended to change over the entire Earth's surface and through all time scales since the formation of the world [7]. Some of these changes are major changes created by individual events, but the majority are regular movements that occur throughout periods of thousands or even millions of years. The climate has undergone a periodic cycle of ice ages and warm periods over the last 400 000 years. In the last 8000 years, the climate of the world has followed a very stable course except for very small temperature fluctuations. However, the increase of anthropogenic activities, rapid population growth and industrialization, distorted urbanization, misuse of land, deforestation and a rapid degradation of the natural environment, with the industrial revolution of the 1860s, has destroyed the stability of natural climate change. The result is a complex process caused by anthropogenic activities, in which "global warming and global climate change" become difficult and even impossible to return from[8].

The source of global warming is the excessive increase in emissions of greenhouse gases such as carbon dioxide (CO2), methane (CH4), chlorofluorocarbon (CFC) and ozone (O3) [8]. The CO2 build-up in the earth's atmosphere has been increasing very rapidly. The average annual CO2 accumulation in the atmosphere, which was about 280 ppmv before the industrial revolution and about 315 ppmv in 1958, reached about 394 ppmv in 2012 and a 400 ppmv monthly average in May 2013 [7].

Global climate change, which has become more evident since the 1980s, has socioeconomic and political impacts such as a scarcity of water resources, food shortages, energy stress, drought, desertification and migration [8]. Floods, landslides, rises in sea level, hot/cold air waves, air pollution, fires, urban heat islands, etc. are also among the likely effects of climate change [4]. In addition, climate change causes the degradation of the natural landscape texture and negatively affects systems that are fundamental parts of biological diversity such as ecosystems, species and genetic resources. The change in water and land ecosystems also causes changes in the biological composition of these ecosystems [8].

Today, many of our cities are under significant risks posed by global climate change, especially drought. It is now known that the losses and damages that may arise through meteorological disasters in our cities can be on very large scales in terms of lives, property and environment. Increased drought means less rainfall, more sunlight, longer and more intense passages of hot air waves, increased insect propagation, exposure to thirst and water scarcity, and more frequent and longer-lasting forest fires. For this reason, today, climate change is one of the most important problems that societies need to focus on, as much as development, hunger and health [9].

Some of the measures that could be taken against drought during the adaptation processes are: growing plants which consume less water; promoting the use of micro-irrigation systems such as drip irrigation instead of wild irrigation, and large sprinkling irrigation systems in agriculture; preventing water loss in water supply; restoring the drainage water with natural treatment; and taking all possible measures to save water in urban water use [9].

3. Grass Formations

The world's plant formations are shaped by certain ecological conditions. Common effects of climate elements such as temperature, humidity, precipitation, wind and light directly influence the spread of plant species around the world and the formations they form [10]. In areas with low rainfall and high temperature conditions, it is not possible to grow trees due to high evaporation, and grass formations occur in these areas [11]. Grass formations are predominantly composed of non-woody species, with less than 25% woody plants, less than 10% low woody plants and more than 10% herbaceous plants [12] [13].

In terms of lifestyle, plants were classified in five main groups by the botanist Raunkiaer, as **Fanerophytes** (trees which had renewal buds 25 cm above the soil surface, prehensile plants, large herbaceous plants and epiphytic plants), **chamaephytes** (plants having renewal buds up to 25 cm from the ground level and buds protected by snow cover in winter), **hemicryptophytes** (plants having renewal buds on the upper layers of the soil or just below the soil surface, usually biennial, with terminal buds protected by dead plants and leaf debris on top of the soil, herbaceous above surface parts and plants which die at the beginning of the critical season), **cryptophytes** (plants whose above soil parts totally die following the vegetation period) and **terophytes** (plants whose lifespan is limited to one vegetation period annually and ephemeral plants, which spend the inconvenient period as seed). **Cryptophytes** are studied in two groups according to the medium in which the buds are found, **geophytess** (bulbous, rhizomatous and tuberous plants with renewal buds in the soil) and **hydrophytes** (plants having their roots, trunks and buds in water or on the surface of water). Grass formations consist of grass chamaephytes, hemicryptophytes, cryptophytes, geophytes, grass hydrophytes and terofits [12, 13, 14, 15].

Grass formations are examined in two parts, savannas and steppes (Fig.1). **Savannas** are seen in tropical regions where summer rains prevail [12]. In areas where annual precipitation decreases or where the annual dry season is prolonged, open tropical or subtropical leafy forests are left to open herbaceous plant communities scattered by rare green shrubs and trees, i.e. savannas [13]. **Steppes** are the type of grass formation situated in the middle belt [12]. The climate conditions in the spreading areas of the tropical areas and the grass formations of the warmer temperate zones are not the same, but precipitation is a common feature that determines the development of the grass formation [11].

Figure 1. Locations of Savannas and Steppes in the World [11].

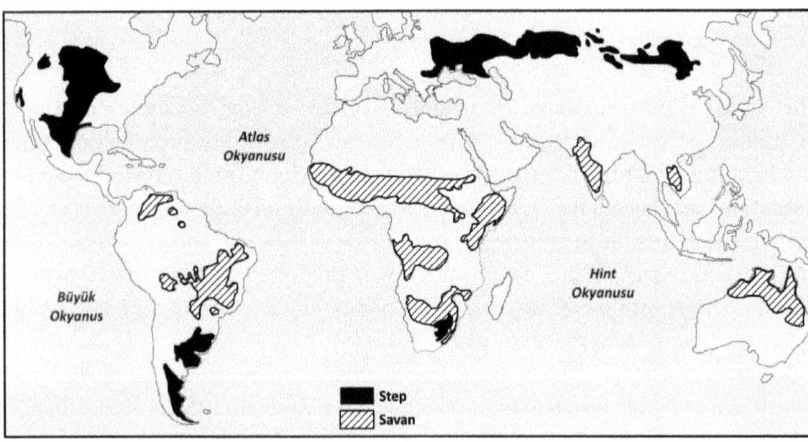

3.1. Steppe Vegetation

Steppe is, according to Trochain (1955), an open vegetation type that dries in summer, spends the winter resting, and is composed of xerophilous plant species which normally grow in non-acidic soil (living in arid environments) or mesophiles (which require water and atmospheric humidity) [16]. The root of the word "steppe" is the Russian word "stepj" [11]. Steppe species grow in the rainy season and die out in the dry season [10]. It is possible to group steppe formations under three groups physiologically:

Malachophile steppes: the type of steppes dominated by large soft-leaved plants. It is thought that these steppes derived from Gramineae steppes due to excessive pasturage in relatively humid conditions and deep soils.

Gramineae steppes: the steppes which largely consist of Gramineae species and are rarely seen today.

Tragacanthic steppes: the steppes dominated by steppes forming pillows such as *Astragalus* L., *Acantholimon* Boiss. [16].

3.1.1. Steppes in the World

The steppe areas are distributed roughly between the forests of the temperate zone and the deserts found roughly between 30–60° geographical latitudes. In these areas where semi-arid climate features dominate, annual precipitation does not exceed 250–350 mm. Grass formations in the temperate zone are known by different names on the continents where they are represented. Prairie in North America, steppe in Europe and Asia, pampas in South America, and veld in south of the African continent are the names given to these groups [11].

The floristic composition of steppe vegetation is very rich and contains many endemic plants. The most characteristic feature of the steppe ecosystem is the predominance of annual or perennial herbaceous plants [17]. Among these, herbs belonging to perennial Euphorbiaceae or Leguminosae families are the most widely observed. In the much drier growing environments of the ecosystem, bunchgrasses growing in clusters become widespread. Some of these grass species exhibit shrubs which reach certain heights from the soil surface, while others grow parallel to the soil surface. Flowering periods also vary [11].

3.1.2. Steppes in Turkey

Steppe ecosystems are spread, especially in Central Anatolia, over the highlands of Aegean and Mediterranean regions, and a large part of Eastern Anatolia in Turkey [17]. A very small part of these areas is natural while a large part is formed of anthropogenic steppe areas.

Natural steppe areas are naturally unsuitable for growing trees. In Turkey, they are located in Central Anatolia around Tuz Gölü, between Konya and Ereğli, in Southeast Anatolia around Akçakale–Ceylanpınar and in limited areas of the Iğdır plateau. The steppe areas outside of these regions in Turkey today are secondary (anthropogenic) steppe areas that have come into being as a result of the destruction of forests for centuries, not based on the characteristics of the natural habitat. Today, forest fragments in the form of greenish spots seen in high mountainous areas (Karadağ, Karacadağ, Erciyes and Hasandağı etc.) within steppe areas or in village gardens and cemeteries continue to exist as evidence of this change [11]. The existence of bushes in steppe areas, whether rare or not, is the indication that these areas are former forests [18].

Evliya Çelebi indicates in 'Seyahatname' that there was a forest consisting of big bushy trees around Ankara and that in 1402 Timur hid his elephants in it during the war of Ankara. Aytuğ (1970) demonstrates findings which indicate that, unlike the present day, there were forests in the region, even if these were not frequent. This was determined through archaeological studies conducted in four different areas in the Central Anatolia region. He concluded his studies using pollen analyses and the assessment of signs in the remains of woody plants such as *Cedrus* Trew, *Abies* Mill., *Pinus* L., *Juniperus* L., *Castanea* (Tourn.) Adans. Woody plants such as *Betula* (Tourn.) L., *Populus* L., *Salix* L., *Tilia* L., *Taxus baccata* L., *Cedrus libani* A. Rich., *Pinus sylvestris* L. *Juniperus foetidissima* Wild., *Fraxinus* L. also existed in the region.

Historically, forests in Anatolia have been significantly reduced and fragmented in some areas due to reasons such as invasions, settlements, wars, forest fires, use of fuel, grazing, field openings and mining activities. The floristic composition has also changed to a great extent. This process of change has led to the deterioration of the growing conditions and in most areas the erosion of the soil surface, resulting in a widening of the spreading areas of the steppe formation [11, 19].

Steppe vegetation is seen in inland areas away from the effects of the sea, where a terrestrial and arid/semi-arid climate is dominant. In arid and semi-arid areas, the rainfall being below 400 mm, the existence of high humidity and high temperature differences between day and night, along with low relative humidity are important factors limiting plant growth [20, 21]. In these areas, soils of considerably different characteristics from the main material have been developed, and because of overgrazing on most of the land, vegetation has been destroyed and the fertile soil has been swept away by erosion [20].

Steppe vegetation is made up of species of the Gramineae family and formations in which annual and perennial herbaceous plants dominate. The parts or even sometimes the whole of bodies of some species, which form the steppe, especially the mountain steppe, and which are the dominant plants of the grass formations, may have become ligneous. Some *Astragalus* L., *Acantholimon* Boiss.

or *Onobrychis cornuta* (L.) Desv. species are examples of this feature [18]. Among the dominant plants of the steppe vegetation are *Stipa* L., *Bromus* L. and *Salvia* L. [17]. Major bushes and tree species that are rarely found in anthropogenic steppe areas are plants such as *Prunus spinosa* L., *Berberis crataegina* DC., *Amygdalus orientalis* Miller, *Pyrus elaeagnifolia* Pallas, *Paliurus spina-christi* Miller, *Rosa sulphurea* Dryand. and *Crataegus orientalis* Pallas ex Bieb. [17,21].

Steppe formation in Turkey is generally divided into "Plain Steppe" and "Mountain Steppe" according to the topographical structure of the area it spreads over [17, 18]. Plain steppe is a type of formation with a generally low height and inclination, widespread in Central, Eastern and Southeastern Anatolia, while high mountain steppe extends over the highlands of Eastern Anatolia, the Black Sea and the Toros Mountains, and in steep and inclined areas [18].

Plain Steppe

Plain steppe is found at heights between 800 and 1200 meters. The characteristic plants of this ecosystem are the salty halophytes around the Tuz Lake, members of the Chenopodiaceae, Juncaceae and Cyperaceae families, and species like *Peganum harmala* L., *Artemisia* L., *Thymus* L. and *Salvia* L. [17].

Plain steppes are divided into many sub-ecosystems. The main ones are; Halophytic (salty) steppes, marly steppes, gypsum-bearing steppes, tuff intensive habitats and serpentine steppes. Each of these sub-ecosystems contains many endemic species of its own, some of which are in the endangered category (Table 1) [17].

Within plain steppe, in places which are hollow and have high ground water, salt concentration increases and a halophytic vegetation type formed of salt-loving plants dominates in these areas. This type of vegetation, usually dominated by Chenopodiaceae taxa and *Frankenia hirsuta* L. and sometimes by *Aeloropus litoralis* (Gouan) Parl., is mostly in the form of dwarf shrubs with small fleshy leaves and unpretentious flowers. In Turkey, halophytic vegetation is prevalent near Tuz Gölü, near Kırşehir Seyfe Lake, Kayseri Sultansazlığı, Burdur Lake and Afyon Acıgöl [17, 18].

There is another group of water-loving plants in areas within plain steppes which are waterfronts, where the ground water is high but the salt concentration is less than that required for the growth of halophytic plants. In these areas, species belonging to the Juncaceae and Cyperaceae families are common. Such groups are also found in similar environments within mountain steppes. These greens are noticed even from long distances, in the form of green belts especially during midsummer or the end of summer, in steppe areas where brown dominates as a color [18].

Table 1. Rare species seen in the sub-ecosystems of plain steppes (prepared with information from The National Biological Diversity Strategy and Action Plan – 2007)

PLANT SPECIES	DANGER CATEGORY	SUB-ECOSYSTEM
Achillea gypsicola Hub. Mor.	VU	Gypsum-bearing
Allium vuralii Kit Tan	VU	Halophytic
Alyssum niveum Dudley	EN	Marly
Asparagus lycaonicus P.H. Davis	EN	Halophytic
Astragalus talasseus Boiss. et Bal.	VU	Tuffaceous
Campanula damboldtiana Davis et Sorger	CR	Marly
Campanula pinnatifida Hub. Mor.	VU	Gypsum-bearing
Campanula ptarmicifolia Lam.	VU	Serpentine
Chrysocamela noeana (Boiss.) Boiss.	EN	Gypsum-bearing
Cousinia sivasica Hub. Mor.	VU	Serpentine
Ferula halophile Peamen	VU	Halophytic
Gladiolus halophilus Boiss. et Heldr.	VU	Halophytic
Gypsophila germanicopolitana Hub. Mor.	CR	Gypsum-bearing
Helianthemum germanicopolitanum Bornm.	EN	Gypsum-bearing
Hypericum salsugineum Robson et Hub.Mor.	VU	Halophytic
Isatis glauca subsp sivasica (Davis) Yıldırımlı	VU	Gypsum-bearing
Kalidiopsis wagenitzii Aellen	EN	Halophytic
Limonium anatolicum Hedge	VU	Halophytic
Limonium tamaricoides Bokhari	EN	Halophytic
Microcnemum coralloides subsp. Anatolicum Wagenitz	VU	Halophytic
Onobrychis germanicopolitana Hub.Mor. et Simon	EN	Gypsum-bearing
Onobrychis paucijuga Bornm.	VU	Tuffaceous
Onosma halophilum Boiss. et Heldr.	VU	Halophytic
Onosma sintenisii Hausskn. et Bornm.	VU	Gypsum-bearing
Physoptychis haussknechtii Bornm.	EN	Serpentine
Reseda germanicopolitana Hub. Mor.	CR	Gypsum-bearing
Salvia vermifolia Hedge et Hub.Mor.	VU	Serpentine

PLANT SPECIES	DANGER CATEGORY	SUB-ECOSYSTEM
Saponaria halophile Hedge et Hub.Mor.	CR	Halophytic
Scorzonera aucherana DC.	VU	Gypsum-bearing
Scrophularia lepidota Boiss.	VU	Gypsum-bearing
Silene salsuginea Hub.Mor.	EN	Halophytic
Tanacetum germanicopolitanum (Bornm. & Heimerl) Grierson	CR	Gypsum-bearing
Taraxacum mirabile Wagenitz	VU	Halophytic
Thesium stellerioides Jaub. Et Spach.	VU	Gypsum-bearing
Verbascum pyroliforme (Biss.et Heldr.) O. Kuntze	VU	Halophytic

Marly steppes of the plain steppe ecosystem, which are very rich in their floristic composition, are largely found in Central Anatolia and the low altitude areas surrounding Central Anatolia. This steppe ecosystem has been constantly diminishing because of anthropogenic pressures and has been undergoing transformation into fields [17].

The gypsum-bearing steppe ecosystem is widely found in Çankırı, Sivas, Ankara-Beypazarı and around the Erzincan provinces. This ecosystem's floristic composition is not very rich, but the rate of endemism is very high. The populations of these plants, which are not found outside the gypsum habitat and whose ecological tolerances are limited, are very low [17]. *Gypsophila* L. is the most notable plant genus in this formation, and is named after it [18]. In Central Anatolia, tuff-intensive habitats, formed as a result of various geological events, cover the large areas in the Ürgüp and Goreme environs and the Ereğli and Karapınar plains. These habitats, where slopes are hardly ever observed, are also considered as part of flat plain ecosystems. Around Sivas, Erzincan and Kırıkkale, serpentine steppe habitats are found on serpentine bedrocks, although to a limited extent [17].

Mountain Steppe

Mountain steppe is generally found between 1300–2500 meters. The characteristic plants of this ecosystem observed above the forest boundary are *Astragalus* L., *Onobrychis cornuta* (L.) Desv., *Acantholimon* Boiss., *Asphodeline* (L.) Reichb. sp, and the *Thymus* L. species. High mountain steppes that are predominant, especially at 2000 meters and even higher levels, are composed of the plant species which are thorny, sclerophyllous and hard stemmed, plants with cushion-formed stems in general, and in some places the *Asphodeline* (L.) Reichb. species, which animals do not like [18]. In the Eastern region of Anatolia, species such as *Ferula* L. and *Prangos* Lindl. have increased in dominance. On the other hand, in higher sections of the Eastern Black Sea mountains and in northern and north-eastern parts of Eastern Anatolia, subalpine and alpine meadows cover large areas. Humidity-loving Gramineae species, Ranunculaceae and Geraniaceae are common in the subalpine and alpine meadows, which have more precipitation than the steppes and shorter arid periods. Subalpine and alpine meadows are still colored in summer when most of the species that make up the steppe vegetation dry up. In Turkey, mountain steppe is very common and its endemism rate is very high. Almost every mountain has a unique endemic plant in its high mountain steppe habitat [17]. Its floristic compositions are more abundant than those of plain steppes, because they grow particularly into thorny and cushion-shaped plants and preserve some species preferred by animals [18].

High mountain vegetation in eastern Turkey differs greatly from that of the west. In some areas, it is recognized that some species belonging to Umbelliferae (*Ammi* L., *Prangos* Lindl., *Ferula* L., *Heracleum* L. etc.) or Compositae (*Helichrysum* Gaertner, *Inula* L., *Centaurea* L., etc.) families are predominant in the general outlook of the formation rather than *Astragalus* L. plants or other cushion-shaped plants. There is also a difference observed in the mountain ranges between the north and south of Anatolia. Particularly in the high mountain steppes of the Black Sea Mountain, one can find endemic plants which are thinlyspread [18].

3.1.3. Pressures on Steppe Areas

The pressures that cause forest areas to turn into anthropogenic steppe areas continue today and cause the steppe areas to turn to different uses, or lead to the deterioration of their botanical composition. In many parts of the world, the extensive areas of grass formations have become the most important cereal production areas for the countries in which they are located. For example, about 70% of North America's tall prairies have now become agricultural land. This has led to the description of the grass formation areas as bread-baskets [11].

The most significant reduction and destruction based on ecosystems is seen in steppe ecosystems in Turkey. Steppe and meadow areas, which constituted 44.3 million hectares in 1935 and 37.8 million hectares in 1950, are now around 21 000 000 hectares. Much of the steppe and meadow areas covering 28% of the country's surface area have been degraded or rendered unproductive because of openings of fields, uncontrolled grazing, unplanned settlements and industrialization, unconscious afforestation and excessive picking up of economic value from natural plants [17]. Because of extreme destruction, the floristic composition of the plain steppe areas in Turkey has been badly degraded and the characteristic steppe plants have given way in some regions to grass species, which animals do not prefer due to their smell [11, 18]. Uslu (1962) found that in areas under grazing pressure, the steppe grass was replaced by *Artemisia fragrans* Willd. and *Festuca sulcate* (Hack.) Beck steppes, annual and bulbous and tuberous plants have increased in number, wait-a-bits such as *Astragalus* L. and *Acantholimon* Boiss., together with poisonous plants like *Salvia* L. and *Euphorbia tinctoria* Boiss. & A. Huet could maintain their presence [22].

As the mountain steppe plants are more resistant to overgrazing and other destructive factors than plants composed of the plain vegetation, and because these areas are not suitable for agriculture, this formation has been degraded less in comparison with the plain steppe [11, 18].

3.1.4. Drought Tolerance of Steppe Plants

As mentioned earlier, annual rainfall is about 250–350 mm in the areas where steppe vegetation is present. It is observed that the arid steppe plants which are found in arid environment conditions resulting from the increase of temperature in the summer period and the decrease in precipitation in this period, are where the plants are in need of water the most but also where these plants have taken some measures in order to adapt to their conditions.

Some steppe species pass through difficult conditions by slowing down their metabolism in a form of stillness period (dormancy) in which they protect the water in their bodies. Summer dormancy is a valuable strategy associated with increased drought tolerance and persistence in perennial grasses. It has been reported that summer dormancy is associated with superior survival and autumn regrowth after severe and repeated summer drought [11].

Steppes' cryptophytes totally hide their bodies, with buds under the soil in the absence of proper climatic conditions. Geophytes, the parts of cryptophytes hiding underground which have bulbous, tuber and rhizomatic groups, store the necessary water, nutrients and building materials in this part. After they foliate, bloom and seed early in life, they form new onions, tubers or rhizomes before the drought starts and they end their life in the soil [11, 23].

The flower-bearing above-ground shoots of the species included in the hemicryptophyte group are alive in a single vegetation period. Hemicryptophytes dry out if the season is not suitable, but the rosette leaves of the plant remain alive at soil level [11, 22].

The short, dwarf shoots and long roots are the most prominent features of perennial steppe plants such as *Astragalus* L. and *Acantholimon* Boiss. The roots of *Artemisia santonicum* L., for example, reaching two meters, help to get the existing water in the soil in the best possible way. Deep-rooted grass is alive even in the summer months; it blooms and seeds quite late [11, 23].

Some perennial steppe plants such as *Astragalus purshii* Dougl. ex. Hook. and *Lomatium geyeri* (S. Watson) J.M. Coult. & Rose have roots which lie dormant for a period of years, if necessary, before sending out shoots with new leaves above the ground [24].

Some of the herbaceous plants have developed succulent structures to adapt to arid conditions; the leaves have fleshy, water-storing forms [11].

Many steppe plants keep leaf surfaces small in order to reduce water consumption due to sweating [23, 24]. In addition, in many steppe plants these small leaves in normal form turn their newly growing leaves into thorns in May-June when it is difficult to get water from the drying soil. For this reason, most of the steppe plants are spiny [23].

Almost all of the steppe plants have feathers on their trunks and leaves. In some plants these feathers become very thick and cover the leaves like a sheath. In this way, loss of water from the plants due to excessive perspiration as well as the overheating of the leaves is prevented [23, 24].

Steppe plants try to prevent the water from evaporating by closing the pores that come out from the leaves on hot summer days [23]. Other plants reduce water loss by having fewer stomata or opening the stomata only at night when it is cooler [24].

4. Results and Discussion

Steppes cover large areas in the world and in Turkey and they are very rich in terms of biological diversity. Steppes are home to numerous endemic plants. However, in every period there are various pressures and threats to these areas and they are becoming narrower. Various work must be carried out to protect steppe areas. First, it is necessary to fully identify steppe biodiversity, to combine the results of the research carried out in this area and fill in information gaps with new research to build a comprehensive inventory.

It is necessary that steppe ecosystems are evaluated as a whole and the threats to these areas and their levels are determined. Thereafter, safeguarding measures should be introduced to reduce these pressures. Natural steppe areas and areas with high endemism rates in particular should be evaluated with the protected area status.

Some of the steppes are used as pastures and the floristic compositions of these areas have been deteriorating due to careless and excessive grazing. For this reason, it is necessary to investigate the bearing capacity of pasture lands, to improve the degraded pasture lands and to investigate alternative feeding sources. The planting of seeds from herbaceous species, together with other conservation and management measures, can be applied for pasture restoration.

Many narrowly spreading endemic plants in steppe ecosystems are in the CR critically endangered category. Ex-situ conservation work is needed for threatened species. Often biological material trafficking for these species is also an important problem to be determined and prevented.

Cultural acquisition regarding species with economic value should be carried out as an important task in terms of conservation, through preventing their from nature and creating income for local people.

Climate change is a major threat to steppe ecosystems, just as it is to all ecosystems. Any work carried out in this regard would also contribute to the protection of steppe ecosystems.

It is envisaged that climate change will have negative effects on water resources and that drought will become a very important threat with the changes in precipitation. Increasing population and unmindful water consumption are accelerating this process. There are many studies that can be carried out to reduce water consumption by managers as well as people at various levels within the scope of adaptation to climate change and the reduction of its impact. Firstly, studies can be carried out to control the heat island effect in cities by increasing the green areas. For green areas, areas such as car parks, refuges, walking paths and pedestrian paths can be evaluated as appropriate, along with large-scale parks. However, it should not be forgotten that a significant part of the water consumption in urban areas is green area irrigation. For this reason, type selection is very important in planting studies. The use of natural plants in green areas, especially those species which are known to be highly resistant to drought in steppe vegetation, by conducting production and aquaculture work, will be an important step in the adaptation process towards climate change. These plants are important not only due to their tolerance to drought, but also because of their values such as providing support for indigenous wildlife populations, their ability to survive with minimal care, and their high chances of success.

References

1. A. ATIL, B. GÜLGÜN and İ. YÖRÜK: *Sürdürülebilir Kentler ve Peyzaj Mimarlığı. Journal of Agriculture Faculty of Ege University*, 42(2):215–226, ISSN 1018-8851 (2005).
2. WCED (World Commission on Environment and Development): *Our Common Future. Brundtland Report.* Oxford University Press, Oxford (1987).
3. R. EMAS: *The Concept of Sustainable Development: Definition and Defining Principles.* Brief for GSDR 2015.
https://sustainabledevelopment.un.org/content/documents/5839GSDR%202015_SD_concept_definiton_rev.pdf
4. U. DUMAN YUKSEL: *İklim Değişikliği Gündeminde Kentsel Yeşil Alanlara Yeniden Bakış. The Journal of Landscape Architecture*, UCTEA The Chamber of Landscape Architecture 2012/2013 (2013).
5. IPCC: *Climate Change 2014: Synthesis Report.* Contribution of Working Groups I, II and III to the Fifth Assessment Report of the Intergovernmental Panel on Climate Change [Core Writing Team, R.K. Pachauri and L.A. Meyer (eds.)]. IPCC, Geneva, Switzerland, 151 pp. (2014).
6. IPCC: *Climate Change 2007: Synthesis Report.* Contribution of Working Groups I, II and III to the Fourth Assessment Report of the Intergovernmental Panel on Climate Change [Core Writing Team, Pachauri, R.K and Reisinger, A. (eds.)]. IPCC, Geneva, Switzerland, 104 pp. (2007).
7. M. TURKES: *Observed and Projected Climate Change, Drought and Desertification in Turkey. Ankara University Journal of Environmental Sciences.* 4(2), 1–32 (2012).
8. A. DEMIR: *The effects of Global Climate Change on Biodiversity and Ecosystems Resources. Ankara University Journal of Environmental Sciences.* 1(2), 37–54 (2009).
9. M. KADIOGLU: *Küresel İklim Değişikliği ve Uyum Stratejileri.* page. 69–94, *Kar Hidrolojisi Konferansı*, 27–28 Mart, 2008 General Directorate of State Hydraulic Works, The 8. Regional Diroctrate Erzurum. (2008)
10. N. GUNAL: *Türkiye'de İklimin Doğal Bitki Örtüsü Üzerindeki Etkileri. Acta Turcica (Online Thematic Journal of Turkic Studies)*, V(1), İstanbul (2013).
11. M. AVCI: *Dünya'da ve Türkiye'de Step Formasyonu.* Ertuğ Öner (Ed.), Prof. Dr. Asaf KOÇMAN'a Armağan. *Ege University Publications*, Faculty of Literature Publication Number:180, pages 111–131, Ege University Press, İzmir (Bornova) (2013).
12. Y. AKMAN and O. KETENOGLU: *Vejetasyon Ekolojisi.* Ankara University Science Faculty Publication Number: 146, Ankara (1987).

13. Y. AKMAN, O. KETENOGLU and F. KURT: *Vejetasyon Ekolojisi ve Araştırma Metodları. Palme Publishing.* Ankara (2011).
14. W.G. SMITH: *Raunkiaer's "Life-Forms" and statistical methods. Journal of Ecology*, 1, 16–26. (1913).
15. E. YURDAKULOL, Ü. BİNGÖL, G. AKGÜL, G.N. TUĞ and A.E. YAPRAK: *Açık Tohumlular Laboratuvar Kılavuzu*, Ankara University Science Faculty Revolving Fund Management Publications Number: 68, Ankara (2003).
16. REPUBLIC OF TURKEY MINISTRY OF ENVIRONMENT AND URBANIZATION, GENERAL DIRECTORATE FOR PROTECTION OF NATURAL ASSETS: *Ihlara Özel Çevre Koruma Bölgesi Biyolojik Çeşitliliğin Tespiti Projesi, Sonuç Raporu*. Ankara. (2012).
17. REPUBLIC OF TURKEY MINISTRY OF ENVIRONMENT AND FORESTRY, GENERAL DIRECTORATE OF NATURE CONSERVATION AND NATIONAL PARKS DEPARTMENT OF NATURE CONSERVATION, NATIONAL FOCAL POINT OF CONVENTION ON BIOLOGICAL DIVERSITY: *The National Biological Diversity Strategy and Action Plan*. ISBN: 978-605-393-030-3 (2007).
18. T. EKİM et al.: *Türkiye'nin Biyolojik Zenginlikleri*, Environment Foundation of Turkey Publication, Ankara,167–195 (2005).
19. B. AYTUĞ: *Arkeolojik Araştırmaların Işığı Altında İç Anadolu Stebi. Journal of the Faculty of Forestry Istanbul University*, A(XX), Number.1, Pages:127–143 (1970).
20. ANONYMOUS: *Protected Areas and Climate Change: National Strategy of Turkey*, (2011). http://www.tr.undp.org/content/turkey/en/home/library/environment_energy/protectedareas.html http://www.kdmp.gov.tr/KAID_strateji.pdf (2011).
21. A. ATALAY: *Türkiye'nin Ekolojik Bölgeleri*. The Ministry of Environment & Forests Publication Number: 163, (2002).
22. F. YALTIRIK and A. EFE: *Otsu Bitkiler Sistematiği*. İstanbul University Institute of Science and Technology Publications, Publication Number: 3, İstanbul. (1989).
23. H. BIRAND: *Alıç Ağacı ile Sohbetler*. The Scientific and Technological Research Council of Turkey, Second Press, Ankara (1996).
24. ANONYMOUS, *Shrub-Steppe Plant Ecology* http://www.cdlandtrust.org/what-we-do/education/plant-ecology (2017).

Aybike Ayfer Karadağ, A. Esra Cengiz, Demet Demiroğlu

Today's Sustainable Placements: Ecovillages

1. Introduction

A quickly developing world population and urbanization continue to be driving planned development. This situation leads to the exhaustion of natural resources and space, and catastrophic events. When we look at the present population issue and the state of urbanization, we can see the immensity of the issue.

In 2016, 1.7 billion individuals (23 per cent of the total population) lived in a city with no less than 1 million inhabitants. By 2030, urban zones are anticipated to house 60 per cent of individuals worldwide and one in every three individuals will live in an urban community with at least half a million inhabitants. Between 2016 and 2030, the population in all classes of city size is anticipated to grow, while the rural population is anticipated to decline a little. While rural regions were home to more than 45 per cent of the total population in 2016, that amount is expected to reduce to 40 per cent by 2030. Of the 1692 urban communities with no less than 300 000 tenants in 2014, 944 (56 per cent) were at high danger of exposure to at least one of six types of natural disaster (cyclones, floods, droughts, earthquakes, landslides and volcano eruptions), based on evidence regarding the occurrence of natural disasters over the late twentieth century. Taken together, urban areas facing high danger of exposure to a natural disaster were home to 1.4 billion individuals in 2014. Around 82 per cent of cities (home to 1.9 billion individuals in 2014) were situated in zones that faced high danger of mortality due to natural disasters. Likewise, 89 per cent of cities (home to 2.1 billion individuals in 2014) were situated in regions that were highly vulnerable to economic losses related towith at least one of the six types of natural disaster [URL-1].

In fact, population density, and the problems that increasingly urbanized living today could bring about were assessed. After the Industrial Revolution, this issue has regularly been brought to our attention. The significance of living in harmony with nature has been particularly stressed in this context. Indeed, even new life and development models have been created. Sustainable development is one of them.

The most widely accepted definition of sustainable development is from the Brundtland Report published in *Our Common Future* by the World Commission on Environment and Development in 1987: "Sustainable development is development that satisfies the needs of the present without compromising the needs of the future." Sustainable development involves three dimensions: social, economic and environmental. It also incorporates the process of bringing these in balance with each other [1]. The overall goal of sustainable development is the long-term stability of these elements [2]. In fact, the sustainability of capital is important at this point. There are several types of capital in sustainable development: social (human resources, skills, and knowledge), natural (forests, air, water, soils and biodiversity and other resources like minerals, etc.), and man-made (manufactured capital and goods, machinery, infrastructure, buildings, etc.) [2; 3]. Actually, the essence of sustainable development is the production of a life that is in balance with nature. Rapidly growing ecological problems are a clear indicator of this situation. As Scheewe [4] said "Man, whether civilised or savage, is a child of nature – he is not the master of nature. He must conform his actions to certain natural laws if he is to maintain his dominance over his environment. When he tries to circumvent the laws of nature, he usually destroys the natural environment that sustains him. And when his environment deteriorates rapidly, his civilisation declines."

Alongside sustainable development, the concept of sustainability has begun to be utilized in various areas of life (such as life, planning, design, management, settlement, space, urban, etc.). Subsequently, along with the concept, the objectives of sustainable development have been moved to these areas. Sustainable placements are one of these areas. Today, sustainable settlements are one of the key aims of urban and rural planning and design.

Ecovillages, which are the subject of this chapter, are examples of sustainable settlements and communities. This chapter deals with the emergence of the ecovillages, their characteristics, and samples.

2. What is an Ecovillage?

Ecovillages have become popular as movements in which people support each other by forming small communities, to create a life compatible with nature with intentions to revive the traditional community life. Ecovillages are also considered to be a reaction to the negative effects of modern lifestyles today [5]. In this context, ecovillages are conscious communities.

In the Sustainable Report prepared by Robert and Diane Gilman, an ecovillage is defined as *"a human-scale, full-featured settlement, in which human activities are harmlessly integrated into the natural world, in a way that is supportive of healthy human development and can be successfully continued into the indefinite future"* [6; 7]. According to another definition, ecovillages are the settlement models which are sustainable in an ecological, economical, socio-cultural and spiritual sense. They aim at diminishing negative effects on nature (ecological foot prints) with voluntary participation and with design applications compatible with the nature. Their objective is to combine ecological design applications and green technologies with participative and supportive social structures. They give the impression of traditional settlements or newly established conscious communities in urban and rural areas, not only for developed countries of the West, but also for underdeveloped countries in the South [8]. Ecovillages are also defined as pioneering human communities that lead to creating a sustainable lifestyle compatible with all living and non-living formations while living in harmony in their own right [9].

The official definition of an ecovillage, agreed upon by the Global Ecovillage Network (GEN) board and used by GEN today runs as follows: *"An ecovillage is an intentional or traditional community that is consciously designed through locally owned, participatory processes to regenerate social and natural environments. The four dimensions of sustainability (ecology, economy, the social and the cultural) are all integrated into a holistic approach"* [URL-2]. A more recent definition of ecovillages by a former GEN-EU president positions ecovillages as: *"private citizens' initiatives in which the communitarian impulse is of central importance, that are seeking to win back some measure of control over community resources, that have strong shared values and that act as centres of research, demonstration and (in most cases) training"* [7] [URL-2].

3. History of Ecovillages

The ecovillage was not a concept that was encountered thirty years ago. Although the ecovillage concept has a much older origin, it is necessary to go to the end of the 1980s to understand the present meaning [7]. With industrialization, a new way of life has emerged in the world. When the quality of life in the industrialized countries is evaluated according to the flow of money available, in the middle of the 1970s it was at the highest level; and when measured by natural resource use, the measurement shows it to be in decline. Towards the end of the 1980s, this decline clearly manifested itself. Perforation in the ozone layer, destruction of forests, depletion of various species, etc. have been the proof of this situation. However, reaction for this case from countries remained weak, the environmental issue was ignored in political negotiations, and business and commercial growth grew rapidly. The emerging political vacuum has led to the development of civil society movements. For the first time in these years, the question of 'how could a sustainable common life look?' has begun to be openly explored. Ecological and social destruction on the one hand, and the political and official circles, which offered limited responses to this destruction on the other hand, have driven people to seek out sustainable life models [7].

The first ecovillage movement was activated in the 1960s, with spiritual movements. These are Findhorn in Scotland, Solheimer in Iceland, Damanhur in Italy, the Farm in Tennessee/USA, Auroville in India, Sarvodaya in Sri Lanka and the NAAM movement in Burkina Faso [URL-2]. In this process, many simultaneous and different movements have found their place in the ecovillage mosaic over time. For example, the hippie and Back to the Land movements of the 1960s and '70s, the common housing movement that started in Denmark and spread rapidly to the international arena, both the environmentalist and feminist movements and some of the nuclear energy protest movements in Germany are examples. [7]. In particular, movements against the negative effects of industrialization and urbanization (green movements, etc.) [10] have become part of this mosaic.

In 1987, the Danish activist Hildar Jackson and the Canadian entrepreneur Ross Jackson (his wife) founded the Gaia Foundation [URL-3]. The foundation started radical work for founding ecovillages and sustainable living. In addition, Robert and Diane Gilman began researching sustainable communities and publishing them in their 'In Context' magazine. In the 1990s, the Foundation recruited Gilmans to conduct field work on 'sustainable communities'. Gilmans, in 1991, prepared a report called 'EcoVillages and Sustainable Communities' and presented the most successful examples on the international scene and gave some suggestions [7].

Alongside all these developments in the northern hemisphere, the pioneers of ecovillage studies also observed the movements in the south on the peninsula and aimed at the unification of the two initiatives. The eco-village movements in both hemispheres also had initiatives aimed at regaining more democratic and populist control over natural resources and revitalizing existing cultures. These initiatives were the key point of the union. The union in ecovillage history is an important step in ensuring the integration of the economic, cultural, technological, educational etc. forces in the fields [7].

After the 'Founders' meeting in 1991, the 'Sustainability Movement' gained momentum. In May 1992, Michael Boddington was commissioned to write the 'Technologies for Life, Sustainable Technologies for, Eco-Villages' report. This report encouraged the Gaia Foundation to establish the Gaia Technologies Company. The company was aiming to lead the development of intermediate technologies for ecovillages from the North and South.. In this process, the first eco-education center was established in The Farm Ecovillage (Tenessee, USA) with support from the Foundation. In 1995, the 'Ecocities and Sustainable Communities: Models for the 21st Century' conference was held at the Findhorn Foundation (Scotland). The conference took place with the participation of 400 people from all over the world. In 1996, GEN was established in the United Nations Habitat Conference (Istanbul, Turkey). GEN presented itself as "a global confederation of people and communities dedicated to sharing ideas, exchanging technology, preparing cultural and educational exchange programs, guides and newsletters, and building sustainable lifestyles", and expressed its main objective as encouraging the evolution of sustainable settlements all over the world. The GEN was primarily divided into three separate zones. These include the Ecovillage Network of the Americas (ENA), GEN for Oceania and Asia (GENOVA) and GEN Europe [7]. Today, GEN regions are classified as GEN Europe, GEN Africa, Latin America (CASA), North America (GENNA) and Oceania + Asia (GENOVA) [URL 4]. While GEN was established as a formal, international network with regional and thematic subnetworks, the ecovillage movement has dependably been a bottom-up movement, conveyed by a variety of single ecovillages. On the one hand, GEN is active in education, networking and the dissemination of information with political organisations such as the European Union (EU) and The United Nations Educational, Scientific and Cultural Organization (UNESCO). On the other hand, it provides a platform for support and exchange for local ecovillages and welcomes not only newly founded villages from the environmental movement but also traditional villages [URL-2]. GEN acts as an advisor to the United Nations Economic and Social Council (UN-ECOSOC). It is also one of the stakeholders of United Nations Institute for Training and Research (UNITAR-CIFAL), which delivers education on sustainable development to the local administrations around the world [8].

4. Ecovillage Features

The idea of the ecovillage depends on a significant understanding which is based on the idea that everything in the universe is in cooperation with everything else and we can influence our environment with our thoughts and behaviors. In this context, it aims at integrating processes of "ecology, design, alternative energy, permaculture, forming community, etc." [11].

At the basis of the reasoning behind ecovillages lies the desire to establish a settlement model that is ecologically the least damaging it could be and integrates with nature in a healthier and more sustainable manner than the average urban environment. Ecovillages see themselves at the service of a greater purpose such as improving themselves ecologically, strengthening community life, reviving local economies, deepening spirituality, etc. Thus, they put into practice actions which will develop and convey their messages. Within these targets, ecovillages aim to design environmentally friendly human settlements, analyze ecological footprints, support sustainable local economies, produce and process local and organic food, repair disruption of the earth's crust, form participatory and community-oriented forms of management, and ensure social participation. With these examples, they lead new models in many fields [7]. The present situation for ecovillages is expressed in the words of Helena Noberg-Hodge, director of the international ecology and culture community. "*The ecovillage movement is perhaps the most comprehensive solution to global economic dependence. People around the world are building ecological village communities in order to get away from pollution and competition*" [12].

The principles and motivations of ecovillages in their four dimensions (ecological, economic, social, and spiritual) are interrelated and interdependent (Figure 1). Therefore, attributes in one dimension can overlap with another dimension, and exert influences on each other as well [6; 13].

Figure 1. Ecovillage cycle [URL-5]

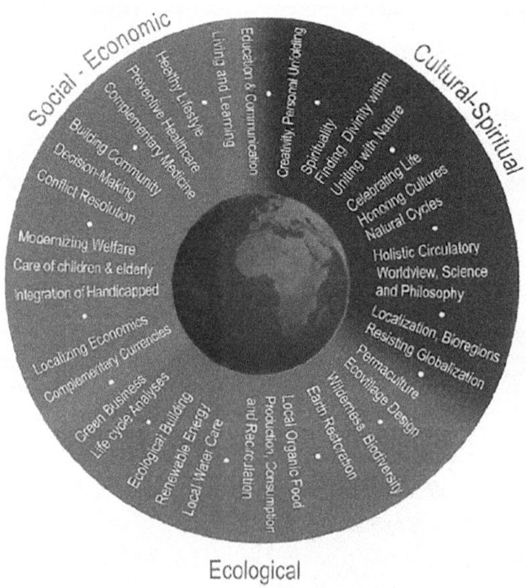

The ecological dimension includes features related to reducing negative impacts on the environment, protecting natural eco-systems, and implementing sustainable practices (planning, design management areas). The ecological dimension imagines a society in harmony with nature, where nature has a critical impact on the physical and psychological well-being of the ecovillage residents. This does not imply that an ecovillage cannot be situated in a metropolitan area. However, if it is, it is of great importance that nature and plants are introduced to the urban environment. The ecological aspect stresses the importance of ecocycles and a sustainable use of resources [14]. The economic dimension includes the prevention of exploitation of business and consumers in ecocities, the creation of legal and reliable service and business management, and the creation of small-scale entrepreneurship [13]. Besides this, ecovillages are protests against the economic activities that weaken cultural values and destroy resources as well as the local economy [7; 13]. This social dimension is concerned with the working and living environment (habitat) in ecovillages. Social interaction and cooperative interdependence and work are rewarded in this environment. Ecovillages follow a well-defined and proper constitution to avoid conflicts and fights [13]. Some parts of the social dimension of ecovillages are their being research centers, both for demonstration and training. They provide opportunities for research into sustainable living and offer education to share this expertize with other people. The fields of training usually include sustainable living, permaculture, energy efficiency, alternative economies or governance [7; 13]. A spiritual dimension is related to the structure of thought and spiritual values in ecovillages. The inhabitants of an ecovillage share common values and goals and believe they will restore, sustain, and protect their environment. Their beliefs are based on various ideologies (ecological restoration, energy efficient buildings or organic food production). In addition, the ecovillages are the quiet builders of a new vision, producing positive approaches to problems rather than criticism [13]. With the change of values towards inner spiritual progress rather than material accumulation and competition, inhabitants feel more whole and content as individuals. Instead of producing and consuming material goods, people will engage in arts, dancing, music, prayer and celebrations [14].

The presence of ecovillages is based on 5 basic principles: social priority (a response to alienation and loneliness in the modern societies), individual initiation (ecovillages are less dependent on governments and official authorities, but are more dependent on imagination, visions of the community members and resources), resource control (ecovillages control their own resources for the control of local resources between communities and the authorities), sharing moral values (showing ecological purpose, belief, humanism and socialism, and sharing spiritual values) and sharing knowledge (sharing ecology based ideas and techniques). Ecovillages also aim at "increasing the feeling of bond and responsibility towards the world; establishing communities that live together in harmony; ending the conflicts with peaceful means; reinforcing the social bonds and providing people with the sense of belonging; developing the act of taking participant decisions; enabling people to commune with nature; using the natural resources productively (controlling the rain waters, etc.); decreasing the carbon emission; using renewable energy technologies; decreasing the need for fossil fuels; building ecological structures; waste management (making compost out of the wastes, recycling, etc.); sharing food, transportation, tools and general heating systems to enable saving; enabling local economical development and decreasing the pressures of the global economy; developing social economies based on cooperation; producing local organic food and growing the community's own food; applying permaculture design methods; giving educations intended for sustainable life; conserving the bio-diversity and promoting it" [10].

5. Ecovillage Examples

GEN is carrying out its efforts to gather and unite under the roofs of ecovillages in the world [9]. GEN today lists more than 1000 local ecovillage projects and networks worldwide, about 130 among them are in Europe (Figure 2). The independent Eurotopia Directory on communities in Europe 2014 lists 430 different kinds of community projects including a hard to define number of ecovillages in Europe [URL-2]. However, not every ecovillage in the world is included in this network since it has no knowledge of this network, it is a member of a different organization, or it does not make any effort to be connected to any network [7]. For these reasons, it is difficult to give a clear figure about the total number of ecovillages in the world.

Figure 2. Ecovillages registered to GEN [URL-6].

Some important examples of ecovillages are given below:

Findhorn (Scotland): Findhorn Ecovillage, in northern Scotland, has changed within twenty-five years from an 'intentional community' in a caravan park into one of the most pioneering and celebrated promoters of sustainable living best practices (Figure 3). In 1982, the Findhorn Foundation hosted 'Building a Planetary Village' and marked the beginning of serious attempts to demonstrate that settlements could be truly environmentally, socially, and economically sustainable. Initially, the focus was placed on the construction of a 75kw wind turbine and the construction of "ecofriendly" buildings. However, they also initiated a biological sewage treatment system, known as the Living Machine, and numerous solar waterheating systems, and are now considering geothermal heating. Findhorn supports traditional organic and biodynamic farming practices and 70 per cent of its fresh food requirements are supplied by the Findhorn community-supported agriculture scheme, established in 1994 – the oldest and largest of its kind in the United Kingdom (UK). In 1998, it was awarded a United Nations Habitat Best Practice Designation. The accomplishment of the ecovillage has carried it to the position of being a major source of inspiration, attracting nearly 14 000 visitors a year. The Findhorn ecovillage also supports a community far wider than the project itself. Many businesses, charities, community bodies and sister communities have come to life in the neighboring fields that are not directly involved with but are connected to the initial project. Along with the 300 jobs it has created, it is projected that the ecovillage and linked activities contribute more than £5 million to the economy of the Findhorn/Forres area of the Highlands [15].

Figure 3. Findhorn

Ithaca Ecovillage, EVI (USA): The Ecovillage at Ithaca (EVI) in New York State is one of North America's most high-profile ecovillages and a growing suburban ecovillage on the edge of a small college town (Figure 4). Consisting of 60 one-to-four-bedroom houses spread over 70 ha and two neighbourhoods, the 160 residents aim to demonstrate a more sustainable way of life for Americans by reaching mainstream culture through education programmes and media coverage. Originally started in 1991, the first homes were completed and the residents moved in after five years. All buildings are passive solar, super-insulated duplexes and some of them have solar electric, grid-integrated systems (14 out of 60 homes). Four also have solar panels for hot water; five have compost toilets; and two are straw-bale. The remaining 90 per cent of the area is preserved as green space, consisting of meadow, woods, streams and ponds, to create a variety of areas for wildlife and natural recreation. There are also two organic community-supported agriculture schemes, which feed about 1400 local people during the growing season. Everyone contributes a few hours a week to keep the community autonomous – including outdoor maintenance, governance, finances and future projects – but being located close to a larger centre means that most people at EVI can access paid work. The community has forged links with Ithaca College, Cornell University and Wells College, where residents offer their village as a working model for study and can build a wider platform to share their experiences [15]. It is a modern-day ecovillage that combines a strong sense of community, small-scale organic farming, land preservation, green building, energy alternatives and hands-on education [16].

Figure 4. Ithaca [URL 7].

Mbam (Senegal, Africa): Located in the Siné-Saloum delta, at the meeting point of two of Senegal's largest rivers, Mbam is home to some 3000 people [15]. The most important feature that distinguishes Mbam from the surrounding communities is the presence of a group of citizens who speak locally at the center of the community. This feature also makes it an ecovillage [7]. Central to the sensitive ecosystem of the delta are the mangrove forests. These came under considerable pressure during the mid-1980s, when a harsh drought not only damaged the plants, but also led to an influx of migrants from surrounding countries, whose arrival resulted in much greater quantities of mangrove wood being extracted to smoke and cook fish. It didn't take long for the mangroves to die back, opening up the paddy fields to the sea. With the salinisation of waterways and soil erosion impacting on their livelihoods, a core group of citizens recognised the need for more sustainable development and approached the Senegalese national branch of the Global Ecovillage Network (GEN). With the villagers' input into all aspects of design and delivery of the projects, GEN has helped the community to realise several achievements. In addition to establishing more sustainable farming techniques and installing solar ovens, the community has been particularly successful in mangrove restoration and management. Mbam has become a centre of excellence, with villagers now training parties from the surrounding area in sustainable mangrove restoration. Aided by GEN, the village is also embarking on a biogas development program sponsored by the Global Environment Fund [15]. They are now obtaining funding from UNIFEM–United Nations Fund for Women, from the Jica-Japanese funds. The Japanese fund assists them in making life jackets for the fishermen. They are the only village in west Africa which produces these life jackets and they sell them throughout Senegal [17]. The Senegalese national branch of GEN is the first network of its kind in Africa. This UN ECOSOC NGO works with 45 member villages to promote sustainable approaches in order to meet the Millennium Development Goals [15].

Auroville – The City of Dawn (India): Auroville (City of Dawn) is an experimental township in Viluppuram district in the state of Tamil Nadu, India, near Puducherry in South India, whose stated purpose is to realize human unity in diversity (Figure 5). It was inspired by the French visionary, The Mother, and by the work and living presence of the great Indian philosopher, yogi, and revolutionary, Sri Aurobindo. Auroville is a place for karma yoga, the transformation of consciousness through action. All life is seen as yoga [16]. The city has about 2000 people [URL–8] from over 35 different nationalities, about one third of whom are Indians (more than 15% are French and another 15% are German). It is a popular tourist destination, and has been described as a "New Age metropolis conceived as an alternative exercise in ecological and spiritual living". The township starkly stands out from the surrounding traditional Indian villages and farms [URL-9].

Figure 5. Auroville/Overview of the ecovillage plan [URL-10].

Crystal Waters Permaculture Village (Australia): Nestled in the hills of subtropical Queensland, 100 km north of Brisbane, is the Crystal Waters Permaculture Village (Figure 6). Covering an area of 259 ha (640 acres), the ecovillage is home to some 200 permanent residents living on 83 freehold plots and running a multitude of environment-orientated businesses. The settlement was formally established in 1986, amid a backdrop of economic and ecological deprivation as local forestry and dairy industries declined. Crystal Waters was designed as an approach to agriculture that takes natural processes into consideration to make the most efficient use of land. At Crystal Waters, 20 per cent of the land is occupied by residential and commercial lots; the remaining 80 per cent is the best land and is owned in common [15]. Two structures cover the legal aspects of the community. The Body Corporate and Community Management Act covers the freehold residential lots and the common land, and is governed by local by-laws. These by-laws are the only criteria for the selection of lot holders – the residents select themselves by accepting that they will abide by these laws when they take over the title [16]. The community can be licensed for sustainable agriculture, forestry, recreation and habitat projects. Land productivity continues to improve today, waterways are clean and community businesses are thriving. It received the 1996 United Nations World Habitat Award for its "pioneering work in demonstrating new ways of low-impact, sustainable living" and is listed in the Top 40 of the UN's Best Practices database [15].

Figure 6. Crystal Waters [URL-11; UR-12].

Sieben Linden (Germany): Sieben Linden near Altmark, Germany, describes itself as a community of communities (Figure 7). The reason why Sieben Linden was established in the first place is to experience a very small ecological trail in the region where it is found. The pioneers of the project began their meetings in 1986, and a decision was reached to establish a new campus instead of improving an existing settlement during the project period. Due to several restrictions such as the planning legislation in force, the lack of sufficient land to buy, and inability to find an open-minded village, the process has been long and difficult [7]. Its residents recognize and celebrate differences as well as common ground in working towards their collective aim of providing "a model for living responsibly with nature" by dividing their ecovillage into neighbourhoods, each with their own way of life. The contrast between the two largest neighbourhoods highlights this point: one, the 99 club, is more radical, being a completely vegan co-operative that does not use machinery, whereas the other came together because its residents' children were roughly the same age. The community has experimented with a range of different governance models such as forums, supervisions and talking sticks. The site is also growing, with the latest land purchase taking the total area of Sieben to 77 ha. This consists mainly of forest and farmland, with approximately 6 ha set aside for housing. Although many residents still live in caravans, there are two modern low-energy houses and one of the first straw-bale houses with planning permission to be built in Germany. Since then, two further straw houses have been built (including the biggest in Europe, with more than 500 square metres of living space) and another two are under construction – Sieben Linden has been lobbying hard for a change in the building regulations in Germany to make it easier and cheaper to build with straw bales. The settlement also has its own water supply, waste disposal and solar heating. There are currently many small enterprises – a jewellery shop and a small publisher, as well as artists, consultants, and tradespeople, including an organic vegetable-grower and joiner. It is hoped that more self-employed craftsmen, artists, architects, and body therapists will soon be able to find ways to earn a living at Sieben Linden [15].

Figure 7. Sieben Linden [URL-13].

6. Conclusion

Sustainable development is an expression born from a period looking for ways to solve problems that developed after the industrial revolution. A wide variety of strategies have been developed and implemented in this period. The concept of sustainability has been introduced and guided in many areas of life. Thus, sustainable life has become one of the main objectives of today's world. In this context, ecovillages are models of lived sustainability. In our world filled with environmental and social problems, ecovillages offer some refreshing solutions. In fact, ecovillages have been created as an answer to the problems of our time.

Ecovillages are sustainable communities based on permaculture (permanent culture). Their main objective is to ensure harmony and sustainability between people and nature. This goal is the main objective of sustainable development. They have shown a remarkable success in implementing sustainable living practices since the theoretical knowledge in ecovillages is transformed into practical action. In a practical way, they show how an ecologically and socially sustainable way of living can work. From this perspective, ecovillages provide live laboratories to test new models and find solutions to many problems relating to sustainable life.

With ecovillages our ecological footprint can be significantly reduced. It is possible to revitalise regional economic and energy cycles; and this constitutes an important counterweight to globalization. Ecovillages enrich regions with ecological business and employment opportunities, cultural activities, and enhanced attractiveness for families. They are, therefore, an active contributor to positive demographic change. Ecovillages worldwide combine modern and traditional knowledge with sustainable innovative initiatives in agriculture, energy, resource management, water, and refuse management, architecture and social participatory processes. They are making a decisive contribution within our society to the quest for local solutions to global problems [URL-14]. In this way, ecovillages have ideas, values, and beliefs in social, economic, ecological and spiritual dimensions. These are strong qualities that must be found in sustainable societies and development.

Ecovillages are now part of a worldwide movement for the social, cultural, ecological and economic revival of both rural and urban areas. In this way, ecovillages are becoming the strongest example of sustainable life. Ecovillages are even a living example of a solution for the challenges we face globally, showing decisive and conscious human societies. Ecovillages are an example of demonstration and teaching centers based on society.

GEN has an explanation for the ecovillages: "*Some will say the vision of ecovillages is utopian, some will say that the world of 2017 has no place for such dreamers. But others, who have lived this vision, know that humans really can live in a way that not only sustains, but even regenerates the social and natural fabric of life. GEN knows that the holistic creativity and innovation which ecovillages bring, when combining the wise use of modern technology and resources with traditional heritage and wisdom, can massively contribute to addressing our global issues of poverty and environmental destruction. We can build a world living within its own means, and at peace with itself.*" [URL-15]. This explanation is far beyond utopian evaluation since ecovillages are the communities that can keep their assets in harmony with nature in today's world. For this reason, they are the most important examples of sustainable life.

References

1. J. H. SPANGENBERG: *Sustainable Development Concepts and Indicators.* Sustainable Europe Research Institute (2000).
2. R. EMAS: *Brief for GSDR 2015 The Concept of Sustainable Development: Definition and Defining Principles.* Available from: www.sustainabledevelopment.un.org/content/documents/5839GSDR%202015_SD_concept _definiton_rev.pdf, Date of Access: 24.5.2017 (2015).
3. H. STODDART: *A Pocket Guide to Sustainable Development Governance, Stakeholder Forum.* Available from: www. sustainabledevelopment.un.org/content/documents/1758GSDR%202015%20Advance %20 Unedited%20Version.pdf, Date of Access: 24.5.2017 (2011).
4. W. SCHEEWE: *Nurturing the Soil Feeding the People.* Rex Bookstore. ISBN: 9712328953

5. A. KIRBY: *Redefining Social and Environmental Relations at the Ecovillage at Ithaca: A Case Study.* Journal of Environmental Psychology, 23: 323–332 (2003).
6. L. JOSEP, A. BATES: *What is an Ecovillage? Communities Magazine,* 117: 1–4 Available from: http://communities.ic.org, Date of Access: 5.5.2013 (1991).
7. J. DAWSON: *Ecovillages-New Frontiers for Sustainability.* Sena Publisher, p. 139, ISBN 978–605-61475-1-7, İstanbul, Türkiye (2006).
8. R. R. ROSENTHAL: *Ecovillage.* Accessed at http://gen-europe.org/uploads/media/leaflet_turkish.pdf, Accessed on 20.01.2013 (2013).
9. A. A. KARADAĞ, D. S. KORKUT: *Sample Societies to Sustainable Lıfe: Ecovillages.* The 4th International Symposium on Sustainable Development (ISSD 2013) (2013).
10. B. GESOTA: *Ecovillages as Models for Sustainable Development: A Case Study Approach.* Master Thesis, Social Sciences. Albert-Ludwigs-Universitat Freiburg i.Br. (Germany) and the University of KwaZulu-Natal, p. 90, Durban (South Africa) (2008).
11. L. SEVIER: *Ecovillages: A Model Life?* Available from: www.gen.ecovillage.org/iservices/publications/articles/ec_ 08_may_ecovillages.pdf, Date of Access: 5.5.2013 (2013).
12. E. BIYIK: *Bu Köyün Kimseye İhtiyacı Yok.* Available from: http://www.aksam.com.tr/cumartesi/bu-koyun-kimseye-ihtiyaci-yok/haber-327374, Date of Access: 29.05.2017 (2014).
13. T. S. COSTA: *Practical Approach Towards Degrowth Transition: a Case Study of an Emerging Ecovillage in Kalmar, Sweden.* Master Thesis. Lund University Lund University, Environmental Studies and Sustainability Science, 53 p., Swedish (2015).
14. S. JANSSON, F. RODHE: 2009. *Challenging Growth Society a Study of Eco-Villages as a Strategy Towards a Sustainable De-Growth Societ.* Bachelor Thesis in Development and International Cooperation and Human Ecology, Gothenburg University, Sweden (2009).
15. L. SEVIER, M. HENDERSON, N. NAIDU: *Ecovillages: A model life?* Available from: www.theecologist.org/investigations/society/268714/ecovillages_a_model_life.html, Date of Access: 22.5.2013 (2008).
16. K. ITONAGA: *Expectation for Ecovillage Conference.* International Ecovillage Conference, Japan. Available from: www.begoodcafe.com/project/ecv/ecvc_tokyo2006.pdf, Date of Access: 2.06.2017 (2006).
17. B. BASSE: *Mbame Ecovillage.* Available from: https://ecovillage.org/project/mbame/, Date of Access: 29.05.2017 (2014).

URL-1: *The World's Cities in 2016 (UN).* Available from: www.un.org/en/development/desa/population/publications/pdf/urbanization/the_worlds_cities_in_2016_data_booklet.pdf, Date of Access: 26.5.2017 (2016).

URL-2: *Social Innovation and the Global Ecovillage Network.* Available from: www.transitsocialinnovation.eu/content/original/Book%20covers/Local%20PDFs/192%20Case_study_report_GEN_FINAL.pdf, Date of Access: 16.5.2017 (2015).

UR- 3: *Partners, Gaia.* Available from: www.ecovillage.org/about/partners/, Date of Access: 11.5.2017 (2017).

URL-4: *GEN Regio.* Available from: www.ecovillage.org/regions/, Date of Access: 11.5.2017 (2017).

URL-5: *Ecovillage cycle.* Available from: www.computingforsustainability.com/2009/11/06/sustainability-pictures-x/, Date of Access: 29.05.2017 (2017).

URL-6: *Ecovillages registered to GEN.* Available from: www.gen.ecovillage.org/ cited in www.degisimicinyaratim.blogspot.com.tr/2012/06/ekokoyler-kitap-tantm-uzerinden-bir.html, Date of Access: 29.05.2017 (2017).

URL-7: *Ithaca.* Available from: www.intuswindows.com/projects/ecovillage-at-ithaca/#; www.ecovillageithaca.org/grow/farming-and-food/, Date of Access: 29.05.2017 (2017).

URL-8: *Auroville.* Available from: www.ecovillagebook.org/ecovillages/crystal-waters/, Date of Access: 2.06.2017 (2017).

URL-9: *Auroville-The City of Dawn.* Available from: www.serreal.ning.com/group/greencommunities/forum/topics/auroville-the-city-of-dawn-in, Date of Access: 29.05.2017 (2017).

URL-10: *Auroville.* Available from: www.auroville.org/contents/51, Date of Access: 29.05.2017 (2017).

URL-11: *Crystal Waters.* Available from: www.//crystalwaters.org.au/, Date of Access: 30.05.2017 (2017).

URL-12: *Crystal Waters Permaculture Village.* Available from: www.ecovillagebook.org/ecovillages/crystal-waters/, Date of Access: 30.05.2017 (2017).

URL-13: *Sieben Linden.* Available from: www.ecovillagebook.org/ecovillages/sieben-linden/, Date of Access: 30.05.2017 (2017).

URL-14: *Ecovillages as a Model of Practiced Sustainability.* Available from: www.gelebte-nachhaltigkeit.de/fileadmin/pdfs/gelebte-nachhaltigkeit/GEN_BROCHURE-EN.pdf, Date of Access: 30.05.2017 (2017).

URL-15: *What is an Ecovillage?* Available from: www. ecovillage.org/what-is-an-ecovillage/, Date of Access: 30.05.2017 (2017).

Tahsin Yilmaz[1]

Age Friendly Landscape Design

[1]*Akdeniz University, Faculty of Architecture, Department of Landscape Architecture, Antalya, Turkey.*

1. Introduction

Studies show that the world population is ageing and the fastest changes occur in less-developed countries [1]. As stated in the reports of organizations such as the World Health Organization (WHO), the Organization for Economic Cooperation and Development (OECD), and the World Economic Forum, ageing all over the world has been at the forefront since the 2000's. The reports state that the elderly population of the world will be two billion by 2050, and reach three times the rate of the year 2000 [1,2,3] The World Health Organization claims that the proportion of elderly people in the world will increase from 11% to 22% over the 45 years between 2006–2050 [4]. In 2015, one in eight people worldwide was aged 60 years or over. By 2030, older persons are projected to account for one in six people globally. By the middle of the twenty-first century, one in every five people will be aged 60 years or over [5]. Future projections of population change can be seen in Figure 1. Beside the organizations mentioned above, the United Nations approaches the two most important issues of the 21st Century within two main themes, "ageing" and "urbanization" [6]. Therefore, studies on ageing have increased recently. The rapid ageing of the world population is one of the priority work areas of the World Health Organization.

2. Ageing

Ageing is defined in the report "A Glossary of Terms for Community Health Care and Servıces for Older Persons" published by the World Health Organization as "the lifelong process of growing older at cellular, organ or whole body level throughout the life span" [7]. However, defining ageing only by the medical aspect will not be the appropriate approach here,because old age does not correspond to a situation where people live longer, healthier, and earn more in their retirement. Along with these aims, old age should be regarded as a category that involves social relations with other groups, is integrated with society, and has suggestions and expectations about the neighborhood, city and country [8]. Ageing is defined as a stage in which physical and mental changes are seen in the later period of life as an extension of adult life. Old age is accepted in definition as a period in which physiological changes, psychosocial factors and chronology are considered [9].

Figure 1. Number of people aged 60 or over: World, developed and developing countries, 1950–2050. (Source: UNDESA, World Population Ageing 2011 (2012; forthcoming), based on UNDESA Population Division medium projection scenario, World Population Prospects: The 2010 Revision.

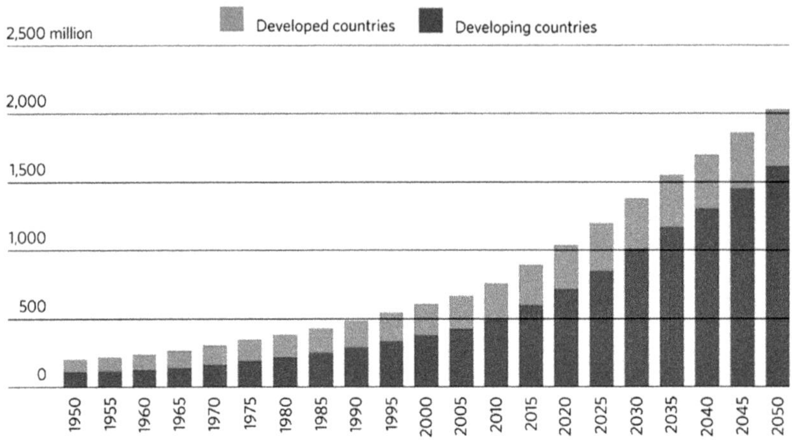

Ageing is a triumph of development. Increasing longevity is one of humanity's greatest achievements. People live longer because of improved nutrition, sanitation, medical advances, health care, education and economic well-being. Life expectancy at birth is over 80 now in 33 countries; just five years ago, only 19 countries had reached this.

An increase in life expectancy, on the one hand, is seen as a great success; it is one of the main causes of the increase in the elderly population and the deterioration of the natural age pyramid from the other side. The increase in the life span is significantly important, but it is also important that the process is carried out comfortably at the same time. Otherwise, unhealthy, anti-social, and living in poor conditions, and more than anything an unhappy "longevity" will be the result of a goal that does not achieve very successful results.

3. Population Change

A population is classified as ageing when older people become a proportionately larger share of the total population. Declining fertility rates and increasing survival at older ages have led to population ageing.

Life expectancy at birth has risen substantially across the world. In 2010–2015, life expectancy was 78 years in developed countries and 68 years in developing regions. By 2045–2050, newborns can expect to live to 83 years in developed regions and 74 years in developing regions. Changes in life expectancy rates in the world during the one hundred year time period between 1950 and 2050 can be seen in Figure 2.

In 1950, there were 205 million persons aged 60 years or over in the world. By 2012, the number of older persons had increased to almost 810 million. This figure is projected to reach 1 billion in less than ten years and double by 2050, reaching 2 billion [5].

According to the report "World Population Ageing" published by the United Nations, the whole world is at the forefront of the ageing phenomenon, but the differences in regions are remarkable. Ageing is particularly noteworthy as being widespread in Africa and Asia. Figure 3 represents the population aged 60 years or over and aged 80 years or over, by region, 1980–2050.

Figure 2. Changes in life expectancy from 1950, with projections until 2050, by WHO

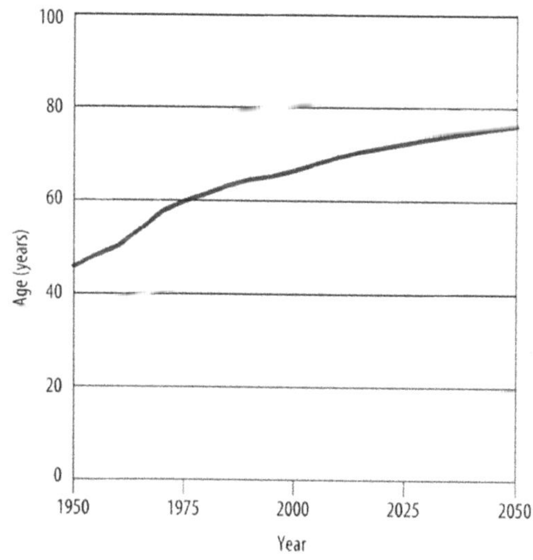

This change in world population has been monitored by many scientists and institutions, especially the World Health Organization. Different scenarios are being produced by scientists and institutions depending on future population change and ageing, and various pieces of scientific research are being carried out in order to prepare the world for the ageing process.

Figure 3. Population aged 60 years or over and aged 80 years or over, by region, 1980–2050 (Source: UN 2015, World Population Ageing Report)

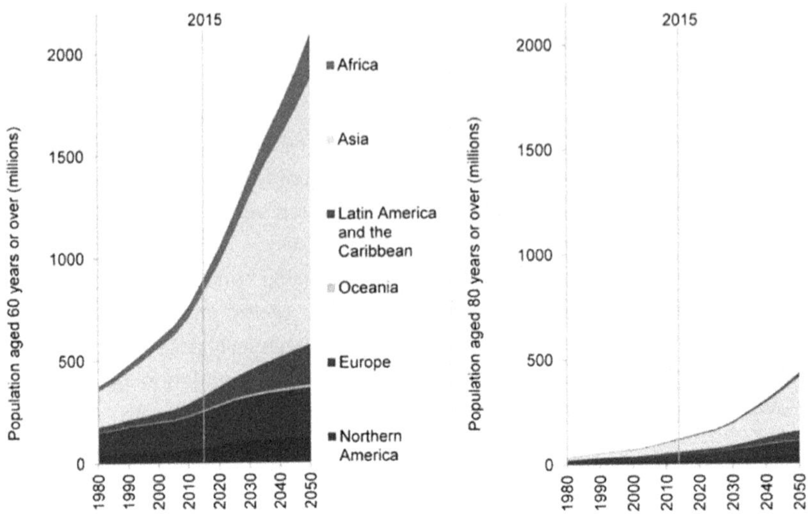

Studies on the "ageing of the world" continue under two main headings. The first is the change in the lifestyle of the elderly in accordance with the "Active Ageing" document published by the World Health Organization in 2002. The other is the "Age-Friendly Cities" project implemented by the World Health Organization in 2006 and based on the principle that the elderly can use their living spaces without encountering any problems [10].

Active ageing is defined by the World Health Organization as "the process of optimizing opportunities for health, participation and security in order to enhance quality of life as people age". It applies to both individuals and population groups [11].

The active ageing approach has a significant importance for the ageing population. The ageing individual naturally experiences various biological losses and their social lives as well as daily lives are shaped and restricted due to these losses.

Beyond these biological losses, older age frequently involves other significant changes. These include shifts in roles and social positions, and the need to deal with the loss of close relationships. In response, older adults tend to select fewer and more meaningful goals and activities, optimize their existing abilities through practice and the use of new technologies, and compensate for the losses of some abilities by finding other ways to accomplish tasks [12].

The application of age-friendly cities is considered a good opportunity to abolish several problems which ageing individuals face.

An age-friendly city offers a supportive environment that enables residents to grow older actively within their families, neighborhoods, and civil societies, and offers extensive opportunities for their participation in the community [13].

An age-friendly community can be defined as a place where older people are actively involved, valued and supported with infrastructure and services that effectively accommodate their needs [14].

The World Health Organization identified eight topics within the Age-friendly Cities framework including social participation, respect and social inclusion, civic participation and employment, communication and innovation, community support and health services, and outdoor spaces and buildings, transportation, and housing (Figure 4) [4]

Figure 4. Age-friendly city topic areas (Source: World Health Organization 2007)

Each of these topics can be further analyzed in terms of service provision, emphasizing the importance of access to services in the development of age-friendly cities and communities [15]. However, when studying each topic, it is important to consider the local characteristics of the place where the work is to be done. On the other hand, even if the age-friendly city approach can theoretically be considered as very appropriate and successful, there can be various difficulties in its implementation.

The implementation of age-friendly policies is being challenged by three major elements: the impact of financial cuts on social programs; pressures arising from the characteristics of urban development; and the shift towards the privatization of urban space [16].

Active ageing and age-friendly city approaches are based on ageing in place. Ageing in place is desirable for older people, policy makers and planners, enabling older people's access to existing social and service networks as well as remaining in the familiarity of their local environment [17]. Successful ageing in place depends on older people having access to relevant service needs in the community. These include access to medical care, retail, recreation, affordable entertainment, social facilities, public transport, housing, home services and environmental "facilities" such as public open spaces [15].

For a well-planned future, it is essential that the relevant professions work in coordination to develop strategies and create the necessary infrastructures. Each profession should provide a guide in its own area of expertise on the subject of ageing and contribute to the ageing process by acting in line with other professionals. This is because the changes in population will cause several problems in many areas such as health, education, social security, architecture etc. One of these areas is the use of public green spaces by the elderly. Within the age-friendly city concept, future green spaces should be designed to be accessible in the direction of the expectations and needs of future users and should be prepared for this transformation process. One of these professions is landscape architecture, with expertise related to the physical design of the outdoor green space in cities.

Green spaces are significant areas that provide several contributions to the city in ecological, aesthetic, functional and economical terms and raise the quality of life, providing social and psychological benefits for the residents [18,19]. They bring significant benefits for individuals and society, such as providing aesthetic experiences and recreational opportunities, contributing to human health and wellbeing by reducing stress and mental fatigue [20,21,22], strengthening social bonds [23,24], and providing learning opportunities for natural processes [25].

Green areas appear set to be used more intensively by the elderly in the future, due to changes in the population. This suggests that green spaces should be shaped according to the expectations and needs of the elderly and should be designed as age-friendly. This approach is also one of the first steps to be taken in becoming an age-friendly city.

Another factor to consider in terms of ageing is disability. According to the World Health Organization, the decline in functional capacity and physiological stress that occur in old age are accompanied by disability. In this context, it appears that the needs of the elderly are related to needs to do with disability [26]. Therefore, designs of green areas, which are used by all disability groups and offer an important social opportunity, should be made considering the criteria of disability and ageing.

While social life is important for every individual's experience, it is more important for disabled and elderly people who have limited social opportunities. Individuals who cannot use green spaces become deprived of socialization opportunities. This process creates social isolation. While social isolation reduces self-esteem and self-respect, it prevents people from integrating into society and participating in life. The ageing person living at a high level of social isolation cannot make independent decisions and make friends. These attitude differences in attitude adversely affect their "hopes and expectations of life" [27]. For this reason, it is important that the green spaces that create a social opportunity for the community should be designed as age-friendly for everyone.

4. Results

Studies show that in the near future ageing will be an inevitable reality. The first condition in order for this process to be properly managed is that each profession should approach ageing and the ageing process with its own expertise, but ultimately the topic should be handled within the framework of vocational cooperation.

There are a variety of knowledge components in the field of landscape architecture that designers need in order to build a successful design. These components can basically be classified under the titles of user expectancy, site analysis, applicable legislation, designer knowledge and managerial expectation (Figure 5).

Figure 5. Necessary components for proper landscape design

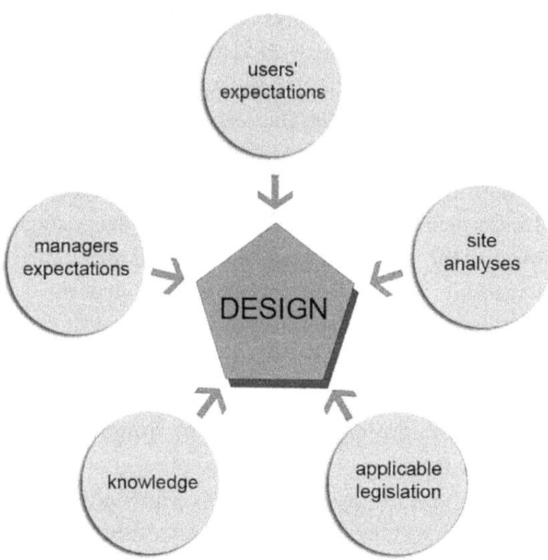

Landscape designs created according to applicable legislation only within age-friendly design cannot reach a successful result because the welfare of the people should be accepted as the main concern in this sort of design. In this context, the first step in design should be to reveal users' expectations.

There are several methods that can be used to reveal the expectations of the elderly. These include questionnaires, surveys, visits to nursing homes, personal observations by a researcher and the results of the studies in current literature.

Places where their expectations are not taken into consideration will not be of interest to the elderly, even if they are built in accordance with design standards. Green spaces are public spaces that provide many activities. The primary responsibility of the landscape architect is to design space, not activities. For this reason, the spaces must meet both standards and offer appropriate opportunities for various activities.

Within the scope of these studies, the aim should not only be that of revealing the expectations of the elderly. In addition to this, all problems related to existing applications should be revealed. This is because, in addition to creating a physical environment that meets the intended standards of age-friendly design, thr right application of design provides opportunities to have a more comfortable life. Further work to be done is that of specifyingthe expectations of managers, non-governmental organizations or decision-makers, if any, who have control on site. Determination of the expectations will be an important source in the course of being able to demonstrate local characteristics and the specific needs associated with them because, in this process, the guidance, restriction and knowledge of the managers will be an important source successful analysis.

One of the main components of the work is assessing the opportunities provided by the current legislation. In this context, the legal situation related to the subject and the design standards produced by the relevant institutions and organizations should be established. The applicable legislation can be a supporting, guiding or limiting component of the study. For this reason, the current legislation must be considered as a fundamental guide towards benefits at every stage of the study.

The fourth component is the site analysis that will allow the local properties of the site to be studied. The climate, topographical structure, social relations, beliefs, economic conditions, etc. of each place are different from each other. In this context, all the features of the site are analyzed and included in the design process. In this respect, while the designs are made in accordance with certain standards, the original designs can be reached by bringing local features to the foreground at the same time.

The last data component on the subject is the knowledge accumulated by the landscape designer. At this stage, landscape architecture creates a general information pool by creating a literature review on the subject and exchanging information with professionals such as psychologists, gerontologists, sociologists, architects, city planners, etc.

By analyzing the data generated by the above-mentioned components, it becomes possible to find the most suitable designs. It is possible in this way that the most suitable environments for the elderly will be created and the cities will be designed as age-friendly. It is necessary not to be afraid of old age but to be prepared for this process.

References

[1] UNDESA, 2007. "World Economic and Social Survey 2007: Development in an Ageing World". United Naitons Department of Social and Economic Affairs.

[2] UN 2012. "United Nations Population Fund". State of the World Population 2011

[3] OECD 2015. Organisation of Economic Cooperation and Development (2015). Ageing in cities. Paris

[4] WHO, 2007. World Health Organization, Global Age Friendly Cities: A Guide, WHO Press, Geneva

[5] UN, 2015. World Population Ageing Report, United Nations, Department of Economic and Social Affairs, Population Division. New York.

[6] Biggs, S. and Haapala, I. 2015. "Age Friendly Environments", Journal of Social Work Practice, 29 (1), 1–3

[7] WHO, 2004. A Glossary Of Terms For Communıty Health Care And Servıces For Older Persons, WHO Centre for Health Development, Ageing and Health Technical Report, Volume 5.

[8] Arun, Ö. 2008. "Yaşlı Bireyin Türkiye Serüveni: Kaliteli Yaşlanma İmkânı Üzerine Senaryolar", Gaziantep Üniversitesi Sosyal Bilimler Dergisi, 7(2), 313–330.

[9] Durak, M. 2004. "Yaşadığı Ortamda Yaşlıya Hizmet Modeli". Ankara: Ankara Büyükşehir Belediyesi, Yaşlılara Hizmet Merkezi Yayınları.

[10] Biggs, S and Carr, A. 2015. "Age and Child-Friendly Cities and The Promise of International Space", Journal of Social Work Practice, 29 (1), 99–112.

[11] WHO 2013. What is active ageing? Retrieved from http://http//www.who.int/ageing/active_ageign/en/

[12] Baltes P, Freund A, Li S-C. The psychological science of human ageing. In: Johnson ML, Bengtson VL, Coleman PG, Kirkwood TBL, editors. The Cambridge handbook of age and ageing. Cambridge: Cambridge University Press; 2005:47–71

[13] Fitzgerald, K.G. and Caro, F.G. (2014). "An Overview of Age-Friendly Cities and Communities Around the World" Journal of Aging & Social Policy, 26:1–18.
[14] Alley, D., Liebig, P., Pynoos, J., Benerjee, T., & Choi, I. H. (2007). Creating elder-friendly communities: Preparation for an ageing society. Journal of Gerontological Social Work, 49, 1–18. http://dx.doi.org/10.1300/J083v49n01_01
[15] Lowen, T., Davern, M.T., Mavoa, S. And Brasher, K. (2015). "Age-Friendly Cities and Communuties: Access to Services for Older People" Australian Planner, 52:4, 255–265.
[16] Buffel, T., Phillipson, C. (2016). "Can Global Cities be "Age Friendly Cities"? Urban Development and Ageing Population" Cities, 55(2016) 94–100.
[17] Australian Institute of Health and Welfare. 2013. The Desire to Age in Place among Older Australians. Canberra: AIHW
[18] Hartig T., Mang M., Evans G.W. 1991. "Restorative Effects of Natural Environment Experience", Environment and Behavior, 23, 3–26.
[19] Chiesura A. 2004. "The Role of Urban Parks For The Sustainable City", Landscape and Urban Planning, 68, 129–138
[20] Ulrich R.S. 1984. "View through a window may ınfluence recovery from surgery", Science, 224, 420–421.
[21] Marcus, C.C., Barnes, M. 1999. "Healing Gardens: Therapeutic Benefits and Design Recommendations", John Wiley & Sons, New York.
[22] Bowler D.E., Buyung-Ali L.M., Knight T.M., Pullin A.S. 2010. "A systematic review of evidence for the added benefits to health of exposure to natural environments", BMC Public Health, 10, 456.
[23] Coley R., Kuo F., Sullivan W., .1997. "Where does Community Grow? The Social Context Created by Nature in Urban Public Housing", Environment and Behavior, 29, 468–494.
[24] Peter K., Elands B., Buijs A. 2010. "Social ınteractions in urban parks: stimulating social cohesion?", Urban Forestry and Urban Greening, 9 (2010), 93–100.
[25] Kahn P.1999. "The Human Relationship with Nature: Development and Culture", MIT Press, Cambridge, MA.
[26] Köse, N., Erkan, N.Ç. 2014. "Kentsel Mekân Örgütlenmesinin Yaşlıların Kentsel Etkinlikleri Üzerindeki Etkisi, İstanbul Ve Viyana Örneği (1)", METU JFA, 31(1), 39–66.
[27] Burcu, E., Türkiye'de Özürlü Birey Olma: Temel Sosyolojik Özellikleri ve orunları Üzerine Bir Araştırma. Hacettepe Üniversitesi Yayınları, Mayıs, Ankara, 2007

Ülkü Duman Yuksel

Key Parameter for Enhancing Sustainable Urban Design: Thermal Comfort in Outdoor Urban Spaces

1. Introduction

Research shows that people spend less than 20% of their time outdoors, and about one-third of their days outside buildings in the developed world [1,2,3]. Therefore, the quality of the outdoor space where people spend their time and conduct their daily activities, social lives, and recreational activities affects the livability and sustainability of a city.

Among many factors that determine the quality of the outdoors, the microclimate is an important factor in contemporary cities. The thermal conditions of the outside directly and indirectly affect the preferences urban dwellers have for the time and duration they spend using the space, their activities, and their health. At a city scale, the thermal conditions of the outdoor benvironment affects the energy consumption and social life of the city.

Since the relationship between thermal conditions and the use of urban open spaces has come to be understood, many thermal comfort studies have been conducted on various open spaces. In the last two decades, outdoor thermal comfort has received great attention.

International organizations such as the International Society of Biometeorology and the World Meteorological Organization have diverted the scientific community working on thermal comfort towards developing an international thermal comfort index. Comprehensive and international research projects have been conducted about thermal comfort.

2. Thermal Comfort

Thermal comfort is a very dynamic and subjective phenomenon; these attributes describing it a complex process. In the relevant literature, thermal comfort has been defined in a variety of ways. Höppe[4] remarks that there are three different approaches used to define thermal comfort according to the principal ways in which it is discussed. These approaches are psychological, thermophysiological, and physical.

The most well-known and used definitions in this literature belong to Olgyay and ASHRAE. Olgyay [5] defines thermal comfort as the conditions wherein the average person does not experience a feeling of discomfort, while in ASHRAE, thermal comfort is defined as the state of mind which expresses satisfaction with the surrounding thermal environment and is assessed by subjective evaluation [6]. According to its thermophysiological definition, thermal comfort is reached when the heat flow to and from the human body is balanced, and the skin temperature and sweat rate are within a comfortable range [4].

In the context of urban design, thermal comfort could be defined as the key indicator of user comfort in outdoor spaces, as it summarizes the impact of the microclimate (sun, wind, air temperature, and humidity) on human climate perception.

A person's sensation of comfort is linked to the heat balance of the human body—the balance of heat generated by metabolic processes and heat lost by conduction, convection, radiation, and evaporation [7]. The heat balance of a person in the summer season (according to the MEMI model) is presented in Figure 1.

Figure 1. The heat balancing of a person in summer [4]

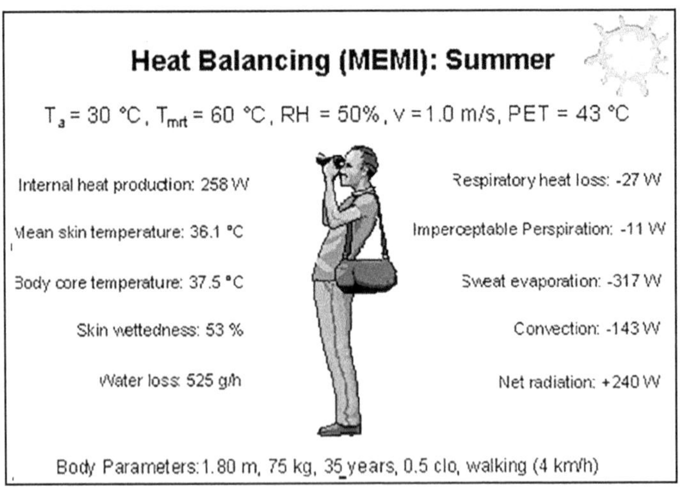

Conventional comfort theories rely on a steady state model where the production of heat is equal to heat losses to the environment, aiming to keep a constant core body temperature of 37°C, so that environmental conditions which provide thermal satisfaction are dependent only upon the activity of the subjects and their clothing level [8].

2.1. Factors Determining Thermal Comfort

Thermal comfort is determined by two main factors; meteorological conditions and personal characteristics.

a. Meteorological conditions
b. Personal characteristics

On the other hand, in outdoor areas, many additional factors participate in and affect thermal comfort either directly or indirectly. For example, landscape materials affect thermal comfort by changing the temperature, and orientation of buildings moderates thermal comfort by changing the wind speed.

a. Meteorological conditions

Meteorological conditions include temperature, humidity, wind, and solar radiation.

Air temperature could be defined as the single most impacting estimator for the thermal comfort. People feel warmer as the air temperature rises and cooler as it falls [9]. Since there are many factors in outdoor areas that interact with temperature and make the process more complicated, it is difficult to predict human comfort simply by taking temperature into consideration.

For example, an air temperature of 20°C could be perceived as uncomfortably warm by someone standing in full sunshine on a humid, windless day, or uncomfortably cool on a windy, cloudy day [9].

Wind is another of the most determinant factors in outdoor thermal comfort. TCOHC [7] describes two effects of wind, the mechanical/direct effect, and the thermal/indirect effect. The indirect effect of wind on human comfort occurs when it combines with air temperature, humidity, and solar radiation.

Humidity—the amount of water vapour in the air—is important because it limits the amount of sweat that can evaporate from the skin, which is the body's main mechanism for heat reduction.

According to Olgyay's study, people feel most comfortable outside when the wind speed is 5 m/sec, the air temperature is 21–27.5°C, and the relative humidity is between 30–65% [10].

The sun, as the main source of heat and light on earth, has an important impact on thermal comfort. The importance of solar radiation on thermal comfort can be expressed very simply: people seek shade in hot weather and sunlight in cold weather.

Solar radiation can caught directly or radiated or reflected from natural or artificial surfaces in urban environments. All forms of solar radiation should be assessed and need to be accounted for in thermal comfort studies. The amount of solar radiation absorbed and/or reflected from these surfaces is determined by various factors such as surface material, colour, roughness, and brightness. Although there are many forms of solar radiation, controlling solar radiation through the use of shade is simple compared to other meteorological conditions.

b. Personal characteristics

Several personal variables influence thermal comfort, such as age, gender, skin colour, height, weight, clothing, level of activity, and psychology. Lai et al. [11] emphasize the significance of subjective human parameters in their research. For instance, age may be a crucial factor because the physiological and psychological characteristics of different age groups may affect outdoor thermal perception and comfort levels. However, Givoni [12] states that clothing and level of activity have the largest impacts on thermal comfort.

The level of activity determines the metabolic activity and internal heat production in the body. The level of activity is often expressed in MET-units corresponding to sedentary activity (Figure 2).

Metabolic heat (M) increases with activity metabolism (M_{act}) due to increased oxygen demands of the muscles (Table 1). M also differs depending on gender, age and fitness level [13].

Table 1. List of activities and corresponding metabolic rates (M_{act}) and activity speeds (va) used for energy prediction (Male -Female) [13]

Activity	M_{act} (W m^{-2})	v_a (m s^{-1})
Running		
Unconditioned	320 (262)	1.5 (1.4)
Conditioned	436 (326)	2.1 (1.7)
Elite	756 (582)	3.5 (3.0)
Cycling		
Unconditioned	320 (262)	4.5 (4.0)
Conditioned	436 (326)	5.1 (4.5)
Brisk Walking		
Average	221 (221)	1.3 (1.3)

Figure 2. Activity levels and their MET

Clothing affects heat and moisture transfer between the skin and the environment; therefore affecting the thermal regulation of the people wearing it. The intensity of this effect depends on the clothing's characteristics, such as colour, fabric, and the number of clothing layers.

Nielsen measured the heat load imposed by the sun's radiation on subjects cycling outdoors for 60 minutes wearing black (albedo=0.28) or white clothing (0.45). They found that black clothing resulted in a 3–4°C higher T_{sk}, a 10bpm increase in HR, and 100 g h^{-1} increase in sweat rate [13].

2.2. A Brief History of Thermal Comfort Studies

The attention paid to the thermal conditions of open spaces in urban areas is relatively new compared with the previous wealth of attention paid to indoor spaces.

The first attempt to represent the human body from a thermokinetic point of view was made by Lefevre in 1911. He considered the body as a sphere with a core producing heat and a shell losing heat to the environment [14]. Following Lefevre's study, many thermal models have been developed.

Nikolopoulou et al., [8] report that Gold in the 1930s, and Siple and Passel in the Antarctic the 1940s, the creators of the wind-chill index, also tackled such issues. In the same period, Büttner postulated, "if one wants to assess the influence of climate on the human organism in the widest sense, it is necessary to evaluate the effects not only of a single parameter, but of all thermal components. This leads us to the necessity of modelling the human heat balance" [15]. In 1963, an architect and city planner, Victor Olgyay, demonstrated in his book *Design with Climate – Bioclimatic Approach to Architectural Regionalism* how an analytical approach to climate management can merge into a harmonious and aesthetically sound design concept.

In 1971 Gehl studied the influence of microclimates on outdoor activities by counting people sitting on sunny and shady benches. He showed that local sunny or shady conditions significantly impacted the desire of people to either stay or leave [16]. Penwarden attempted a more systematic approach for thermal conditions outdoors in 1973 by adding a term for solar radiation to the steady state model [8].

In his book Landsberg [17] focused on the physical relations that create climatic differences in urbanized areas, and presented an assessment of urban energy fluxes, the urban heat island, and the urban wind field on both a synoptic and local scale. In the 1980s, a team of researchers at Berkeley worked on thermal comfort outdoors, particularly on the implications for design solutions for the microclimate of San Francisco, which led to legislation for solar access and wind protection. Again, a mathematical model of the thermoregulatory system was utilized to calculate comfort conditions [8].

One of the important information sources for designers was written by Brown and Gillespie in 1995 and focused on ways to create comfortable outdoor spaces in a wide variety of climatic conditions. Through the use of formulas, they determined human thermal comfort, estimated solar radiation absorbed by a person, and estimated wind in a given landscape.

Another designer named Emmanuel [18], in his book *An Urban Approach to Climate Sensitive Design,* focused on thermal comfort in the tropical urban outdoors. Emmanuel developed the concept of the shadow umbrella for radiation reduction outdoors during the day.

The final book, entitled *Urban Microclimate: Designing the Spaces between Buildings,* was written by Evyatar et al. [19]. They provided an exploration of the energy balance of a pedestrian in the street and how it differs from a person in an indoor environment. This book is devoted to practical issues, such as building density, street orientation, and the extent of landscaped areas with respect to their effect on pedestrian thermal comfort.

When the literature is chronologically evaluated, is becomes clear that thermal comfort research is explored in a variety of disciplines, by professionals such as climatologists, engineers, and urban designers. These were the first studies that examined the physical and physiological aspects and components of thermal comfort. Today, this knowledge has been translated into an instrument and guide for urban design.

2.3. Assessment of Thermal Comfort

As mentioned above, human thermal comfort is difficult to measure because it is highly subjective, and the conditions outdoors are very dynamic and complex. Assessments of outdoor microclimates requires the use of models and indices which combine meteorological elements with personal parameters. There are different methodologies used to measure thermal comfort because of its complex nature and its various principal discussions.

Fanger defined three parameters necessary for a person to be in thermal comfort [20]:

- The body is in heat balance,
- Sweat rate is within comfort limits
- Mean skin temperature is within comfort limits.

To measure thermal comfort, these three parameters are converted into measurable values by taking meteorological, personal, and environmental factors into account. Since the perception of these values differs according to personal factors such as gender, age, psychology, and activity, new models have been developed which take additional mechanisms and factors into account. But selecting the most appropriate assessment methodology (index, model etc.), and knowing the limitations and accuracy of each assessment is the main task in thermal comfort research.

There are different types of assessment methodologies and approaches in thermal comfort studies literature. One of the main differences in assessment methodologies is the use of steady-state models or dynamic models, which are based on behaviorpeople's behavior in the outdoors.

Comfort models range from simple, one-dimensional, steady-state simulations (e.g., Fanger model, Gagge two-node model) to complex, transient, finite element codes with thousands of nodes [13].

In this study, thermal comfort assessment approaches are grouped into four categories:

- **Empirical thermal comfort indices:** Only a few meteorological parameters are considered. Though they are useful and easy to calculate, they have many limitations. They don't assess thermal physiology, and have usually been developed for specific climate zones.

Discomfort index, wind chill index, and humidex are well-known and the most widely used empirical indices.

- **Energy balance equation-based models:** Since Lefevre first represented the human body from a thermokinetic point of view, a number of other entire-body thermal models have been developed [14]. After the 1960s, heat-balance models of the human body have gained more acceptance in the assessment of thermal comfort. The basis for all of these models is the heat balance equation for the human body [15].

The first models to assess thermal comfort in outdoor areas were developed mainly from indoor and steady-state condition models. Due to the complex structure of outdoor areas and the variety of activities conducted there, more complicated models have been developed (Table 2).

Energy balance equation-based models can be classified into three groups: one-node models, two-node models, and multi-node and multi-element models[3]. In one-node models, the human body is represented by a single node (PVM index and COMFA+). Two-node models divide the entire human body into two concentric shells, an outer skin and a central core. Gagge's model, developed in 1971, and MEMI are the best-known two-node models. Multi-node models divide the human body into more than two concentric shells, as in the UTCI, AUB Bio Heat Model, and Fiala-model.

- **Psychological studies:** Nikolopoulou et al. [8]'s study is one of the first thermal comfort studies to address people's subjective perception of meteorological parameters. Their study proved the large discrepancy between the actual thermal comfort sensation people experience, described as the Actual Sensation Vote, and the theoretically predicted thermal comfort condition described as PMV [16]. Their study showed the interrelationships between outdoor thermal comfort and psychological adaptation.

Table 2. Main thermal comfort indices and energy balance equation-based models proposed by thermal environment literature, in chronological order

Cold Environments	Moderate Environments	Hot Environments
IS, Sharlau Index[6] (1950)	IS[6], Sharlau Index (1950)	P4SR[21], Predicted 4 h sweat rate (1947)
DI, Discomfort Index[7] (1959)	DI[7], Discomfort Index (1959)	IS[6], Sharlau Index (1950)
T_{eq}[8], Equivalent temperature Index (1984)	HD[14, 15], Humidex (1965, 1979)	HSI[22], Heat Stress Index (1955)
IREQ[9, 10], required insulation (1993, 2007)	ET*[16] (1971)	DI[7], Discomfort Index (1959)
WCI, wind chill index[9] (1993)	T_{eq}[8], Equivalent temperature Index (1984)	RHS[23], Relative Heat Strain (1964)
THI[11], temperature humidity index (1994)	SSI[17], New Summer Simmer Index (1987)	ITS[24], Index of thermal Stress (1976)
PT[12], Perceived Temperature (1997)	RSI[18], Relative Strain Index (1992)	HD[14, 15], Humidex (1965,1979)
PET[13], Physiological Equivalent Temperature (1999)	THI[11], temperature humidity index (1994)	HI[25], Heat index (1979)
T_{wc}[10], wind chill temperature (2007)	**PMV[2, 19, 20], predicted mean vote (1994, 2005)**	T_{eq}[8], Equivalent temperature Index (1984)
UTCI[30] (2009)	PT[12], Perceived Temperature (1997)	SSI[17], New Summer Simmer Index (1987)
	PET[13], Physiological Equivalent Temperature (1999)	**SW_{req}[26], Required sweat rate (1989)**
	UTCI[30] (2009)	RSI[18], Relative Strain Index (1992)
		WBGT[27], Wet bulb globe temperature (1989)
		THI[11], temperature humidity index (1994)
		PT[12], Perceived Temperature (1997)
		PET[13], Physiological Equivalent Temperature (1999)
		PHS[28], Predicted heat strain(*) (2004)
		STI[29] (2005)
		UTCI[30] (2009)

In bold letters international standardised indices. (*) In place of SW_{req} index.

- **Comprehensive models:** A comprehensive model has been suggested by Chen and Ng [16] under the concept of behavioral assessment. In this concept, since the perception of outdoor thermal comfort consists of at least four levels—physical, physiological, psychological, and social/behavioral (Figure 3)—all these levels should be taken into account.

Vanos focused on bioclimatic design in her model. The idea behind her model was the interdisciplinary nature of bioclimatic design (Figure 4). Thermal comfort modelling with this methodology comprises of going beyond simple methodologies, perspectives, and concepts in order to corroborate ideas and collect data from multiple disciplines, advancing our fundamental understanding [13].

Figure 3. Behavioral assessment model [16]

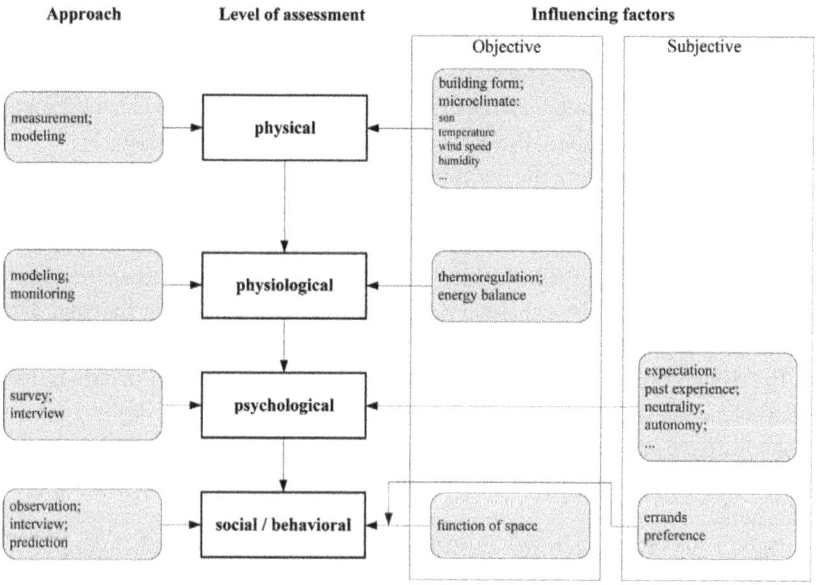

Figure 4. Interdisciplinary framework for advancement of thermal comfort research [13].

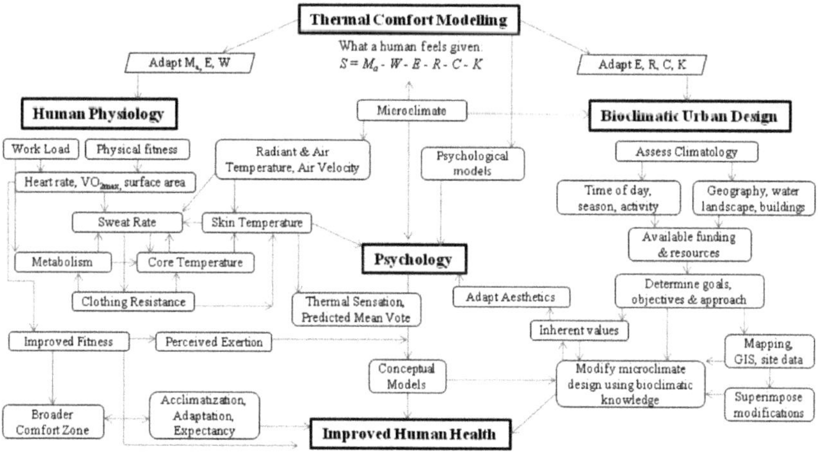

4. The Importance of Thermal Comfort Studies in Urban Design

Information about the thermal properties of the outdoors revealed by comfort studies is very important in terms of quality of life and human health. Therefore, the transformation of this information into usable qualities for disciplines engaged in spatial studies has been discussed frequently in recent years. Problems in the urban environment, such as the effects of climate change, air pollution, and urban heat islands, are the main reason for this growing debate.

For urban areas, where the outdoor thermal environment is shaped by stakeholders' design decisions in the urban design process (such as architects, urban planners and landscape architects), an analysis of the urban built environment is of great importance when making design decisions. At this stage, measurements, surveys, models, and interviews are used to provide the thermal data which is to be used in outdoor planning and design.

The design process of outdoor thermal comfort studies often occurs in two different stages: the pre-design stage, and after application of the design because problems have arisen. In both stages, the maps and simulations that make the spatial analysis and presentation of the data possible are essential for urban designers.

There have been many site-specific studies in recent years for different climate regions which use and produce spatial thermal comfort data using different models and indices. In this chapter of the book, three of them will be reviewed.

The first one is Katzschner's methodology for thermal comfort mapping, a methodology for mapping comfort conditions needed for open space planning and design using measurements, calculations, observations, and questionnaires.

Katzschner derived an easy-to-use and easy-to-understand tool/methodology which takes into consideration three main influential issues; morphology of the site (geometry of buildings), surfaces (ground / façades), vegetation, meteorological parameters, and time parameters [22] (Figure 5).

Figure 5. Thermal comfort mapping process [22]

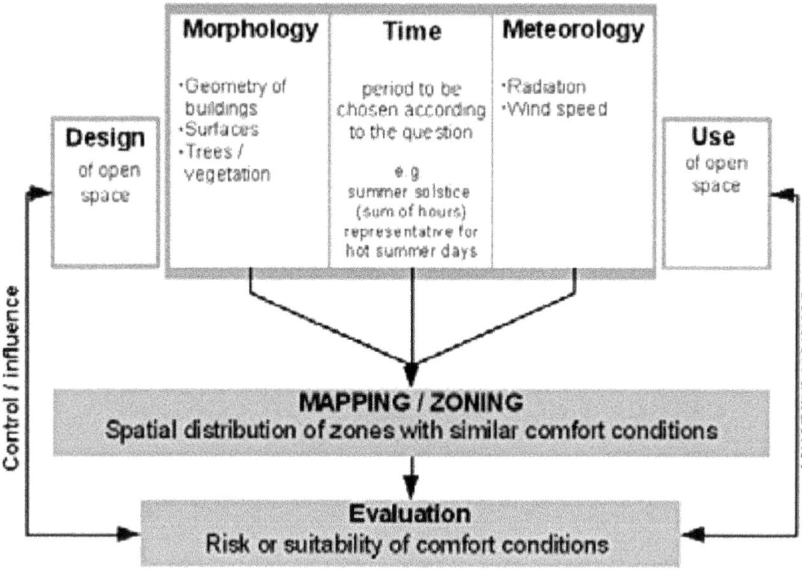

PET, which was developed by Höppe, has been used to calculate the thermal index in Katzschner's methodology. In the end, design proposals are made based on land use patterns, observations, questionnaires, and microclimatic mapping results.

The second tool is the numerical microclimate model ENVI-met. ENVI-met is a three-dimensional non-hydrostatic model for the simulation of surface-plant-air interactions in urban environments designed for microscale simulations with a typical horizontal resolution from 0.5–10 m and a typical time frame of 24–48 h. (Figure 6). This resolution allows for a fine reading of the microclimatic changes which is sensitive to urban geometry and relevant for comfort issues and the analysis of small-scale interactions between individual buildings, surfaces, and plants [23]. Vegetation is handled not only as a porous obstacle to wind and solar radiation, but also as a contributor to the physiological processes of evapotranspiration and photosynthesis. Various types of vegetation with specific properties can be used to great effect [24].

Figure 6. ENVI-met model

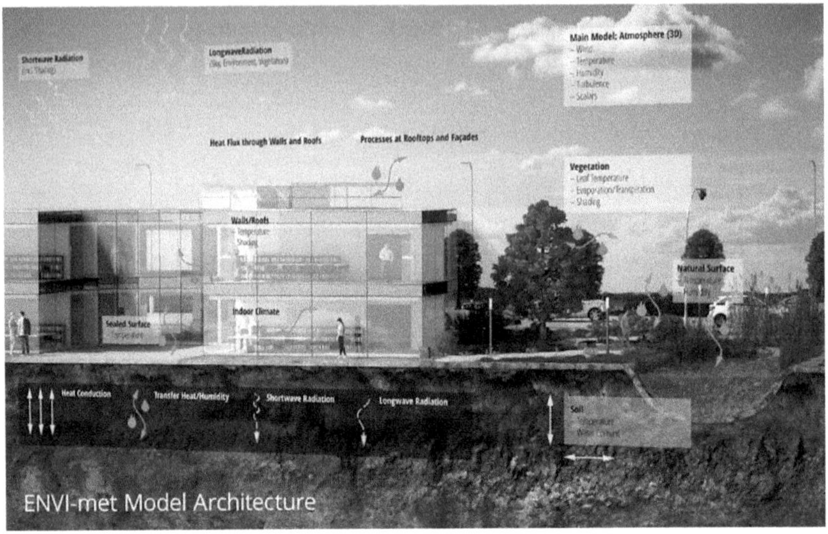

Through the use of ENVI-met, the three-dimensional microclimate model Ali-Toudert and Mayer [26] used PET patterns to map improved urban design aspects for shading, such as trees and overhangs, which caused marked improvements in climate quality on an urban street.

4. Results

Issues of human comfort and quantitative indices generate valuable information for urban designers, helping them to increase the well-being of an urban population by planning a suitable and healthy environment [27]. But the integration of the thermal comfort dimension in the design process is lacking because of poor interdisciplinary work and a missing common language between urban climatology and urban design. Improving comprehensive models with simulation tools could help to accomplish this integration.

References

[1] J. A. LEECH, W. C. NELSON, R. T. BURNETT, S. AARON, M. E. RAIZENNE: *It's about time: a comparison of Canadian and American time-activity patterns. Journal of Exposure Science and Environmental Epidemiology*, 12(6), 427 (2002)

[2] S. BRASCHE, W. BISCHOF: *Daily time spent indoors in German homes-baseline data for the assessment of indoor exposure of German occupants. International journal of hygiene and environmental health*, 208(4), 247–253(2005)
[3] G. KATAVOUTAS, H.A. FLOCAS, A. MATZARAKIS: *Dynamic modeling of human thermal comfort after the transition from an indoor to an outdoor hot environment. International journal of biometeorology*, 59(2), 205–216 (2015)
[4] P. HÖPPE: Different aspects of assessing indoor and outdoor thermal comfort. Energy and buildings, 34(6), 661–665 (2002)
[5] O. VICTOR: Design with climate. Climate Responsive Building, 1963.
[6] American Society of Heating, Refrigerating and Air-Conditioning Engineers (ASHRAE), A Comprehensive Glossary of Terms for Built Environment, https://www.ashrae.org/resources--publications/free-resources/ashrae-terminology, (Accessed 17/04/2017)
[7] Task Committee on Outdoor Human Comfort (TCOHC), ASCE Outdoor Human Comfort and Its Assessment http://ascelibrary.org/doi/abs/10.1061/40700%282004%2949 (2005) (Accessed 12/05/2017)
[8] M. NIKOLOPOULOU, N. BAKER, K. STEEMERS: *Thermal comfort in outdoor urban spaces: understanding the human parameter. Solar energy*, 70(3), 227–235 (2001)
[9] R.D. BROWN, T.J. GILLESPIE: Microclimatic Landscape Design: Creating Thermal Comfort and Energy Efficiency, Wiley, (1995)
[10] M. TOPAY, B. YILMAZ: Biyoklimatik Konfora Sahip Alanların Belirlenmesinde CBS'den Yararlanma Olanakları: Muğla İli Örneği, 3.CBS Bilişim Günleri, Fatih Üniversitesi, İstanbul, 425–434 (2004)
[11] D. LAI, D. GUO, Y. HOU, C. LIN, Q. CHEN: *Studies of outdoor thermal comfort in northern China. Building and Environment*, 77, 110–118 (2014)
[12] B. GIVONI: Climate considerations in building and urban design. John Wiley & Sons, (1998)
[13] J.K. VANOS, J.S. WARLAND, T.J. GILLESPIE, N.A. KENNY: *Review of the physiology of human thermal comfort while exercising in urban landscapes and implications for bioclimatic design. International journal of biometeorology*, 54(4), 319–334 (2010)
[14] L. YI, L. FENGZHI, L. YINGXI, L. ZHONGXUAN: *An integrated model for simulating interactive thermal processes in human-clothing system. J Therm Biol* 29:567–575 (2004)
[15] P. HÖPPE: *The physiological equivalent temperature–a universal index for the biometeorological assessment of the thermal environment. International journal of Biometeorology*, 43.2: 71–75 (1999)

[16] L. CHEN, E. NG: Outdoor thermal comfort and outdoor activities: A review of research in the past decade. Cities, 29(2), 118–125 (2012)

[17] H.E. LANDSBERG, The urban climate. Academic press, (1981)

[18] R. EMMANUEL: An Urban Approach To Climate Sensitive Design, ISBN: 9780415334099 (2005)

[19] E. ERELL, D. PEARLMUTTER, T.J. WILLIAMSON: Urban Microclimate: Designing the Spaces Between Buildings, Routledge, 2011

[20] Y. EPSTEIN, D.S. MORAN: *Thermal comfort and the heat stress indices. Industrial health*, 44.3: 388–398 (2006)

[21] G. SCUDO: Thermal comfort in green spaces. Proceeding COST C, 2002, 11.

[22] L. KATZSCHNER: Open Space Design Strategies Based on Thermal Comfort Analysis, Proceedings of PLEA, vol.1, NL Technische Universitat Eindhoven (2004)

[23] ENVI_met. A holistic microclimate model, ENVI-met Model Architecture, http://www.model.envi-met.com/hg2e/doku.php?id= intro:modelconept

[24] G. LATINI, R.C. GRIFONI, S. TASCINI: Thermal comfort and microclimates in open spaces. In Proceedings Building XI Conference, Vol. 79, 5–9 (2010)

[25] M. BRUSE: Simulating human thermal comfort and resulting usage patterns of urban open spaces with a Multi-Agent System, in: Wittkopf, St. and Tan, B. K. (eds.): Proceedings of the 24th International Conference on Passive and Low Energy Architecture PLEA, 699–706 (2007)

[26] F. ALI-TOUDERT, H. MAYER: *Numerical study on the effects of aspect ratio and orientation of an urban street canyon on outdoor thermal comfort in hot and dry climate. Building and Environment*, 41, 94–108 (2006)

[27] Á. GULYÁS, A. MATZARAKIS, J. UNGER: Comparison of the urban-rural comfort sensation in a city with warm continental climate, 7. BIOMET Conference, Freiburg,Germany.http://www.urbanclimate.net/matzarakis/papers/BIOMET7_Gulyas_Matzarakis_Unger_473_479.pdf (2010)

Timur Kaprol[1]

Postafter Republic Housing in the Edirne City Center

[1]*Asist. Prof. Dr.Namık Kemal Üniversity Faculty of Fine Arts, Design and Architecture, Department of Architecture*

1. Introduction

In the history of our post-Republic cities, we can read the evolution of the changing Turkish family structure by looking at the plans of those buildings which were built as products of the economic, cultural and social structuring of the eras they were built in. Buildings that reflect the social acclaim, economic opportunities and cultural formations of the eras they were built in are the architectural documents of their eras.

Three historical elements of the architecture of the Edirne City, **'Necdet Koman', 'Recep Zıpkınkurt', and 'Mustafa Birik' Buildings'** are traces of the socio-cultural and socio-economic development of the renewal and modernization of the Turkish Republic. In this article, the socio-cultural and socio-economic values and traces of socio-cultural facts in architecture relating to these three houses which were built in Balıkesir mid-way through 20th Century were researched, and common architectural features of these buildings were studied with reference to information and data about these houses collected using 'Structure Analysis', a sociological method used in studying the architectural identities of houses.

1. Application of the Structure Analysis Method in the Study

As there has been a mutual and interconnecting relation between social structure/life and sociological institutions, with new formations that have occurred in social structures providing opportunities for emergence and development of new theories, these theories on the other hand have contributed to the explanation and restructuring of social structure (Kızılçelik, 2000). Within this motion, the plans of the houses in which people spend significant parts of their lives can, in particular, be considered as data in architecture. Vital necessities of the times define a distinctive spatial pattern in domestic architecture as a result of their interaction with the social structure.

In sociology, *"basic research searches whether certain theories are correct or wrong and aims to develop new theories with reference to the acceptance that basic research has to be "theoretically" valid"* (Çelebi, 2001). While scientists approach the subject with the purpose of "knowing for knowing's sake" in basic research, they approach the subject with the purpose of "knowing to benefit" in applied (needs) research (Arslantürk, 2001). However, the issue where these two research strands combine is in attaining knowledge of socio-cultural reality and as a matter of fact, these two research strands feed each other in social structure analysis. This is a basic-applied research strand in sociology which might be helpful in architecture if we refer to the idea that space setups have their own identities. This article aims to add depth to the subject through analysing of the impacts of social facts relating to these eras on the approaches to building design, by examining three of the houses of the era.

In "Bir Analiz Modeli Denemesi" (An essay on the Analysis Model) by B. Dikeçligil, features of the structures in terms of their types were determined before establishing a model and thereby these structures were interpreted sociologically. In the study, which was about basic forms of research, the application of quantitative and qualitative methods was discussed. Five features of the study are presented below. Although these five features aim to review the behavior of society on a macro scale, they also help to explain the social facts that are determinant in the building's planning, particularly when they are applied to building analysis (Dikeçligil,1997).

Pieces or elements that have no common features whatsoever don't build a structure.

The areas of a building and the parts that form it affect the formation of all parts. In the definition of space, these elements provide help both in a functional sense and in the architectural design of the buildings in three dimensions, and such elements appear before us with their identities both in city walls and in the city silhouette.

Structure comprises the (relatively) steady relations between a part and the whole.

In terms of inner spaces, it is observed that there is a steady relationship between spaces belonging to the house and the areas shared by the houses. In order to develop an analysis model in accordance with the features mentioned, a basic acceptance of the harmonizing approach has been determined as below (Dikeçligil,1997).

First Basic Acceptance: Between the individual and society there is a mutual interaction rather than a one-sided relationship in which one determines the other. As for houses, they have macroscaled interactions with the city they are in, the space they occupy and the social life they integrate into the city, while they have microscaled interactions with the city and the spatial array formed by the needs of the users as well as in terms of the third dimension setup.

Second Basic Acceptance: Socio-cultural facts have interconnected, intertwined three-dimensional structures.

Changes in society, (women entering business life, for example), have had a particular effect on the operating characteristics the houses assumed in previous centuries and as a result of lives in which products manufactured abroad were consumed more than domestic products, and in which lives were spent more outside than inside houses, meaning that spatial designs required modification for the time spent in houses, and, given the intertwined life setup, it became essential that these spaces be identified. Wet areas started to be designed within the houses; spaces that were situatedin cupboards inthe rooms, as bathing places started to find their place inside houses as parents' bathrooms. Cellars, and storerooms which were previously placed near kitchens, started to appear within house spaces. On the basis of this development, it can be said that people formed their new houses within the context of customs and traditions which came from the past.

Third Basic Acceptance: principle of impartibility of social and cultural things. A sense of privacy, which is particularly prevailing in Turkish society, is a social and cultural return in building planning. Guest room solutions with entrances from the outside are the indicators of social and cultural impartibility in the building plan.

Fourth Basic Acceptance: In a sociological analysis, "existing/happening" and "should exist/should happen" should be separated. Achieving the contemporary living conditions that should be achieved in the time the house is built is ***what should happen***, while what ***happens*** is that the house is designed within the context of socio-cultural data, on the basis of customs and traditions.

Fifth Basic Acceptance: In order to obtain scientific knowledge, the world of socio-cultural facts should be understood and explained. While the structuring of the Turkish Republic continued right up to mid-way through the 20th century, both economic and cultural lives were formed. All these formations have affected life and brought new setups of space.

3. Urban Development of Edirne in the Post-Republic Era

In the same way that Thrace was on a prominent route which connected the Balkans to Istanbul before the Republic, so was Edirne on the same route during the same era. Buildings which were constructed during the Republic era were situated in Edirne, such as Kaleiçi which was formed in the pre-Republic era. The buildings which are analyzed in this study, **'Necdet Koman'**, **'Recep Zıpkınkurt'** and **'Mustafa Birlik'**, are residential buildings to be found in Kaleiçi.

4. Reflections of The Part/Whole Relation of Social Life in the Residence

Similar characteristics to the features of structures, which bear the same characteristics, are the patterns of interactions that form them. Judging from the discourse that this pattern involves steady relations between the part and the whole, the lifestyle in Turkish society has changed as a result of the transformation from the patriarchal family to the nuclear family, and the land -oriented agricultural society has given way to an industrial society **[8]**. The transformation of the social life and a reduction in family size provided the opportunity to live together **[9]**. Since a grouping of a multitude of various houses brought about communal life, both public domains and private areas have emerged in the buildings.

Figure 1. Positions of the analyzed buildings in Edirne

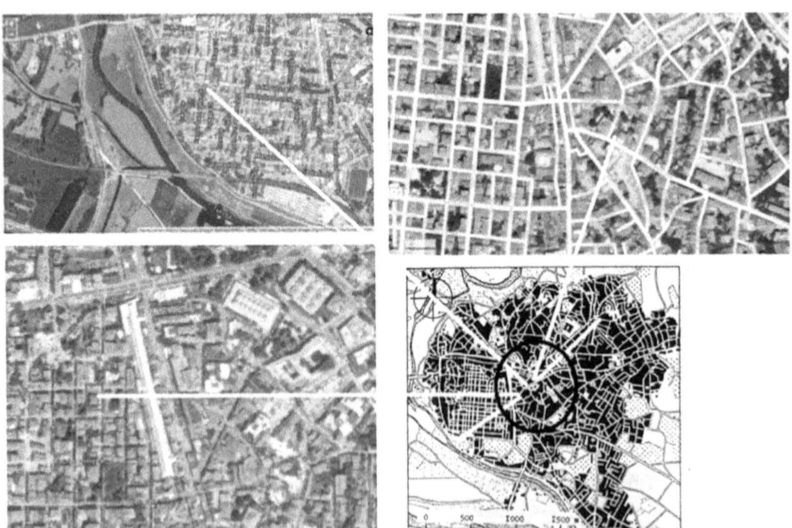

4.1. Areas within the Buildings

With the transformed social structure during the Turkish Republic, residence architecture required solutions for wet areas such as the kitchen or toilet, which were located outside the building in traditional architecture. The spatial organization of the place in which the function is set, and the relationship between the spaces, have affected the spatial organization of the residence. In order for the collected data to be analyzed in the new structure, inner space units that possess certain features were analyzed separately.

Entrance hall: If we consider the sofa, which was seen in traditional architecture, with the rapid and new structuring that started after the Turkish Republic, we see that the inner sofa became narrower and transformed into a corridor while the middle sofa also narrowed, but became a hall that enabled the passage between spaces **[10]**. The entrance hall in buildings in Mustafa Birlik and Recep Zıpkınkurt is placed in the middle and is connected to the living spaces with a glass door. The entrance hall, whose size is determined according to the building it is placed in, serves for additional functions as well as a travelling area. The functions of halls have been to provide a separate area for going to other rooms and eating. In this context, the traditional sofa has been used in the contemporary residence via modernization. It can also be seen that the entrance hall in Necdet Koman building was designed as a connection area as well as being a continuing part of the entrance to the eating area.

Room connections: One of the three buildings analyzed uses doors for passing from one room to another. This door provides for privacy between spaces. While the door is a construction unit that increases the functionality, it provides to a passage to move from one room to another in the Necdet Koman building, and connects the entrance hall to the room in the other two buildings.

Cellar: The cellar, which is situated in the post-Republic buildings and serves as a storage unit, is not found in the analyzed buildings. This space, whose origins are in traditional architecture, can be seen in the residence designs made by the important architect of the era, Sedad Hakkı, who modernizes the traditional residence.

Wet Areas; as a part of a transformed life, the kitchen and toilet are located inside the house. The toilet, which is located in the garden in the traditional residence, is given a space inside the house, at the same level with the rooms and in relation with them. The bathing function, which was located in built-in closets in traditional houses, is taken to the bathroom in the new residence. In addition to serving for bathing, these bathrooms also provided for washing clothes.

Pan closets are used in the toilet, and they are situated in the bathroom. Mosaic pavements can be seen on the floor of wet areas, corridors and entrance halls. Counters with extensive potential usage are placed in kitchens. Fume hoods are located above ovens. Due to the usage of oil stoves for cooking, cast counters have been widely used.

Figure 2. Storey plans of the analyzed buildings

'Necdet Koman' House **'Recep Zıpkınkurt' House** **Mustafa Birik' House**

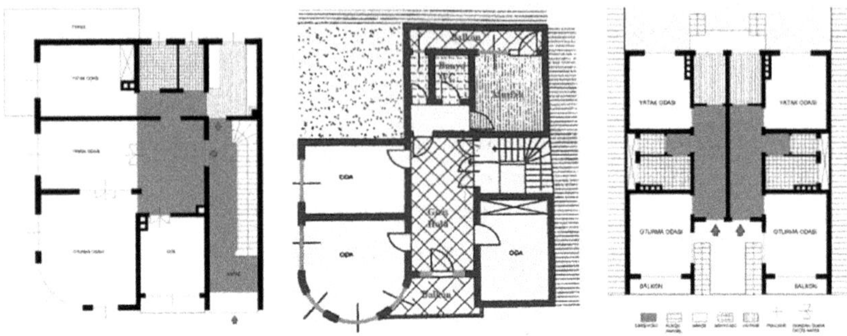

4.2. Features of Façades

Regarding the façades of buildings, coves and consoles that provide dynamism to the façades can be observed. While coves can be seen in the Mustafa Birlik building, projections are seen in the Necdet Koman house and Recep Zıpkınkurt residence. The coves and console façade formations are seen as open projections in the Necdet Koman house, open-and-closed consoles in the Recep Zıpkınkurt residence, and as entrance niches on the façade in the Mustafa Birlik building. Projections show two features on the façade: firstly a balcony which is an open console that allows the room to have a projection on the façade, and secondly as balconies to the rooms which open out through the façade. The open console is situated in front of the entrance of the hall in the Recep Zıpkınkurt residence while it is situated in front of the living spaces in the Necdet Koman house, allowing façades of the living spaces to be projected to the outside. Unlike these two buildings, there are two open consoles on either sides of the stairs at the entrance in Mustafa Birlik building. This residence building has four apartments on the first floor and upstairs there are are twin houses.

Figure 3. Façades of the analyzed buildings (Necdet Koman', 'Recep Zıpkınkurt', 'Mustafa Birik' buildings

Necdet Koman residence building: The corner of the building, which was built on a corner parcel, is in a circular form. This form can also be identified in the upstairs open projection. Next to the entrance door which is structured as a niche, there are French windows on the façade of the room which opens to the entrance hall and part of the ground floor apartment. There are corniche window jambs in the form of vhamfered plaster ending around the windows. Forged bands embroidered as circular forms are used for balcony railings. **Recep Zıpkınkurt residence building:** the building, attached to another on one side, is situated on the corner parcel. The supporting units under the closed-and-open console on the lateral façade of the building are references to the buttresses in traditional architecture. A modernist approach can be seen in the façade formation of the building. The influence of the Bauhaus style from the era is visible on the façade. **Mustafa Birlik residence building:** The era which was known for the application of Bauhaus and International Styles and employed the idea of "two eyes side by side, the skyline continuing on a horizontal line, and the outer view entering the inner space without interruption" introduced window openings with horizontal proportions to the architecture. The open spaces on either sides of the entrance axis emphasize the entrance axis on the upper floors. The "full+empty+full" application which almost resembles a loggia, is limited with the living spaces situated in the entrance of the building.

4.3. Façade Units of the Period

4.3.1.Consoles: two approaches can be seen regarding the open and closed consoles. **Type 1 – Open Consoles:** these are generally situated in front of the living space and the façade of the space is opened to the outside. Thus, the space assumes another dimension with the usage of the balcony. **Type 2 – Closed-and-open Consoles:** this is consisted of collaboration of both kinds (open consoles and closed consoles). While the open consoles can be seen as located on either side of the closed consoles, there are also examples of the open console only on one side.

Figure 4. Scheme regarding the assembly of the closed and open consoles

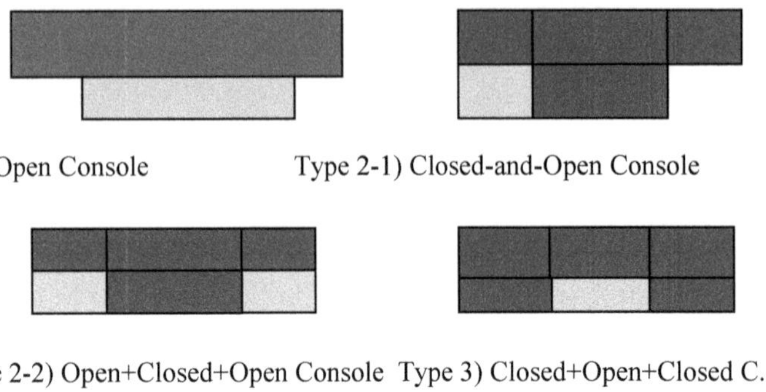

4.3.2 Interaction between the Basement and the Ground Floor on the Façade: The basement in some of the analyzed buildings is constructed as an example of Sokl. If the storey which bears the features of Sokl is constructed by retracting from the parcel line, a garden floor is situated which serves as a residence on the precipitated platform. An architectural solution was introduced to this storey that bears Sokl features by using semi-storey features that were used in historical Edirne houses. The vestibule in front of the entrance door and the organization of stairs are observed to be taken as references in the new architecture

Figure 5. Visual regarding the window structuring of the Sokl featured storey in Necdet Koman house

4.3.3. Doors and Transoms: Doors were structured in metal. Simplicity was the main feature. However, doors with ornamental patterns can also be seen. Among the analyzed buildings, in the Necdet Koman house, doors are designed by retracting from the surface into a niche. They are situated with transoms and door niches are framed. The transom is separated from the door with iron features and is a single piece in the door niche. Single-sash windows can also be encountered on either side of the entrance door in the building mentioned. In the door leaves which are made of iron, chilled cast iron is used, such as when heating and cutting the iron without forging, bending, welding and clinching. Transparency is significant for these doors, so the glass parts can be opened for ventilation. There are decorative iron frames around the glass panes.

Figure 6. Visuals of metal joinery on doors

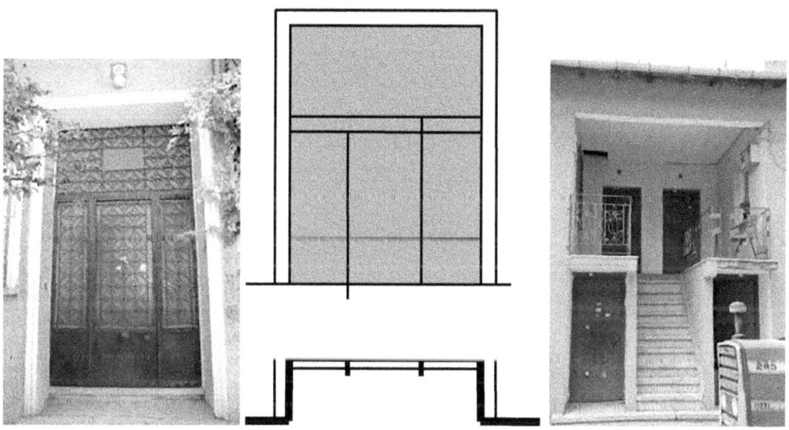

4.3.4. Windows: The idea of "two eyes side by side, the skyline continuing on a horizontal line, and the outer view entering the inner space without interruption' which is dominant in Bauhaus and International Styles has introduced horizontal proportions to the acchitecture of the the window openings.. The examples in the countries where these styles were developed introduced the solution of using metal joinery for the widening of the window openings. In order to avoid a dense appearance because of the joinery incisions, iron frames were used. Horizontally proportioned windows can be seen in Edirne and our country in this period. Wood materials were preferred to iron for window joinery due to their being cheaper. Concrete lintel and joist limit the window opening which allows windows to expand horizontally. Improvements in the carrying system have led to a variety of window kinds and window openings. Regarding the positioning of the façade, the cornet window shows features as can be seen in Figure 7.

Figure 7. Horizontal windows in Necdet Koman house

4.3.5. Balcony Railings: A variety of formations in railings have been observed during the study. Upon examining the details of the railings, geometrical forms as well as botanical patterns can be observed. A variety of forms have been applied by using iron profiles that are products of forged iron works. Permeability is of importance in this kind which can be seen in many of the surviving examples in the residences b analyzed.

Figure 7. Balcony railings in the residences analyzed

4.3.6. Coating and Window Frames: Edelputz (Edel coating), which is a kind of rugged exterior coating, is used on all façades. The coating of the era is generally colorless, but its colored usage on buildings often features light green and pink tones. The smooth and plain façade which is the characteristic of Bauhaus and International Styles is seen on buildings using the mentioned coating.

For window frames; a frame which starts on the windowsill and gradually continues by covering the lintel can be observed.

Figure 8. Example of the frame used around windows.

5. Conclusion

The buildings of the era analyzed in this study are documents of the social structuring and the **cultural continuity** of the era they were built in. The tendency to create a singular space in the middle spaces functioning as central or the inner sofa can be seen during the planning of the relationships of these building with the outside spaces. Window openings with horizontal proportions, chilled iron entrance doors, balcony railings, and handrails provide a perspective which combines past and present as reflections of the social acclaim of the era. The distance between what 'should have been' at a cognitive level which can open up new horizons to designers and the socio-cultural 'happened' level at which these functions are realized could only be explained by a structural analysis model which was developed with a holistic approach.

Referenses

[1] Kızılçelik,S.; (2000), **Sosyoloji Yazıları**, Anı Yayıncılık, ISBN 975695633X, Ankara,

[2] Çelebi,N.; (2001) **Sosyoloji ve Metodoloji Yazıları,** Anı Yayıncılık, ISBN 9756956542 Ankara

[3] Arslantürk. Z.; (2001) **Araştırma Metot ve Teknikleri,** İstanbul :Çamlıca Yayınları, İstanbul,

[4] [5] [8] Dikeçligil B "Bir Analiz Modeli Denemesi 'Sosyal Yapı' ve 'Toplumsal Yapı", **Yeni Türkiye**, cilt no 3, Sayı15, 447–666. 1997

[6] Durkheim E **Sosyolojik Yöntemin Kuralları,** Dost Kitabevi, ISBN 9789752984127, Ankara 2010

[7] Yüksel **'Konut Mekanı Kavramanın Tipolojik Temelleri'** İTÜ Mim.Fak. Baskı Atölyesi, İstanbul 1995

[9] Stewing, **''Bursa, Nordwestanatolien''** IM Selmverlag des Geographischen Inlstaius der Universitat Kiel, 1986

[10] Eruzun, C **''Kültürel Süreklilik İçinde Türk Evi''** Mimarlık Dergisi IV. İstanbul 1985

[11] Eldem, S.H. **''Plan Tiplerine Göre Karakteristik Örnekler''** Türk evi Osmanlı Dönemi I: 35, Türkiye Anıt ve Çevre Değerlerini Koruma Vakfı Yayınları, İstanbul 1984

[12] Kaprol, T., Meral A **''Necdet Koman Evi''** Türkiye Mimarlığında Modernizmin Yerel Açılımları X.Bildiri Özetleri Kitabı, Atatürk Üniversitesi Yayınları no:1064, syf:80 Erzurum 2015

[13] Kaprol, T., Meral A **''Recep Zıpkınkurt Evi''** Türkiye Mimarlığında Modernizmin Yerel Açılımları X.Bildiri Özetleri Kitabı, Atatürk Üniversitesi Yayınları no:1064, syf:83 Erzurum 2015

[14] Kaprol, T., Meral A **''Mustafa Birik- Orhan Birik Evi''** Türkiye Mimarlığında Modernizmin Yerel Açılımları X.Bildiri Özetleri Kitabı, Atatürk Üniversitesi Yayınları no:1064, syf:82 Erzurum 2015

Aylin Salici[1]

The use of Sustainability in Landscape Design

[1]*Mustafa Kemal University, Faculty of Architecture, Department of Landscape Architecture, 31040 Antakya, Hatay, Turkey*

Sustainable development, sustainable improvement, sustainable energy, sustainable agriculture, sustainable fashion, sustainable production, sustainable consumption, sustainable life.

The concept of "sustainability", which is a widely used concept of today, is a concept that was on the agenda for the first time in 1972 at the United Nations Environment Conference in Stockholm. This concept was introduced internationally in the report *Our Common Future* (Brundtland) in 1987. Sustainability can be defined as providing today's needs without compromising the ability of future generations to meet their needs.

Hübler Hermann (2001) pointed out that these rules existed in Europe in the twelfth century, and according to these rules, forests could only be cut back by as many trees as they grow. The starting point of the concept of sustainability, which came to the fore in the context of the Environmental Development Conference held in Rio in 1992, focuses on protecting the ecosystem and avoiding environmental problems that arise in parallel with economic and technological developments. (Korkut et al., 2017).

Sustainability is a concept that connects usage and makes today's usage responsible for that of the future. Once more, this concept reminds us that the world, which is inherited from our ancestors, is not entirely ours, is only entrusted for a period of time, and will be left to future generations.

The concept of sustainability, dealing with landscapes, (urban) planning and design, is a rather new phenomenon. At the 1996 Habitat II Urban Summit declaration, it was stated that it is possible to make people living in urban areas healthy, safe and happy, with a high quality of life, by establishing settlements where a sustainable life is possible. Hence, the tendency towards sustainability in today's cities has become progressively more important.

The concept of sustainability has been associated with the discipline of landscape architecture and has been maintained as one of the main subjects of the discipline due to the changing world order and the increase of environmental pollution, the depletion of raw material resources, the inability to renew nature itself and so forth.

Landscape architecture is the design of outdoor public areas, landmarks, and structures to achieve environmental, social-behavioral or aesthetic outcomes. It involves the systematic investigation of existing social, ecological, and soil conditions and processes in the landscape, and the design of interventions that will produce the desired outcome. The scope of the profession includes landscape design; site planning; stormwater management; environmental restoration; parks and recreation planning; visual resource management; green infrastructure planning and provision, and private estate and residence landscape master planning and design, all at varying scales of design, planning and management. A practitioner of the profession of landscape architecture is called a landscape architect.

From this perspective, it is understood that all the work related to environmental protection, which is inevitable in sustainable cities, such as ecosystem and resources analysis and management, rural and urban spaces planning, the coordination of environmental impact assessment works, recreational areas, urban open spaces, pedestrian zones, highways, industrial and agricultural areas planning, designing and land use decision are among the tasks of landscape architects (Yaşar and Düzgüneş, 2013).

Landscape design, which is a study area of landscape architecture on a subscale, is a process in which outdoor spaces are shaped in the direction of planning decisions. Landscape design takes place in an environment directly in outdoor spaces. Therefore, natural conditions have a direct impact on landscape design work (Dong, 2009).

Landscape design, trying to find some space under the pressure of intense urbanism when it comes to daily life, can be described as a tired, deteriorated, sometimes impossible effort to find the life of nature. At this stage, an attempt to achieve more protection and the survival of existing natural values rather than trying to reconstruct something that is gone is a current effort in landscape design (Erdoğan and Oktay, 2010). From this point of view, it is understood that landscape design is based on sustainability and design principles in revealing the best possible spatial composition relating to an area.

Sustainable Landscape Design

The concept of sustainable landscape design can be defined as the art of using natural, structural, cultural and scientific data to create outdoor spaces in line with resource conservation and development principles, within the framework of aesthetic rules. (Seçkin et al., 2011).

This concept is not only a space design process consisting of technical, architectural, social or cost constraints, but also an understanding that aims to produce solutions that will strengthen the importance given to the long-term perspective (Ender and Bağcı, 2010). Accordingly, the concept aims to develop a sustainable system that can be a part of a self-sufficient urban ecosystem. This system is based on an understanding that takes nature as a model, brings systematic solutions to the structural and ecological characteristics of the area and natural processes (Korkut et. al., 2017), respects natural resources and embraces cultural and historical differences (Ender and Bağcı, 2010).

Targets of Sustainable Landscape Design

Most studies focus on the following targets to maintain sustainable landscape design:

Adaptation of existing topography

It is common practice to obtain flat areas in building construction. For this purpose, more excavation or filling is done and can cause very costly results. Most compatible solutions and techniques must be applied to topography for sustainability. Landscape design should be shaped with and by slopes and their characteristics.

Minimizing existing landscape intervention

The second important point among targets of sustainable landscape design is minimizing effects on the environment during construction. If not, the land quality of large construction sites deteriorates after excavation. It is important to recapture it in terms of sustainability. Building a protected zone around the construction sites especially helps in the form of natural corridors and protection of natural plants. In addition, constricting floor areas in architectural spaces and increasing building height will help in creating open wide spaces between buildings, so it will help to maintain the existing landscape.

Effective use of natural resources

Preservation of natural land leads to increased aesthetic qualities. It also helps to form suitable floors for inter-building uses between buildings. It is easy to create solutions for recreational and sportive activities in open green areas. Preservation therefore provides spaces for human uses and natural activities such as surface water flow, wildlife movement, and ensuring the continuity of the natural habitat. In this way natural vegetation, water flow, flora, fauna, etc. will be preserved and open green areas will be increased.

Provision of excavation load balancing

Using natural forms of land in landscape design is understanding the provision of minimum excavation-fill and land- forming balance. This understanding aims at the least destruction of nature and thereby protects it (Thompson and Sorvig, 2007).

In excavation work the following solutions should be avoided;

Damaging existing flora

Extreme excavation

Drainage pits in flat areas around buildings

Erosion because of slope areas

Land solutions such as walls and stairs that require expensive landscaping

In addition to this, fertile soil during excavation should be moved into green areas within the scope of sustainability.

Use of natural and local resources and materials

Using local natural materials in the landscape design applications which include firm ground and 3D elements causes a minimization of the damage to the environment and energy consumption, especially during phases of production, transportation and application.

Besides the color and texture effects of the plants in sustainable landscape design, it is important to use natural or exotic plants, water consumption, regional climate compliance and drought tolerance in plantation.

Protection of existing plants

Protection of existing plants in the landscape area makes the design more effective and reduces the cost of implementation. Existing plants, especially old varieties and trees which have more ecological importance, have shown certain improvements and will not require additional maintenance. Furthermore, storage and removal of excess water in surface drainage will be protected by existing plants. In addition, the identification of existing plants is also important in terms of the detection of plants that will subsequently be planted on the outside and the elimination of plants that do not fit in the center (Jim, 2003).

Use of natural plants

Use of natural plant material in urban landscape design increases the success of the plant, provides integrity among urban and rural areas. and especially provides an economical approach for arid cities. Besides, it helps to recover fragmented or destroyed ecosystems and the harmony between natural and urban areas will be realized. Thus, urban areas where natural species are used will be able to ensure the sustainability of natural ecosystems and serve for the rehabilitation of degraded parts of the ecosystem. Using natural plants adds richness to plant combinations by removing the exotic species which are very well known and frequently used, and have a uniform appearance. Natural plants have a positive economic contribution for the city because of low costs due to over-maintenance and a minimum workforce (Bilgili et. al., 2014).

Reducing grasslands

It is important to reduce grassland areas for sustainability because of the high requirement of irrigation. For this reason, grass or ground cover plants should be used instead of grass species. This causes a reduction in irrigation which is very important not just for sustainability but also for the economy. For instance, using ground cover plants such as "*Sedum* sp." and "*Mesembryanthemum cordifolium*" will reduce water consumption because of low irrigation requirements.

Water and energy saving

The use of water in landscape design can be discussed in terms of sustainability under 3 headings;

- *Reducing watersurface in landscape design*

In addition to causing excessive use of water resources, stagnant water surfaces need extensive maintenance to prevent the accumulation of algae. For this reason, reducing the scale of such areas which were created for aesthetic purposes in design is important for water and energy consumption.

- *Use of plants that require less irrigation*

Another important feature is classifying plant groups according to their water consumption and preparing an irrigation program which relates to these classifications. For this point of view, water-loving plants should be planted in the south, while plants that do not need irrigation should be planted on the north side of the park area.

– *Use of water and energy saving systems*

It is important for green areas to have irrigation with a system that stores rainwater underground after filtering. The system provides water and energy saving. This kind of irrigation is the best solution for regions with seasonal drought. Water on pathways in landscape areas or other spaces with appropriate slopes should be collected; after that it could be used for irrigation (Hilaire, Rolston St, et al, 2008). It is known that sink and bathroom waters can be used for irrigation in green areas after being filtered. This kind of building is known as a "green building" (Gary et. al., 2009).

Energy efficient landscape design

Cities are the areas where energy consumption and resource use are at their most intense. Especially in urban areas with high population density, there is a high potential for energy consumption. Components such as interaction between buildings and outdoor weather conditions, solar radiation, temperature, precipitation regime, and wind should be considered during the design process (Karaca and Varol, 2012)

Sustainable landscape design applications have the potential to significantly reduce resource utilization either inside or outside energy consumption.

With landscape design, it is possible to save energy easily especially in buildings, to increase the aesthetic value of the building and its surroundings, and to confidently provide desirable areas. If the selected plants are placed with a particular strategy around a building, it can be protected from the harsh winter winds and the burning summer sun. In this way, both inside and outside, the comfort of organisms will be developed, and the amount of energy consumed for heating and cooling will be also reduced (Dikmen, 2011).

Deciduous trees and prehensiles can be used for shading in the indoor and outdoor living areas in the summer, while they can allow sunlight to pass through the twigs in winter. The use of low-bush shrubs and ground cover plants around the building can prevent heat and sunlight reflections on buildings. Frequently, textured trees and shrubs can be used as barriers to strong winds and to create a corridor to take advantage of light summer breezes (Motloch, 2001).

With the landscape design of the area, the temperature can be changed inside and around the building. Shading with roof, walls and windows made of wood can change the indoor temperature by 6–12 °C in summer. However, with proper plant selection, the entry of the winter sun into the windows provides a significant reduction in the energy used for heating. The temperature outside the building can be reduced by 6°C by the use of ground cover plants or lawns instead of floor coverings. These temperature changes both improve "bio-comfort" and reduce the amount of energy used for heating and cooling. Birişçi et al. (2012) indicated that the temperature of a concrete surface exposed to an entire solar ray is 42 °C,the surface of the grass is 36 °C and the measurements made in the same area in the shade show that the temperature of the concrete surface decreased to 38 °C and the temperature of the surface of the grass decreased to 34 °C.

In terms of energy savings, the designer has to make proper arrangements such as providing access for the winter sun to a house and protecting the smooth surface of this house from the impact of the summer sun. Ideally, to ensure that the interior is cool and the outside is comfortable, or comfortable for human use, the house should be protected from all effects of the summer sun on its south-facing side. It is acceptable that winter rays enter through the house windows and expose the house's south-facing side to their rays. Thus, when the temperature falls below the human comfort level, heat is stored so as to be useful to the thermal mass (Seçkin, 2007).

Additionally, energy-saving outdoor lighting elements and proper planning will reduce energy consumption during everyday life processes. At the same time, light pollution will be prevented and the lighting pressure on the flora and fauna will be reduced. Standards for illumination projects in all recreational and walking areas in urban parks are not considered. Generally, it is common to use abundant lighting on pathways and large projectors in large areas such as squares and sport areas. In addition, park equipment, architectural items, orientation and information sheets are illuminated by colorful and effective LED lighting. This light pollution in parks causes too much energy consumption and also creates adverse effects in terms of aesthetics.

Therefore:

- Unnecessary areas shouldn't be lit, but minimum power sources use should provide night vision and prevent reflection from illuminated surfaces.
- Solar energy systems should be used.
- Some selected varieties of plant instead of all plants in the park should be lit, emphasizing these plants .
- To help strengthen the silhouette effect, some plants should be in darkness and some plants in light.

- Lighting elements should be placed at a certain distance from trees. This is an important point because these elements will form light spots on the trees if they are close to or on the body of the tree.
- When flower beds are illuminated, short-term flowers should light up conveniently, and this lighting should shift in time, so that attention is given to the size of the plants and the flower beds.
- The items around artificial surfaces such as lakes and ponds should be lit to show their reflection on the water surface, thus also producing the light of their reflection, and proving economical for lighting in this way. All these items are important means of taking precautions to maintain sustainability in landscape areas (Alpay et al., 2013).

References

Motloch, J.L. (2001) Introduction to Landscape Design. John Wiley& Sons, Inc., USA.

Jim, C. Y. (2003) Protection of urban trees from trenching damage in compact city environments, *Cities*, 20 (2), 87–94.

Seçkin, N.P. (2007) Solar and Wind Control in Ecological Design. Ecological Architecture and Planning National Symposium Proceeding Book, 89–94. 27–28 April 2007, Antalya.

Thompson, J. W., and Sorvig, K. (2007) Sustainable landscape construction: a guide to green building outdoors. Island Press.

Hilaire, R. S., Arnold, M. A., Wilkerson, D. C., Devitt, D. A., Hurd, B. H., Lesikar, B. J. and Pittenger, D. R. (2008) Efficient water use in residential urban landscapes. HortScience, 43(7), 2081–2092.

Dong, D. Z. 2009. Study on Sustainable Landscape Design of Abandoned Quarries An Example: Zhushan Ecological Park in Xuzhou. Procedia Earth and Planetary Science, (1):1107–1113.

Gary L. Wade, Midcap J T, Coder K D. Landry G, Tyson AW, Weatherly N Jr (2009) A Guide to Developing a Water-Wise Landscape. University of Georgia Environmental Landscape Design Department, Georgia 30602, 44.

Ender E. and Bağcı, Ö., (2010) The Effectiveness of Landscape Architecture on Urban Design. Union of Chambers of Turkish Engineers and Architects 4 th. Landscape Architectures Symposium Proceeding Book. 21–24 October 2010. 321–331 Selçuk-İzmir.

Erdoğan, R., Oktay, H.E., (2010) Investigating Natural and Cultural Terms at Landscape Architecture by The Design Context. Union of Chambers of Turkish

Engineers and Architects 4 th. Landscape Architectures Symposium Proceeding Book, 21–24 October 2010, 393–400.

Dikmen, Ç.B., (2011) Sample Study of Energy Efficient Building Design Criteria. *Journal of Polytechnic*, 14 (2), 121–134.

Seçkin, N.P., Y.Ç., Seçkin, Ö.B. Seçkin, (2011) Sustainable Landscape Design and Application Principles. Literature Publishing.

Birişçi, T., Güney, M.A., Türel, H.S., Kılıçaslan, Ç., (2012) Planting Design. Universities Offset, Bornova.

Karaca, M. and Varol, Ç. (2012) Energy Efficiency in Residential Areas in Mass Housing; Chairmanship of Mass Housing Administration (Housing Development Administration of Turkey) An Assessment of the Mass Housing Project. METU Publications. Doi: 10.430; 127–14, Ankara.

Alpay, C.O., Kalaycı, A., Birişçi, T., (2013) City Park Improvement Model According to Ecological Design Criteria: Example of İzmir Kültürpark. Union of Chambers of Turkish Engineers and Architects İzmir 2 nd. Urban Symposium: Take Ownership of the City, Nov, 2013, İzmir.

Yalçın Yaşar, Y.,. Düzgüneş, E. (2013) Integration of Sustainability Concept to Landscape Design: A Studio Work. İnönü University Journal of Art and Design, 3 (7), 31–43. ISSN: 1309–9876.

Bilgili, B.C., Öner, N., Aytaş, İ., (2014) Determination of Natural Tree Species Used For the Plant Landscape Design in the Parks of Çankiri Province. 3rd International Non-wood Forest Products Symposium Proceedings Book 786–795. 8–10 May 2014, Karamanmaraş-Turkey.

Korkut, A., Kiper, T., Topal, T.Ü., (2017) Ecological Approaches in Urban Landscape Design. Journal of Artium, 5 (1), 14–26.

Nurgül Arisoy

Energy Efficient Landscape Planning and Design Approach for Sustainability

1. Introduction

The rapid growth of cities with the Industrial Revolution, which has been perceived as a source of environmental problems in the nineteeth century, continues to increase in the twenty first century. Rapid urbanization has increased the number of motor vehicles, population density, noise, poor quality of life and a deterioration of environmental quality. One of the issues which is overemphasized on the literature of urbanization and is at the center of sustainable urbanization is that of providing energy efficiency.

Energy efficiency refers to different policies, technologies and strategies that aim to solve issues related to energy use, whether for residential, commercial, industrial and national purposes, so as to minimize the emission of greenhouse gases which cause global warming, whilst also reducing financial costs [1].

An approach to energy efficiency in landscape planning and the design process is emerging as a necessity when we consider the environmental problems caused by today's conditions and thescarcity and finite nature of resources.

Energy efficiency demands a series of decisions, starting from the design process and continuing throughout the application process, to be made in line with a holistic planning and design concept.

The emergence of "sustainable development" as a popular concept has revived a discussion about the form of cities and has given a major stimulus to the question of the contribution that certain urban forms might make to lower energy consumption and lower pollution levels [2,3,4]. In addition, sustainability can be described as raising the quality of life of communities without exceeding the carrying capacity of global ecosystems.

Energy efficiency has been demonstrated through providing sustainability in landscape architecture, planning and design in a period in which the phenomenon of sustainability has become even more important at all levels of life.

This chapter identifies the intertwined principles for achieving energy-efficient planning and design.

2. Sustanability and Energy Efficient Planning and Design

The concept of sustainability is the basis for an energy-efficient planning and design approach as well as for all environmentally-conscious approaches. Sustainability from the point of view of designers and planners can be defined as upgrading the communities' quality of life in the repair and sustainable use of existing structured urban landscapes, reducing the energy consumption starting from the housing environment scale and not exceeding the carrying capacities of ecosystems, for the possibility of living in a healthy, climatically comfortable environment [5,6].

There are great benefits of energy efficiency, such as microclimate regulation, reducing water consumption, reusing treated wastewater, using environmentally friendly and economic material, controlling noise and air pollution, reducing $CO2$ emissions, increasing quality of life, reducting and preventing losses in the ecosystem [7].

An energy-efficient landscape planning and design approach requires detailed data collection regarding the analysis and evaluation of direct natural and climatic data and energy use. In this context, a number of principles should be taken into account in ensuring sustainability. The main planning and design principles that affect sustainability in energy efficiency in terms of landscape architecture are as follows:

1. Effective use of climatic and microclimatic data: The external character of a geographical location is prescribed by its climate, which refers to the long-term weather patterns over a specified time frame for a specific area.

The climate consists of elements such as sunshine, temperature, humidity, wind, and precipitation, and is a natural factor that has a great impact on the ecological, social and economic dynamics which have the greatest effect on integration, natural processes and socio-cultural processes, and is of topical concern to planners and designers.

It is possible to reduce energy consumption through landscape design and planning which ensure correct and conscious energy conservation. For this purpose, regional, local and microclimatic properties need to be taken into consideration [8].

It is said that "understanding the microclimate may provide the necessary tools for creating a comfortable thermal habitat for people and energy efficient landscapes for buildings" [9].

We cannot control the weather, but with careful planning and design we can moderate the effects of the weather. Vegetation creates a microclimatic space where the humidity is controlled by transferring the water to the atmosphere, and reducing the evaporation of the water in the ground, holding and filtering the sunlight, and cutting and controlling the flow of wind.

Trees are associated significantly with improved thermal comfort and relief from heat stress, particularly during hot seasons and times of day.

The first step in any energy-efficient design is to make observations about where the sun comes from over the course of the day, to block the sunlight according to the climate zone, to allow the light to enter, or both to block and allow it. Then the design can be made [10].

The wind can be considered as an element of the microclimate which will be guided by landscape planning and design processes in the passive ventilation and cooling of cities [11]. An energy-conserving landscape slows or prevents air leakage in a building by reducing wind velocity. Firstly, a prevailing wind direction should be determined for wind control, and the position of the current vegetation and neighboring structures should be taken into consideration.

Urban settings experience higher temperatures than rural areas due to a lack of vegetation, properties of urban materials that have a greater thermal storage capacity, the geometry of urban areas, a release of waste heat (e.g. from vehicles and buildings) and a city's size [12]. This is known as the urban heat island (UHI) effect [13].

Furthermore, water surfaces and the characteristics of materials (artificial materials) are also effective climate regulators and serve in energy-efficient landscape design and planning. Water surface absorbs a large amount of heat, converts this type of heat to the form of water vapor, and reduces the temperature.

2. Effective use of topographic data: One of the materials to be considered when analyzing the land to be planned and designed is the topography that has an effect on its microclimate, water drainage patterns, and soil properties.

Microclimate, temperature, noise and wind patterns can be controlled and directed through design by coordinating features of the surface created by changing the topography and architectural components, vegetation, structures and solar collectors [8].

The topography of the location is important because of the effect of the angle of incidence of solar radiation, slopes, and the orientation of the land in terms of the use of daylight and natural ventilation, and solar radiation. Therefore, the slope of the land, the amount of incoming solar radiation and the latitude are very important parameters.

Topographic properties have an indirect effect on the control of humidity. The amount of rainfall over the planting elements is filtered to form the areas that provide humidity sources, primarily by accumulation in the pit areas on the ground surface. Therefore, in hot weather conditions, the air on these areas creates a cooler environment than the air on dry areas [14].

The locations of the hemisphere, slope, and aspect are important design parameters. The aspect shows how the area is affected by the sun and how it influences the field conditions according to the amount of sun it receives. A slightly inclined slope receives the sunlight at a convenient angle for warmer summers, and the steep slopes are the best locations for winter.

3. Effective use of natural resources: It should not be forgotten that natural resources are not unlimited. The existence and conscious use of natural resources will be possible if there is a healthy environment.

Water: It should be known that landscapes that conserve energy also conserve water at the same time.

In a sustainable landscape approach, wastewater is considered as a valuable resource. Irrigation requirements can be reduced or even eliminated by proper design and plant selection.

It is also possible to use water for irrigation purposes, such as rainwater collecting units and surface water flows that disappear, are mixed with underground water sources, or are evaporated from concrete asphalt.

The techniques used in this sense are also called "Low Impact Development – LID", and have been recommended as an alternative to traditional stormwater design.

Another type of recycled water is "gray water", which is reusable wastewater from residential, commercial and industrial bathroom sinks, bath tub/shower drains, and drains from clothes washing equipment.

To catch and store water in the ground in order to bring it above the ground level is a technique that decreases green water levels in the aquatic ecosystem and also soil humidity [15].

In addition, we can consider tactics to slow the movement of water and retain water on site where plants can absorb it.

There are many systems for catching, directing, and purifying rainwater [16]:

- Downspout disconnection
- Rainwater harvesting
- Rain gardens
- Planter boxes
- Bioswales
- Permeable pavements
- Green streets and alleys

- Green parking
- Green roofs
- Urban tree canopy
- Land conservation

One of the ways to reduce unnecessary water and fertilizer consumption in planting design is to group plants according to their needs [17].

An energy-efficient method is to plant groundcover and small shrubs instead of turf grass for water and maintenance if an area is not to be used actively.

4. **Efficient use of waste:** Energy can also be conserved by composting plant waste. Excess plant material can be recycled with the nutrients it contains going back into the landscape by converting it to compost or mulch. The best-known materials for preparing organic mulch are tree bark chips, wood shavings, pine needles, grass clippings, shells of hazelnuts or walnuts, leaves, small-sized gravel and finely chopped pruning waste [18].

Mulching helps energy efficiency by minimizing evaporation, reducing weed growth, moderating soil temperature, preventing erosion, maintaining surface permeability, increasing soil biological activity, and protecting soil health [10, 17].

Returning these organic materials to the land perpetuates the natural biological cycle, improving the health of soil and plants. Healthy plants are better able to tolerate pests and environmental extremes, reducing the need for irrigation, fertilizer, or pest control. Ecologically, this is a sensible means of reusing organic wastes.

Creating and using compost on-site not only reduces the energy needed for transporting the material to a landfill, but also reduces the energy used in producing and transporting synthetic fertilizers and pesticides. Adding compost to a soil also increases permeability, water-holding capacity, and the stormwater retention capacity, thereby reducing the need for irrigation [19].

The use of manual and solar powered tools to minimize or eliminate the fossil-fuel energy used for maintenence ought to be considered.

This can be used in the production of alternative energy by the systems for the collection of solid waste city-wide.

5. Effective use of renewable energy sources

Renewable energy sources (sun, wind, biomass, biogas, geothermal energy, hydro, wood, ocean thermal and flow, wave, sea flows) are the energy resources that are accepted as inexhaustible. It is possible to benefit from renewable energy resources by passive and active methods. Solar and wind energy are the most effective renewable energy sources that can be used for lighting in the landscape area.

Biomass is the most common form of renewable energy; it is a term for all organic material that stems from plants (including algae, trees and crops) [20].

Suitable landscape areas are determined for energy forestry and energy crops. Both woody and herbaceous plant species have specific growing conditions, based on soil type, soil moisture, nutrient balances, and sunlight, which determine their suitability and productive growth rates for specific geographic locations [21].

Biogas can be produced from industrial waste containing organic matter as a result of urban activities, garbage from human waste, and sewage waters.

6. Effective use of Materials: The materials used in energy efficiency are divided into two: planting (natural) and structural (artificial). The ecological approach requires the use of local resources as much as possible in the planting and structural design of the area. Thus, while on the one hand durable materials that can easily adapt to the conditions are used, on the other hand low-cost applications that minimize natural resource consumption in the installation and maintenance of these areas can be included [22].

Plants have a tremendous potential for energy-efficient landscape design and planning. Well-designed landscaping not only creates an aesthetic effect around the building, but also greatly reduces the building's heating-cooling costs.

Each plant group has a different function in energy-efficient landscape design.

Groundcovers prevent the reflection of heat rays and help reduce the effect of urban heat islands (UHI). Dense trees allow strong winds to be avoided and the summer breeze to reach the desired area in the appropriate directions. Trees are also significantly associated with improved thermal comfort, and relief from heat stress at the street level and neighbourhood scale, particularly during hot seasons and at specific times of day.

Studies have shown that evapotranspiration, alone or in combination with shading, can help reduce peak summer temperatures by 2–9°F (1–5°C) [23, 24]

Grass surfaces consume a lot of water, so local plant species should be chosen which can be an alternative to energy efficient landscape design. Native plants work best when the functions required for remediation are tied to the plants' effect on soil fertility, microbial populations, water consumption, or to their effectiveness as ground cover, since native plants often have a positive effect on these attributes [25].

Unlike planting materials, walls, fence walls, pavements, lighting elements, roof surfaces, pavilions, earth sets and terraces can affect the microclimate of the land by changing the wind velocity and temperature.

Climatic factors also need to be taken into consideration when selecting materials (stone, brick, wood, concrete, plastic) and color (hot and cold).

Absorbent and reflective materials: Materials such as asphalt, light colored gravel or rock that store heat in the daytime and continue to radiate heat through the night, increasing the HIE.

Impenetrable materials: Sites with high ratios of impermeable surfaces increase ambient air temperatures and require more energy for cooling due to the heat observing quality of paving materials such as asphalt, concrete, brick stone.

Pavements comprise about one third of urban surfaces. Hot pavements can raise the temperature of storm water runoff and aggravate urban heat islands by warming the local air, and contribute to global warming by radiating heat into the atmosphere [26].

Reclaimed and Recycled materials: A large amount of energy is used in manufacturing many building materials.

In the manufacture of construction material, using recycled sources instead of those sources which are not newly processed material provides a considerable preservation of raw material and also a considerable amount of energy saving.

Recycled materials lessen the need for virgin feedstock and avoid sending useful materials to the landfill; however, significant energy and other resources are often required during the recycling process [27]. Because of this, recycled materials should be considered after options to reduce or reuse materials have been fully explored.

Landscapes can support the use of recycled materials in two ways: firstly, by purchasing materials with recycled content. Typically, the higher the recycled content the better. And secondly, by purchasing materials that can be easily recycled in the local area. http://landscapeforlife.org/materials/reduce-reuse-and-recycle/

Minimally processed material: Generally, the energy content of natural materials is lower than that of artificial materials since these materials are manufactured with less energy and lower labor costs. Such kinds of materials which are can be easily provided locally are wood, bamboo, reed, straw, sunflower stalk, and mushroom.

The existing structural landscape elements (existing stones, concrete slabs, and natural stones) in the area can be used for the construction of new pedestrian paths or vehicle paths. In terms of energy efficiency, this is saved from the energy which is to be consumed and the carbon emissions that will be needed to reach the application area by using the existing material. In addition, when the existing material is disassembled to dispose of it, an unnecessary stack of garbage is formed and moving this also requires energy.

Using the existing material on the field to recycle is both energy-efficient, and protects the genius loci so that it harmonizes with the physical environment surrounding it [17].

Local material: In the total energy consumption of construction work, the amount of energy spent for transportation of the construction materials to construction sites is considerable and also affects the constructions' energy efficiency and economical cost. For this reason, if the construction materials are local and are manufactured in places near to the construction site as much as possible, energy consumption in transportation will decrease and that saving in transportation will give construction an important ecological quality.

Long-distance transportation of construction materials (walls, floors, pergolas) causes intensive energy consumption and an accompanying environmental pollution. Therefore, construction materials should be selected, if possible, using locally sourced materials.

The constructions we have designed should not be such that they force the functional capacities of natural resources.

The five most important components that must be considered in landscape lighting design are security, safety, usability, aesthetic value and energy efficiency.

Today, the concept that stands out in lighting design for landscape in the world is energy-efficient lighting. Organizations such as The International Dark Sky Association (IDA) are looking for ways to be energy-efficient while lighting an area and avoiding undue lighting in cities because of light pollution [10].

LEDs ***(Light Emitting Diodes)*** allow more savings with less energy; they are the leading source in energy-efficient lighting options.

Another reference for energy efficient landscape lighting are the LEED ***(Leadership in Energy and Environmental Design***) standards [29]. These LEED standards have three main components concerning lighting. The first insists on limiting light levels; the second aims to enhance the natural skyglow by prohibiting the use of luminaires with lumen output greater than 3500. Lastly, LEED standards concerning landscape lighting forbid light trespassing [30].

They also contribute to the design of energy-efficient lighting by reducing false and upward-directed lighting on pathways and controlling the lights with a time-sensitive and motion-sensitive automatic system.

7. *Effective land use planning*: Land use regulations and built-up environments directly impact energy needs.

Transportation: Transportation planning is closely related to land use decisions at the planning stage, and they should be tackled together to develop priority transportation facilities for pedestrians and cyclists and to offer as many public transport options such as buses and trains as possible. Green corridors should be designed in accordance with pedestrian and bicycle routes.

Energy-efficient transportation needs must be encouraged on three different levels (vehicle efficiency, travel efficiency, system efficiency) [31].

Corresponding to these three levels of energy efficiency in transport, three basic strategies exist to improve energy efficiency:

- avoiding increased transport activity and reducing the current demand for transport;
- shifting demand to more efficient modes of transport;
- improving the vehicles and fuels used.

Land-use planning should therefore optimise the positioning of settlement and production structures to avoid traffic or to reduce travel distances [32].

Limiting the maximum parking capacity at particular sites helps to discourage the use of inefficient cars and promote the use of efficient public transport.

Settlement: The basic principles to be considered for planning energy-efficient settlements are [33]

Schools and other service buildings: must be within walking distance to reduce fossil fuel consumption.

Trade and equipment / activity areas around the neighborhood: Small trading units should be included to meet daily needs.

Recreational and social facilities: Open and impact areas that reveal a positive effect in terms of insolation should be included.

Housing: Firstly, characteristics such as housing and user type, income, and age group should be specified, dwellings should be formed as clusters, and energy efficiency should be provided.

Transportation: Alternative public transport systems should be developed.

Local energy systems: Opportunities on a local scale such as wind, solar energy and biomass should be evaluated and used in heating and cooling systems.

Water management: The infrastructure system should be arranged so as to reuse wastewater in open and green areas.

3. Conclusion

Planning and design approaches should at all times endeavor to harmonize the needs of the community with the natural layout and resources of the environment, to ensure an effective and sustainable design.

In paticular, data about climate and microclimate, topography, natural vegetation, the existence of local materials and resources, alternative energy sources and potentials in order to provide sustainable development in terms of planning and design are of capital importance for minimizing the use of energy.

In this chapter, we tried to explain the principles of energy efficient landscape planning and design. Besides this, we looked at planning and design tools to ensure energy efficiency. These tools are xeriscape, roof gardens and vertical gardens, and the planning tools are green urbanism, new urbanism, new green urbanism, landscape urbanism and smart growth.

We have demonstrated that caring about issues such as planning and design compatible with topography and climate, planting, sustainable material use, renewable resources used passively and actively, natural resource use, waste use, the use of construction elements, and eco-friendly transportation bring sucessful results in terms of energy efficiency.

As a result, it appears that the "energy efficient landscape planning and design" approach is important in terms of contributing to the sustainability of spaces, minimizing pressure on natural resources and environmental problems caused by energy production and the consumption of natural resources, and providing energy efficiency and conservation.

References

[1] Ganda F., Ngwakwe C. C. 2014. Role of energy efficiency sustainable development. Environmental Economics, Volume 5, Issue 1, 2014,

[2] U.K. Department of the Environment (DoE). 1996. Greening the city: A guide to good practice. London: Crown.

[3] Breheny, Michael, T. Gent, and D. Lock. 1993. Alternative development patterns: New settlements. London: HMSO.

[4] Jabareen, Y. 2004. A knowledge map for describing variegated and conflict domains of sustainable development. Journal of Environmental Planning and Management 47 (4): 632–42.

[5] Oktay, D. (2001). Planning housing environments for sustainability: Evaluations in Cypriot Settlements, Istanbul: YEM.

[6] Sert E. 2013. Enerji etkin kentsel peyzaj tasarımında yağmur suyu, Yüksek Lisans tezi, Peyzaj Mimarlığı Anabilim dalı, İstanbul Teknik Üniversitesi, Fen Bilimleri Enstitüsü.
[7] Seabrook, L., Mcalpine, C. A., Bowen, M. E. (2011). Restore, repair or reinvent: options for sustainable landscapes in a changing climate. Landscape and Urban Planning, Sayı 100, Sf. 407–410.
[8] Moffat, A. S., Schiler, M. ve The Staff of Green Living (1994). Energy Efficient and Environmental Landscaping. Appropriate Solutions Press, ISBN: 0–9638784-0-9.
[9] De Robert D.B., T.J. Gillespie, 1995, Microclimatic landscape design: creating thermal comfort and energy efficiency, John Wiley & Sons, Inc., New York, 17.
[10] Yeang K, 2006. Ecodesign: a manual for ecological design, Wiley academi, London.
[11] Erbaş, E. 2002. Enerji Kullanımı ve Yerleşme Alanları Planlaması, Mimar Sinan Üniversitesi Şehir ve Bölge Planlama Bölümü, Planlama 2002/2–3, İstanbul.
[12] Hajat &Kosatky, 2010 Hajat, S., & Kosatky, T. (2010). Heat-related mortality: a review and exploration of heterogeneity. Journal of epidemiology and community health, 64(9), 753–760.
[13] Zupancic T., Bulthuis M. 2015 The impact of green space on heat and air pollution in urban communities: A meta-narrative systematic review, David Suzuki Foundation.
[14] Goulding, J.R., Lewis J. O., Steemers T.C. (Ed.), Energy in architecture – the European passive solar handbook, B.T. Batsford Ltd., London, 1992.
[15] Kinkade-Levario, H. (2007). *Design for Water- Rainwater Harvesting, Stormwater Catchment and Alternate Water Reuse.* New Society Publishers, ISBN: 978–0-86571-580-6.
[16] https://www.epa.gov/green-infrastructure/what-green-infrastructure
[17] Cook, T.W., VanDerZanden, A.M., 2011. Sustainable Landscape Management Design, Construction, and Maintenance. John Wiley & Sons, Inc., Hoboken, New Jersey, ISBN 978–0-470-48093-9
[18] Barış, M. E. (2007). Kurakçıl peyzaj. Bilim ve Teknik, Sayı 478, Sf. 22–27.
[19] Chollak T., Rosenfeld P., 1998. Guidelines for Landscaping with Compost-Amended Soils, City of Redmond Public Works, Washinton.
[20] McKendry P. 2002. Energy production from biomass (part 1): overview of biomass, Bioresource Technology 83 (2002) 37–46
[21]] P. McKendry / Bioresource Technology 83 (2002) 37–46 39

[22] Cranz, G. ve Boland, M. 2003. The Ecological Park As An Emerging Type, PlacesMassachusetts, 15(3), pp 44–47, Design History Foundation, USA.
[23] . Akbari, H., D. Kurn, et al. 1997. Peak power and cooling energy savings of shade trees. Energy and Buildings 25:139–148.
[24] Huang, J., H. Akbari, and H. Taha. 1990. The Wind-Shielding and Shading Effects of Trees on Residential Heating and Cooling Requirements. ASHRAE Winter Meeting, American Society of Heating, Refrigerating and Air-Conditioning Engineers. Atlanta, Georgia.
[25] Negri, M. C., 2004. Phytoremediation Using Native Plants, Exploring the Environmental, Social and Economic Benefits Conference December 6–7, 2004
[26] Akbari H, Rose LS, Taha H. 1999. Characterizing the fabric of the urban environment: A case study of Sacramento, California. Lawrence Berkeley National Laboratory.
[27] Calkins, M., 2009. Materials for Sustainable Sites. Hoboken. NJ: John Wiley and Sons.
[28] http://landscapeforlife.org/materials/reduce-reuse-and-recycle/
[29] Şermet R., Özyavuz M. 2017. Uluslararası yeşil bina sertifika sistemlerinin değerlendirilmesi, TMD, Uluslararası Hakemli Tasarım ve Mimarlık Dergisi 4 (10) p:290–303
[30] Gates C. 2008. Energy efficient landscape lightning options for commercial and residential sites, A senior project, University of California.
[31] GIZ (ed.) 2012: Urban transport and energy efficiency module 5h, sustainable transport: a sourcebook for policy-makers in developing cities.
[32] Ercoşkun, Ö. 2007. Sürdürülebilir Kent İçin Ekolojik-Teknolojik (Eko-Tek) Tasarım Ankara- Güdül Örneği, Gazi Üniversitesi Fen Bilimleri Enstitüsü Doktora Tezi, Ankara.
[33] Tuğaç, Ç. 2003. Enerji Etkin Planlama Yaklaşımının Kentlerin Ekolojik Planlaması ve Tasarımı Süreçlerine Etkisi – Güney Ankara Ahiboz Koridoru Örneği, Gazi Üniversitesi Fen Bilimleri Enstitüsü, Yüksek Lisans Tezi, Ankara.

Ahmet Benliay[1], Orhun Soydan[1]

Evaluation of Sustainable Lighting Systems in the Context of Landscape Design

[1]*Akdeniz University, Faculty of Architecture, Department of Landscape Architecture, Antalya, Turkey*

1. Introduction

Human survival depends upon adapting ourselves and our landscapes – cities, buildings, roadways, rivers, fields, forests – in new, life-sustaining ways, shaping places that are functional, meaningful, and artful. We shape contexts that acknowledge connections to air, earth, water, life, and each other [1].

Landscapes are dynamic and change over time due to a combination of human and natural processes [2]. Human interaction can alter this process and landscapes change faster than is natural. This can cause a dramatic effect on environment and ecology. This is mainly characterized by the rapid growth of the human population and its associated activities, the depletion of both non-renewable and renewable resources, and extensive and intensive damage caused to ecosystems and biodiversity [3]. The growing awareness of local and global environmental decline gave rise to an appreciation of ecology and its implementation in design and planning works, especially in terms of sustainability [4].

Ecological studies focus on protecting and enhancing natural processes, such as succession and water flow; biodiversity, including rare species; fish and wildlife populations, and landscape elements, such as wetlands and stream/riparian corridors [4].

Urban environments such as cities which can support significant biodiversity in the form of both resident and migratory species can be called urban ecosystems. This may be surprising if ecosystems are considered to be strictly self-maintaining, homeostatic, and essentially closed entities [5,6].

Today the way of understanding nature, the environment and landscape has changed to form a more dynamic, complex and integrated view which explains the relations between human systems and natural systems. This leads landscape designers to review their design approaches in order to respond to current environmental trends. This implies a shift in design methodology in which ecology forms a fundamental link between creativity and science. Now more effort has been given to understand and interpret the living systems and their dynamics in landscape design process [4].

The first studies of sustainable design were based on small-scale experiments [7]. Energy efficiency, alternative building materials, conservation and recycling have been widely adopted in a piecemeal fashion. Later studies in sustainable design began to realize that the integration of all these factors can produce the best results for sustainable design. This includes the synthesis of all the various ideas and strategies of the past into complete and large-scale sustainable design theories and practices. The term sustainable design for the built environment is used interchangeably with green design. However, there is a subtle difference. The Green Design Education Initiative states that green design often implies an interest in design that protects people's health and well-being, while sustainable design also protects the global environment and the world's ecosystems for future generations [8].

These two terms, often used interchangeably, reflect the concept of creating buildings that can be constructed and operated in ways that enhance the impact of the buildings on the environment and occupants. The U.S. Green Building Council (2003) defines "green design" as design and construction practices that significantly reduce or eliminate the negative impact of buildings on the environment and occupants in five broad areas. These are: site, water, energy, environmental quality, and material and resources [8].

Opportunities for achieving this goal occur throughout the lifecycle of the environment. In other words, it is a cradle-to-grave approach that recognizes ecological consequences of the entire lifecycle of the design environment. The Minnesota Sustainable Design Guide (MSDG) (Center for Sustainable Building Research, 2002) provides instructions and strategies for each phase of the building lifecycle. The design strategies for the MSDG fall into eight environmental topics: conservation, site, water, energy, indoor environmental quality, materials, waste, and innovation. Integration of these factors produces the most sustainable results, and is high performance in terms of building a healthy environment, whether it is called green design or sustainable design [8].

Sustainable landscaping is low impact, low maintenance, low resource use and frequently low-cost landscaping that fits each particular site and climate – virtually taking care of itself. The landscape is unique among the construction elements because it involves a living and therefore perishable finished product. A significant part of the built environment is landscape, not structures. As land itself becomes scarce and ever more precious, outdoor spaces need to be designed to deliver value in as many ways as possible, i.e., increasing land values, rewarding the senses, promoting environmental quality, and enhancing mobility. Sustainable landscapes incorporate and balance the human desire for beautiful and functional landscapes with the imperative to preserve valuable resources [9].

Sustainable landscapes incorporate and balance the human desire for beautiful and functional landscapes with the imperative to preserve valuable resources. Benefits of sustainable landscaping are that it [9]:

- reduces/prevents pollution and waste disposal; reduces environmental harm;
- requires fewer inputs (e.g., fertilizers, pesticides) and conserves natural and financial resources;
- maximizes ecological function and benefits wildlife;
- looks attractive and provides seasonal interest;
- provides safer environments for our families;
- provides quieter neighborhoods (from reduced use of power equipment);
- provides water conservation that benefits the homeowner and community;
- reduces flooding and costs for stormwater management;
- provides greater opportunities to enjoy nature;
- reduces landscape maintenance labor/ gives more free time;
- reduces landscape maintenance costs;
- puts less strain on municipal waste collection and water treatment;
- provides cleaner water bodies for fishing, swimming, drinking;
- produces lower heating and cooling bills;

A sustainable landscape requires informed and thoughtful decision-making about every aspect of the development site: topography and layout, hardscape and lighting, vegetation and irrigation systems [9].

Figure 1. Pillars of Sustainable Landscape Design [10]

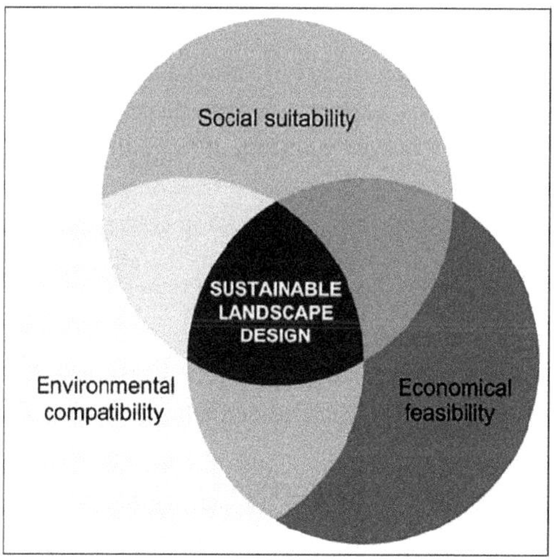

2. Outdoor Lighting Design

Lighting is the application of light to object, urban areas or their surroundings for seeing [11]. Extensive outdoor electric lighting is a recent phenomenon. Thomas Edison commercialized the electric light bulb in the late 1880s, and outdoor use was largely limited to cities until well into the 1900s. Significant outdoor lighting spread with the rural electrification programs of the 1930s and 1940s. Other sources of artificial night lighting have proliferated as well [12].

2.1. Lighting Terms

Quantifying and qualifying the nature of light within a space calls for a basic understanding of certain lighting terms. To explain these terms:

- Candela are the units of luminous intensity of a light source in a specific direction. A single ray of light has its luminous intensity value, measured in candelas. One candela directed perpendicularly to a surface one foot away generates one foot candela of light.
- Lumen is a quantitative unit of measurement referring to the total amount of light energy emitted by a light source, without regard to the direction of its distribution.
- Footcandle (fc) is a U.S. unit of measurement referring to incident light. Footcandles can be derived from lumens (1 fc = 1 lumen/sq. ft.) or candelas (fc = candelas/distance squared).
- Lux (Ix) is the International Standard (SI) measure of incident light. It is equal to one lumen uniformly distributed over an area of one square meter (10.7 Ix = 1 fc).
- Illuminance is incident light, or light striking a surface.
- Luminance is light leaving a surface, whether due to the surface's reflectancy, or because it is the surface of a light-emitting object. Luminance is the measurable form of brightness.
- Efficacy is a measure of how efficiently a lamp converts electric power (watts) into light energy (lumens) without regard to the effectiveness of its illumination.
- Light depreciation is the effective life of lamp.
- Color is the apparent color of a light source.
- Glare: A point or surface of luminance that is above one's current state of adaptation.
- Cutoff light distribution is a term used in reference to the optical design of some fixture types [13, 14, 15].

2.2. Lamp Characteristics

Selection of a lamp involves trade-offs between lamp size, optical control, efficacy, appearance, color temperature, color rendition, lamp life, costs, and maintenance [12]. The most used lamp types are

incandescent lamps, fluorescent lamps and induction lamps.

The fluorescent lamp is energy-efficient and has a long life. It has a good color rendition, whitish appearance, and superior life. Although they have good efficacy, their light output may be severely diminished by very cold weather. The incandescent lamp is the second most used lamp in the world today following fluorescent lamps. It is great for small area lighting, good color rendering, is cheap to produce, has no quantity of toxic materials to dispose of and is easily used in strobe or dimming circuits. But it has several disadvantages. It is not energy efficient (90% of the energy goes to heat, 10% makes visual light), and traditional incandescent light bulbs are not useful for lighting large areas. Induction lamps are relatively new, high-frequency sources that have no filaments or electrodes, but rather use the electromagnetic spectrum to directly energize a phosphorescent coating on the bulb. These lamps have a light quality similar to fluorescent lighting in a 90–115 mm (3 ½-4½ in) spherical lamp envelope, but with a significantly longer average rated life (up to 100 000 hours), and with almost no sensitivity to the ambient temperature [12,16].

Common light sources include low-pressure sodium, high-pressure sodium, metal halide and light emitting diodes (LEDs). Low-pressure sodium is very energy efficient, but emits only a narrow spectrum of pumpkin-colored light that some find to be undesirable. High-pressure sodium is commonly used for street lighting in many cities. Although it still emits an orange-colored light, its color ing is more "true to life" than that of low-pressure sodium. In areas where it is necessary to use white light, two common choices are metal halide and LEDs. One of the advantages of LED lighting is that it can be dimmed. Because of their reported long life and energy efficiency, LEDs are rapidly coming into widespread use, replacing the existing lighting in many cities [16]

2.3. General Principles

The objectives of outdoor lighting are: improving the legibility of critical nodes, landmarks, and circulation and activity zones in the landscape; facilitating the safe movement of pedestrians and vehicles; promoting a more secure environment; minimizing the potential for personal harm and damage to property; and helping to reveal the salient features of a site at a desired intensity of light in order to encourage nighttime use. The main focuses of general principles of outdoor lighting design are:

- Orientation: Driver and pedestrian orientation can be aided by providing a hierarchy of lighting effects that correspond to the different zones and uses of a site. Recognizable distinctions can be made between major and minor roads, paths, and usage areas by varying the distribution and brightness of the light and by varying the height, spacing, and color of lamps. Clear optical guidance can be provided with the alignment of light fixtures positioned in consistent, recognizable, and unambiguous patterns. A staggered layout of road and pathway lights tends to obscure rather than reinforce the direction of circulation and the location of the intersections.
- Identification:Intersections, decision points, crossings, bus stops, steps, arrival points, etc., should be articulated in a manner that signals their presence, shape, and nature. The illumination pattern should serve as a visual cue to what conditions may lie ahead. Spacing, height, and distribution of luminaires should avoid foliage shading, provide uniformity and vertical surface illumination. Lower mounting heights and closer spacing between fixtures may create a more uniform distribution of light, promoting the pedestrian's sense of security.
- Safety: In addition to poor lighting layout and design, glare and underlighting can create hazardous conditions for pedestrians and vehicles. Glare is more of a problem when exposed light sources, such as lamps or lenses, can be seen directly. The location and mounting height of lighting, as well as the fixture type and lamp intensity, must be carefully selected to optimize light distribution and minimize glare.
- Security: Darkness, together with unfamiliar surroundings, can incite strong feelings of insecurity. To provide a sense of security, possible hiding places and dense shadows should be minimized by the placement of appropriate light fixtures. Walkway lights should have enough peripheral distribution to illuminate the immediate surroundings. Vertical light distribution over walkway areas should cover or overlap at a height of 2.10 m so that the visual recognition of other pedestrians is maintained. For surveillance needs, lighting requirements should permit the detection of suspicious movements rather than provide for the recognition of definitive details.
- Atmosphere and Character: The clarity with which an object is perceived is influenced by its context. Exterior spaces should have a well-defined sense of background. Background spaces should be illuminated as unobtrusively as possible to meet the functional needs of safe circulation and protect people and property. Foreground spaces or objects may be major elements and should

be treated accordingly. Foreground spaces should utilize local lighting which produces maximum focus, minimum distractions, and no glare. Objects of interest and activities can be brightly illuminated while the background produces only minimal distraction. The direction of the light source is important for the perception of three-dimensional objects. The ability to perceive volumetric form is influenced by the gradient of light and shadow falling on the object. Differences in lamplight color are often used with great effect in public lighting to color-code roadways or to clearly delineate one area from another. As the general illumination level rises in a given situation, the preference usually shifts away from a warm appearance towards a cool range. Accurate color rendition will aid recognition and improve the perception of outdoor environments [13, 17].

Overall, there are standards and recommended levels of illumination for landscape area or activities considering these design principles of outdoor lighting. These are given in Table 1.

Table 1. Recommended levels of illumination [13]

Outdoor Area / Activity	**Lux (lx)**
Active use building entry or Vital locations or structures	50
Locked or infrequent use building entry	10
Building surrounds	10
Monuments on bright surroundings (Residental – Commercial)	150–500
Monuments (Dark surroundings) (Residental – Commercial)	50–200
Bikeways (Along roadside) (Residental – Commercial)	2–10
Bikeways (Distant from roadside)	5
Bulletin, posterboard, sign on bright surroundings (Light-dark)	5–1000
Bulletin, posterboard, sign on dark surroundings (Light-dark)	200–500
Express roadways (Residental – Commercial)	9–14
Major roadways (Residental – Commercial)	9–17
Collector roads (Residental – Commercial)	6–12
Local roads (Residental – Commercial)	4–9
Along roadside (Residental – Commercial)	2–10
Walkways	5
Pedestrian tunnels	20

Outdoor Area / Activity	Lux (lx)
Pedestrian overpasses	2
Pedestrian stairways (Light-dark)	200–500
General lighting	5
Path, steps away from home	10
Backgrounds, fences, walls, trees, shrubbery	20
Flower beds, rock gardens	50
Focal points (Small-Large)	100–200
Parking areas (Self – Attendant)	10–20
Recreational Basketball, Tennis, Volleyball	100
Football (Class III to Class I)	300–1000

2.4. Ecological hazards of urban light

The disruption of natural patterns of light and dark, which have been more or less reliable for millions of years, has a range of adverse consequences for wildlife across taxonomic groups and landscape types [12]. Light is a vital requirement for plants: it is necessary for photosynthesis, for information (i.e. when to bud, flower, germinate, etc.) and for their growth form. Natural light varies daily and seasonally. Artificial light sources interrupt that natural cycle — and may disrupt both the plants and the ecology they support. These manmade light sources are concentrated along road verges and hedgerows, in gardens and in urban environments, which may represent a significant and potentially hitherto overlooked threat [18].

There are three elements of light pollution: light trespass, glare and sky glow. Light trespass is light cast where it doesn't belong or isn't wanted or needed. Glare is light that is so bright that it creates a nuisance or a hazard, or inhibits our ability to see effectively. Sky glow is caused by light projected up into the night sky, either directly or by reflection, which lights up particulate matter in the air, thereby obscuring our view of the stars [19].

Artificial light sources vary in intensity, duration and spatial distribution. Plants have different behavioral responses to different wavelengths of light – with flowering, germination and photosynthesis being associated with exposure to different parts of the visible light spectrum. Photoreceptors in plants use light to sense information about the season and even the time of day, therefore controlling germination, growth and shade avoidance. Studies have shown that agricultural crops such as soya (Glycine max) and maize (Zea mays) are affected by their proximity to artificial light, which leads to changes in their development, for example growing rapidly but failing to flower. The wild plant species Lotus pedunculatus produced 10% to 25% fewer flower heads under simulated street lighting, which in turn led to reduced numbers of the aphid Acyrthosiphon pisum [12].

Figure 2. Bortle scale – light pollution visualition [20]

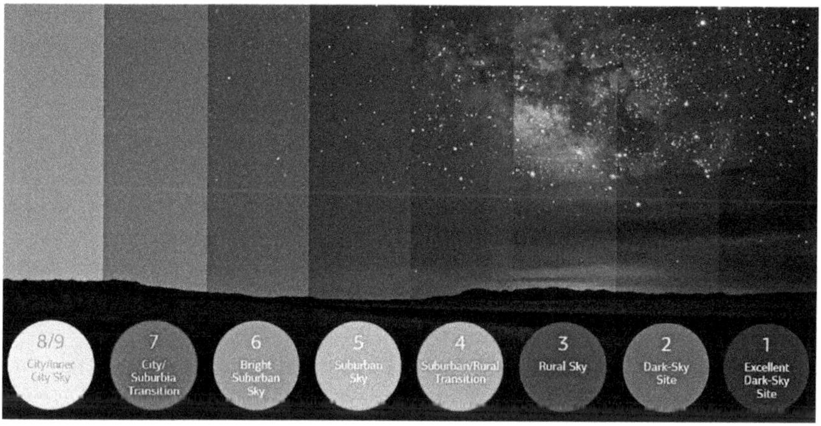

The effects that the light creates vary according to the regions, but the effects are the same. When examined according to the region;

- Coastal dunes, beaches, shorelines, deserts and scrublands are open habitats with few barriers to light transmission. Many nocturnal species prefer low illumination levels and have good visual performance under the faint light of the darkest nights. The movement of species into and out of wetlands and streams is influenced by lights, as is the movement of animals, such as fishes or aquatic invertebrates, up and down rivers and streams. Downwelling light mediates most predator-prey interactions in the water column.
- Islands, oceans, and reefs are increasingly influenced by lights from onshore sources, hydrocarbon extraction platforms, fishing vessels, and all manner of ships. Lighting associated with oil and gas development illuminates large terrestrial and offshore regions. Similarly, light-assisted fishing operations illuminate oceans in many regions and ocean-going freighters and passenger ships introduce mobile light sources along oceanic routes. Together, these and other light sources introduce novel lighting conditions that have no historical precedent in natural ecosystems. Natural patterns of darkness are lost or endangered globally.

- Grasslands are also open habitats with few barriers to block lights. Research shows the influence of lighting on the nesting behavior of birds, distribution of predators, and signaling of bioluminescent organisms such as fireflies. Deciduous and evergreen forests can block light and reduce its influence, but also contain communities of forest floor species adapted to lighting levels much dimmer than in exposed habitats. Therefore, even low levels of light can influence foraging times or the timing of reproductive activity.
- Alpine and tundra habitats are well represented in protected lands. Many species have annual rhythms designed to avoid the harsh winter that are potentially disrupted by lighting cues.
- Finally, urban environments have many artificial light sources, but still can support significant biodiversity in the form of both resident and migratory species. Migratory birds are attracted to illuminated structures at night and collide with windows during the day. Some bat species are attracted to insects found under city lights, while others avoid them [12,21,22,23].

3. RESULTS AND DISCUSSION

Creative solutions are often available to avoid the use of lights where they are not absolutely necessary. Especially in natural areas, managers should exercise discretion in limiting the lighting infrastructure. Various suggestions have been developed to minimize the effects of light on ecosystems. These are:

1. Landscape designers and planners must create broad-scale plans considering nocturnally-active organisms and the beneficiaries of nocturnally-active organisms which aim to protect existing dark landscapes, restore those currently exposed to moderate levels of lighting, and control the lighting environment where new development is anticipated.
2. The lighting design should not be identical. Different design for all design areas must be created. Light sources are needed for seeing our surroundings after the sun goes down. But more lighting is not always suitable. Excessive lighting causes light pollution and can annoy the user of the place [24]. Poor lighting can not only affect the health of people and cause symptoms like eye strain, migraine and headaches, but can also cause lethargy, irritability and poor concentration [25].
3. Reducing the intensity of lights can often improve visibility for humans by reducing the contrast between light and shadow, allowing people to see a larger area than they might otherwise be able to discern.

4. Although no color of light is benign in all situations, designers should avoid lights that have shorter wavelengths and in general use lights with red and yellow hues.
5. Both LED and metal halide fixtures contain large amounts of blue light in their spectrum. Lighting with lower color temperatures has less blue in its spectrum and is referred to as being "warm." Higher color temperature sources of light are rich in blue light. The IDA recommends that only warm light sources be used for outdoor lighting. This includes low-pressure sodium, high-pressure sodium and low-color-temperature LEDs.
6. Lights should be shielded to avoid glare and light pollution such that they only cast light where it is needed, and never be directed upwards.
7. Timers and motion detectors can reduce the time a light is on and may therefore reduce its impact. Curfew hours for lights can also enhance visitor experience. Instead of always lighting an empty street or parking lot at full brightness, LEDs can be turned down, or even off, when they aren't needed and then brought back to full brightness as necessary. This feature both saves on energy and reduces light pollution during the night.
8. Lighting fixtures must require less maintenance. Lamp output (lumens) will depreciate over its effective life. Illumination will be reduced further due to an accumulation of dirt and grime on the lamp and fixture. Adjustments should be made to compensate for this depreciation when determining the average values of illumination maintained over time.
9. Designs must require fewer inputs and conserve natural and financial resources, which maximize ecological function and benefit wildlife.
10. Outdoor lighting materials must be recyclable or if possible biodegradable, to place less strain on municipal waste collection and water treatment.
11. Decisions must be made in terms of what needs to be lighted, what doesn't need to be lighted and what shouldn't be lighted. Outdoor lighting is good when the right amount of light is provided for the seeing task, from a source that is shielded and aimed so it cannot readily be seen and is turned on only when needed. Determine how much light is needed to comfortably perform the task.

In any case, studies must focus on minimizing the spectrum, intensity, direction, and duration of lighting. When these four aspects are addressed, mitigation can be effective in reducing ecological disruption from artificial night lighting. With careful planning and collaboration, usually with nearby jurisdictions, designers of landscapes and other protected lands can be leaders in the control of light pollution, and increase the enjoyment of natural land, from inner city parks to wilderness areas.

References

[1] A. W. SPIRN: *Ecological Urbanism: A Framework for the Design of Resilient Cities,* http://www.annewhistonspirn.com/pdf/Spirn-EcoUrbanism-2012.pdf *(2012).*

[2] J. PRIMDAHL, S. R. SWAFFIELD: *Globalization and rural landscape change – key concepts, development trends and some implications for policy. Portuguese Journal of Regional Studies*, 20, p. 87–99. 2009.

[3] F. ÇELİK: *Ecological Landscape Design, Advances in Landscape Architecture, InTech,* DOI: 10.5772/55760. P.325–50. 2013.

[4] M. ERDEM: *Revaluating ecology in contemporary landscape design, A|Z ITU Journal of the Faculty of Architecture,* 9(1), p. 37–55, 2012.

[5] T. LONGCORE, C. RICH: *Artificial Night Lighting and Protected Lands, Ecological Effects and Management Approaches, Natural Resource Report U.S. Department of the Interior National Park Service Natural Resource Stewardship and Science Fort Collins, Colorado* (2016).

[6] M. L. CADENASSO, S. T. A. PICKETT: *Urban Principles for Landscape Design and Management, Cities and the Environment.* 1(2): Article 4, 16 pp. 2002.

[7] V.S. YRN, and D.S. COWAN. *Ecological Design.* (ISBN:978–4-559633-89-5) England: Island Press. p:216. 1995.

[8] *Implications: A Newsletter by Informe Design. A Web site for design and human behavior research.* Vol:1 Issue: 06 http://www.informedesign.umn.edu/ 2016.

[9] *Creating Sustainable Communities: A Guide For Developers And Communities. New Jersey Department Of Environmental Protection.* http://www.state.nj.us. 2014.

[10] F. AŞILIOĞLU. *Sustainable Landscape Design in Contemporary Residential Gardens. Environmental Sustainability and Landscape Management, Chapter: 51, Publisher: St. Kliment Ohridski University,* pp. 703–720. 2016.

[11] S. SAZİ,. *Aydınlatmanın Geçmişi*, Lighting Design Türkiye, Sayı 7, p. 64. 2006.

[12] T. LONGCORE, C. RICH: *Artificial Night Lighting and Protected Lands, Ecological Effects and Management Approaches, Natural Resource Report U.S. Department of the Interior National Park Service Natural Resource Stewardship and Science Fort Collins, Colorado.* 2016.

[13] C. W. HARRIS, N. T. DINES. *Time Saver Standards for Landscape Architecture: Design and Construction Data. Second Edition.* QKP: 9021098. McGraw-Hill Publishing Company. USA. 1998.

[14] O. RANSEN. *Candelas, Lumens and Lux.* http://www.ransen.com/ photometric/Candelas-Lumens-And-Lux-Chapter-Samples.pdf/. 2017.

[15] P. D. HISCOCKS. *Executive Summary, Introduction to Photometry and Measurement of Glare: Measuring Light. City of Toronto Light Pollution Abatement Guideline.* Ryerson University. Canada. 2008.

[16] *How it works. Edison Tech Center.* http://www.edisontechcenter.org/ Fluorescent.html#howitworks/. 2017.

[17] *Outdoor Lighting Basics. International Dark-Sky Association.* http://www.darksky.org/lighting/ lighting-basics/ 2017.

[18] J. BENNIE, T. W. DAVIES, D. CRUSE, K. J. GASTON. *Ecological effects of artificial light at night on wild plants. Journal of Ecology.* 104(3): 611–620. DOI:10.1111/13652745.12551. 2016.

[19] C. S. STUBBE. *Outdoor Lighting Tips to Avoid Light Pollution. Pennsylvania Outdoor Lighting Council. Pennsylvania Department of Conservation and Natural Resources.* 610.326.1402.http://www.dcnr.state.pa.us/cs/groups/public/documents/document/d_001050.pdf. 2017.

[20] Bortle Scale. Sky Glow Project. Light Pollution Web page. https://skyglow project.com/#light-pollution. 2017.

[21] J. BENNIE, J. P. DUFFY, T. W. DAVIES, M. E. CORREA-CANO, K. J. GASTON. *Global trends in exposure to light pollution in natural terrestrial ecosystems. Remote Sensing* 7:2715–2730. 2015.

[22] J. P. DUFFY, J. BENNIE, A. P. DURÁN, and K. J. GASTON. *Mammalian ranges are experiencing erosion of natural darkness. Scientific Reports* 5:12042. 2015.

[23] M. MARCANTONIO, S. PAREETH, D. ROCCHINI, M. METZ, C. X. GARZON-LOPEZ, M. NETELER. *The integration of artificial nighttime lights in landscape ecology: a remote sensing approach. Ecological Complexity.* 22:109–120. 2015.

[24] D. GÖKÇE. *Aydınlatma Kavramı. Yakın Doğu Üniversitesi. Fen Bilimleri Enstitüsü, Yüksek Lisans Tezi. Lefkoşa / Kıbrıs.* 2014.

[25] *Lighting at Work. Her Majesty's Stationery Office, St Clements House, 2–16 Colegate, Norwich NR3 1BQ.* (ISBN 0–7176- 1232–5) England. 68 Pages. 2010.

Assoc. Prof. Dr. Deniz Dokgöz[1]
Assist. Prof. Dr. Ferhat Hacialibeyoğlu[1]

Sustainability as a Design Strategy: a Case Study of Three Public Buildings

[1]*Dokuz Eylül University, Faculty of Architecture, Department or Architecture, İzmir, Turkey*

1. Introduction

Production and consumption phenomena refer to the creation of an integrated system, which essentially contains opposites and interactions. When this system with its aforementioned nature is considered as a balanced whole, every production-oriented process has to consider the phenomenon of consumption. In this sense, architectural practice as a form of intervention in the natural environment takes place in a dominant and important position when we consider the resources and the amount of energy consumed in the processes of producing and using the built environment. It is a known fact that natural environmental values and energy resources have suffered / are suffering a great deal as a result of the consumer attitude which has been introduced, disregarding environmental awareness through the forms of construction / production which, especially, accelerated in the aftermath of the Industrial Revolution. In the face of this reality, it is seen that architectural practice is struggling to transform itself by questioning the ongoing built-in attitude within the action field, which is that of design and construction oriented by nature and by rediscussing its position within the long-existing concept of 'sustainability'. In this sense, the concept of sustainability has become one of the most frequently seen concepts on the agenda in recent years. According to McDonough, within the general framework (1992), the concept is expressed as meeting the current needs without harming future generations' opportunities to meet their own needs [1]. Sustainability is conceptually new, but it has been a lasting approach since ancient times especially in the context of using climatic data in design and observing local data [2]. When the definitions of the concept's relation to design and architecture are examined; according to Yeang (2001), the main feature of sustainable design is the harmonization of construction systems with life cycles; construction systems are able to create the least destructive effect and the most positive effect on natural systems [3]. Taking a similar approach, Foster (2001) emphasizes the importance of passive design decisions which reduce the use of energy resources rather than mechanical system-dependent approaches, while defining sustainable design as achieving the most with the least and expressing that the quality of sustainable architecture is much more about the quality of a design idea than the materials used [4]. Sev (2009), on the other hand, defines sustainable architecture as all the activities aimed at creating structures that give priority to the use of renewable energy resources, are environmentally conscious, use energy, water, material and space in an efficient manner, and protect people's health and comfort whilst also taking into consideration future generations, in all conditions of their existence and every period of their existence [5]. Ekim (2004) states that making the concept of sustainability visible in architecture can be evaluated based on the two basic approaches of space and technology. Place has a phenomenological meaning in the context of the continuity of local culture and identity for a sustainable society, while it has an ecological meaning in the context of the continuity of local resources. When understood in this form, place has an importance in the context of cultural and climatic values, along with topography and the use of landscape. Technology on the other hand, emphasizes the universal, which transcends the boundaries of the Earth and therefore works towards the continuity of the whole world ecosystem [6].

For today's developing engineering technologies, the importance that the construction industry gives to ecology and recycling, as well as the sensitivity that the public has in this regard, includes sustainability solutions that have met certain standards for the structures to be built, especially in terms of public resources. What needs to be addressed through a design should be how the sustainability criteria will be translated into architecture through space by way of the material, detail, and engineering solutions suggestions. Therefore, the identification of the location of a structure, the creation of the requirements program and the identification of the areas and functions are parts of this process. At this point, the way in which the concept of sustainability is addressed is of great importance. Many parameters such as the program parts that a structure contains and the way in which these program parts come together, tectonics, the relationship with the environment and the context are the basis of the sustainability concept.

This study covers the design frameworks examined in the context of sustainability [7], which won the first prize in architectural project competitions of the Kayseri Chamber of Commerce, Bornova Municipality Service Building, Caferağa Sports Hall and Cultural Centre that opened in 2012, 2015 and 2016. The design idea framework of these three buildings which have been constructed in three different geographical areas with different climatic and environmental data in Turkey, such as the Central Anatolia region, the Aegean Region and the Marmara Region, along with the contextualization of this framework on sustainable terms, will be evaluated with a focus on the design process and the final product.

2. Three Geographies, Three Structures, Three Programs

The main project subjects of the designs which are located in three regions of Turkey that have differences in climatic and environmental data are the Chamber of Commerce, the Municipal Service Building and the Sports Hall. The common factors directly affecting the design in the three examples are the functions which we can expect to be expressed in detail in the specification. These will then be reflected in the requirement program and will serve to strengthen publicity. When the specifications, the jury's expectations and the requirements of the programs of each example are analyzed, the main features that shape the design shall emerge.

Here the targets are: the architectural ideas of the Kayseri Chamber of Commerce (KTO) 's service building, which is thought to be constructed in an area whose boundaries are set in 1/5000 and 1/1000 zoning layouts; the research of the economic and modern solutions that are appropriate to today's architecture; engineering; city planning; and the encouragement of fine arts. In this context, functionality, flexibility (to accommodate the solution potentials that enable the development / diversification of requirements within the lifecycle), feasibility, affordability, persistency, valuing the environment, compatibility with the natural environment (showing an approach that is sensitive to local climate, local ecology and local social life), openness to the public / invitation / participation, transparency, disabled access (access in universal principles in open, semi-open and closed spaces), sustainability (the assessment of the sensitivity that the structure shows to location, climate and context rather than the technological equipment within the structure as the design input) are the most important criteria in the KTO's service building project competition. [8] The prominent topics in the building requirement program are the service departments of the Chamber of Commerce, the administrative departments, service departments and a commonly used council hall, a meeting room and a restaurant.

The basic expectation in the design competition of the Bornova Municipality Building and its surroundings, which will be situated in an approximately 21 720 m^2 area in the Bornova province of Izmir, is not limited to the acquisition of a qualified local government building but to the creation of characterized common living areas that urbanites need more and more to enrich the everyday life of the city. Therefore, economic, genuine and qualified architectural approaches which will have a place in the city's collective memory and will serve as a model for the future work towards achieving a better quality environment in Bornova, functional and innovative solutions; approaches that are respectful of nature; are sustainable; and reveal a contemporary environment and architectural understanding are expected. At this point, it is important that the new municipal service building of Bornova defines an environment which is part of the urban area, easy to reach, and used in an adapted way and where there is sustainability within contemporary urban life and the city. The architectural language of the newly created surroundings and structures should determine the strength of the relationship established between the city culture and contemporary architecture and the artistic environment. One of the significant topics of the competition is that of introducing open space into city life, to transforming it into a social space and integrating it with the municipality building. In the framework of green, sustainable and smart building (a feature which is related not only to automation systems but also to architectural design, natural lighting, heating-cooling, acoustics, interior comfort etc.) concept in design, criteria such as the use of alternative clean energy sources, the maximum efficiency from the ground floor usage, the setup of a transparent, flexible and open office system, the spatial configurations that the municipality employees may use and socialize within (floor gardens, terraces, etc.), the differentiated and good analysis of the circulation (document circulation between archives and units, protocol, parliament halls, differentiation of exhibition halls and citizen entrances, freight elevators required for services, etc.) and the suitability for the disabled are prominent [9]. The most prominent topics in the building requirements program are municipal service building units, underground parking and open space arrangements.

The area in which the Caferağa Sports Hall and Cultural Centre will be built, in other words central Kadıköy, has recently become a busy entertainment centre due to some functional shifts resulting from developments in the metropolis and some new transportation possibilities. At this point, the Caferağa Sports and Cultural Centre is expected to become an attractive socialization centre that brings together different age groups in the framework of culture and sport as an alternative to an eating, drinking and entertainment-oriented socialization, whose dynamics are mostly created by the younger population. In designs, the setup of open / semi-open public spaces with flexible usage that are able to directly relate to the "street" at all the possible levels but primarily on the ground, beyond the stated program items, is considered important. Designs in which the transitions of the public areas to one other is provided, the elevations are used fluently and the usages associated with the city square are emphasized, are expected. In the requirements program for the newly proposed building, it is anticipated that the capacities of all functions in both existing centres (Caferağa Sports Centre and Baris Manco Cultural Centre) should be protected as much as possible, increased and diversified in a moderate manner, and that the spatial qualities and equipment should be improved. The Caferağa Sports Hall and Cultural Centre should be interpreted as a centre that strengthens social communication, where residents of all ages would like to be for more than one reason, ascribe different meanings and use the functions in the building in an effective way. The consideration of the structural density in the environment and the sensitivity to mass and height issues are other criteria put forward by the jury. [10] The most prominent topics in the building requirements program are the gymnasium, cultural centre, swimming pool and parking lot.

3. Design Setup in the Context of Sustainability

The design process involves a creative acquisition which is dependent on natural surroundings, towns, neighbourhoods and regions and which emerges through the handling of different parameters for each building. In this frame, the way the structure program is handled and interpreted determines the situation in which the building can be compatible with the area. A good and powerful design carries a setup that has managed to belong to the place where it is and holds contextually strong expressions. From this point of view, we can define setups that stand out in the design process through specific titles. These titles can be counted as; tectonic/formation/form setup that describes the formation of building envelope; open space setup in which publicity and public area concept come into existence; planimetric setup in the context of functional integrity and correct solutions; section setup expressing the development of the predicted life on the vertical axis; selected material setup, including the predictions regarding the construction process of the building; and climatic setup that leads to the handling of the building in the context of the geographical region in which it is located. The fact that the design decisions and the strategies that are carried out in accordance with them are correctly put forward means that they also embody the sustainability principles of the building. The correct formation of the setups, which are addressed in the design process and determine the design, will create the context of the sustainability concept which exists with the design itself.

KAYSERİ CHAMBER OF COMMERCE

Fig. 1: The main approach perception of Kayseri Chamber of Commerce

Tectonic / Formation / Form Setup *Urban context / City / Texture / Environment*

As the general urban texture of the city of Kayseri is examined, it becomes evident that it contains a fragmented texture preserved as a single mass on the basis of a traditional texture. At this point, the location of the Kayseri Commerce Chamber building is an area that is unrelated to the urban texture and is withdrawn by creating a wide gap over the main boulevard. The area of construction is in a discrete position off the surrounding texture and creates an area that is deprived of urban references within the gap, and this has led to the structural tectonic formation being directed to an inwardly focused scheme. At this point, the form setup was created through a courtyard typology that has a place in the urban memory within the traditional texture of Kayseri. This typological approach was strengthened in the third dimension and a layered, vertical courtyard setup was established. Among the prominent components in structural tectonics, the perception of the building from the main transport artery and its exhibition of an identity that reflects the Chamber of Commerce's institutional identity and prestige have been given a lot of importance. In this direction, a strong reception for public activities around the courtyard has been set up vertically and the gap created in the north-western front, together with this reception, has created the social living courtyard structure tectonics.

Fig. 2: The mass formation logic of Kayseri Chamber of Commerce

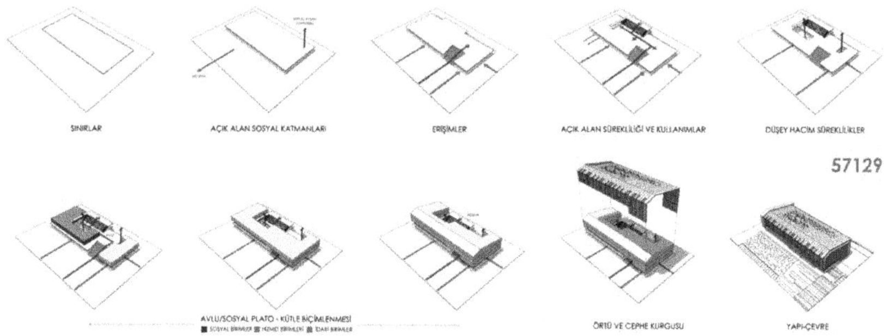

Open Space Setup *Publicity*

Chambers of Commerce buildings generally have an inward-oriented function when their usage is reviewed. This inward-oriented state of being embodies exteriorisation potentials with social and cultural actions such as clubs, exhibition halls, etc. However, it is thought that the aforementioned public relations are insufficient due to the irrelevance of the internal-external field and the visibility and accessibility problems of the social strata. With these thoughts, design has been handled by regarding the open-closed area relations created by the interpretation of traditional courtyard typology and their continuity. In this setup, which integrates the service units of the Chamber of Commerce and the social spaces included in that program, both inward and outward-oriented, multi-layered social spaces have been created, vertical-horizontal modes of handling and functional and vital continuities have been enabled among these spaces. In the formation of forms of public usage, easy access is taken into account and new grounds are created by increasing the ground floor usage on the vertical axis. In this direction, the areas that are heavily stepped on and intensively used are located on the ground floor and on the upper ground floor.

Planimetric Setup

The planimetric setup of the design is designed in such a way that the ground floor planes are increased, along with the space created on the upper ground floor. The public activity areas are organized and the chamber of commerce units can produce visual, physical, sensory relations with public spaces from each levels of elevation. The assembly hall, meeting room, foyer and exhibition halls are accessible from the square (+ – 0.00) elevation of the building, and these units can be opened to the square elevation by enabling the use of the outside area. Furthermore, the service units of the Chamber of Commerce are reached from the square (+ – 0.00) elevation. Restaurants and intensively used service units can be accessed from the social courtyard created at an elevation level of +4. The reception square in front of the building and the "social life courtyard" created on the +4 elevation provide common use for service units and social units, taking into account an accessible unity from open and closed spaces. In addition, these social open spaces are continuous, through the activity-show amphitheatre, up to the level of +8. Service units continue at +8 elevation, and from this level, the other layer of the social life courtyard can be accessed from the refectory. Administrative units are located at +12 elevation.

Section Setup

Spatialisation of the gap in the courtyard typology, which continues on the vertical, has created the section setup of the design. The highlighted sub-floor and top-floor relationship for public use has evolved into a gap surrounding use, and this use has enabled both visual and physical access between the floors.

Fig. 3: Sections of Kayseri Chamber of Commerce

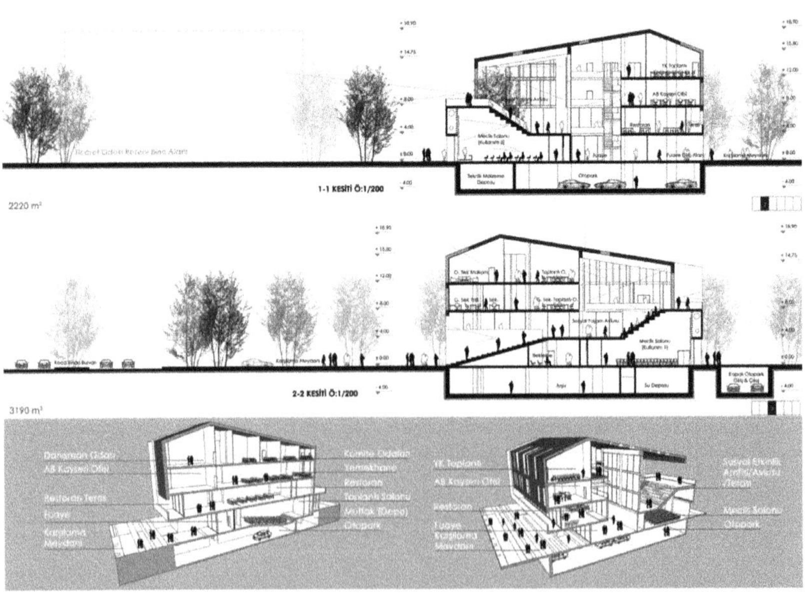

Apart from today's solutions and the social life courtyard created in the design, it has also been made possible to make a vital and visual communication with the Chamber of Commerce Reservation area located right next to the building site. This setup can be described as an interpretation of the different layers of social use, which have been fastened onto the gap, in the vertical continuity of the traditional courtyard structure typology.

Material Setup

The material selection of the building is made up of materials that are easily accessible and comply with the conditions of the existing geographical area. The wrapping of the white reinforced concrete surface with the wooden shell elements describing the units of the Chamber of Commerce is a reference to the main colour tone of the material setup in the Central Anatolian climate.

Fig. 4: The social life courtyard of Kayseri Chamber of Commerce

Climatic Setup *Direction / Sun / Wind / Northern front*

An inclined mass action is a way to form an emphasis on the roof in this region, where the terrestrial climate is intense and heavy snowfall is dominant in winter. In terms of the mass formation of the structure, it is a more filled formation; instead of an intense use of glass on the façades, it is a layout plan which surrounds the courtyard and is weatherproof during summer, and where winter use is preferred. Service units and administrative units are placed in the south-eastern and north-eastern façades of the social life courtyard and natural light-heat values are provided for the working units through the gap created in the north-western façade of the courtyard and from this point, a formation has been created in which climatic data are adapted. In particular the massive modelling that forms the structural tectonics of the northern façade has been set up with the service spaces and in this way, both mass movement and functional settlement have been supported.

BORNOVA MUNICIPALITY SERVICE BUILDING AND ITS SURROUNDINGS

A series of research questions, which are related to the structural formations to be added to the urban environment in the light of the construction plan data, constitutes the main approach to the design. How the Municipal Service building and the public open space in front of it will be connected, the relation of the municipal building that will create a dense settlement due to its interaction with its surroundings, that the public open space due to the large size of the area will have a new square scale and its relation with the scale of Bornova, the potential of the public open space to be transformed into an active recreation area and its use at night, along with the relation of the streets surrounding the area with the settlement are the most prominent of these research questions. Apart from these problems, the qualification of the possibility of interfering with the situation, caused by the addition of the space which is formed due to a large inward-oriented space on the scale of a stadium becoming outward-oriented, is suggested to be an important problem. As a result, the new situation that the possibilities of this solution create on an architectural scale will create in the urbanite a perception which has begun to turn into a number of architectural decisions and suggestions for solutions that lead design.

Fig. 5: Bornova Municipal Services Building and its location in the city

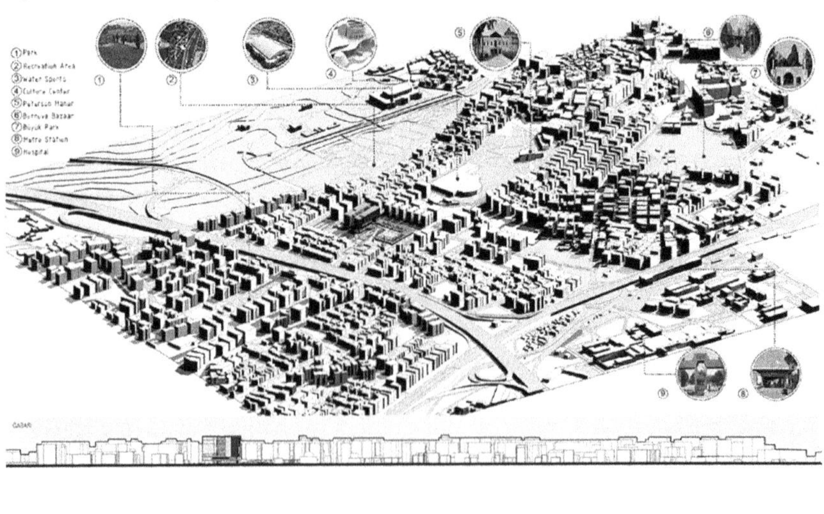

Tectonics / Formation / Form Setup *Urban Context / City / Texture / Surroundings*

The tectonic setup of the building has been arranged in order to create an intermediate space in the municipal service building.

The massive formation of the municipal building, which occurred as a result of the division of its intensive program into two along the 555th street axis, extending from the area that includes the historical bazaar and the Bornova square on the ground and upper floors, is defined with the vaulted space composed of the surrounding area of the canopy, and defining the main street and the park in front of the building. The intensively-used public functions of the building are set up in this vaulted space in an inviting formation that will welcome the urbanites. Although the transportation from the area where Bornova Square and the historic bazaar are located is provided via Mustafa Kemal Street, the 555th street appears as a pedestrian axis that leads directly to the Municipal Service Building. The revitalization with the necessary equipment and the part-pedestrianization of this axis present a significant decision which will connect the municipal building to the St. Peterson mansion and the Bornova Cultural Centre, from the historical Bornova Bazaar. As mentioned at the beginning, the division of the Municipal Building's intensive program into two along the 555th street axis on the ground and upper floors causes the axis, which continues as an inner street, to end up with a forum area enriched with cultural functions. The surfaces obtained by the articulation of the mass composition brought by the formation setup aim to create elements that will increase spatial and vital richness. Floor gardens, terraces and balconies created in this direction are designed with an approach which questions the concept of public space with open and semi-open space uses, embracing the city and the urbanites and providing opportunities for vital richness, rather than the effect of administrative buildings which strengthen managerial positions, and are strict, monumental, with general characteristics that do not include the traces of public life.

Fig. 6: Bornova Municipal Services Building and the general perception of its surroundings

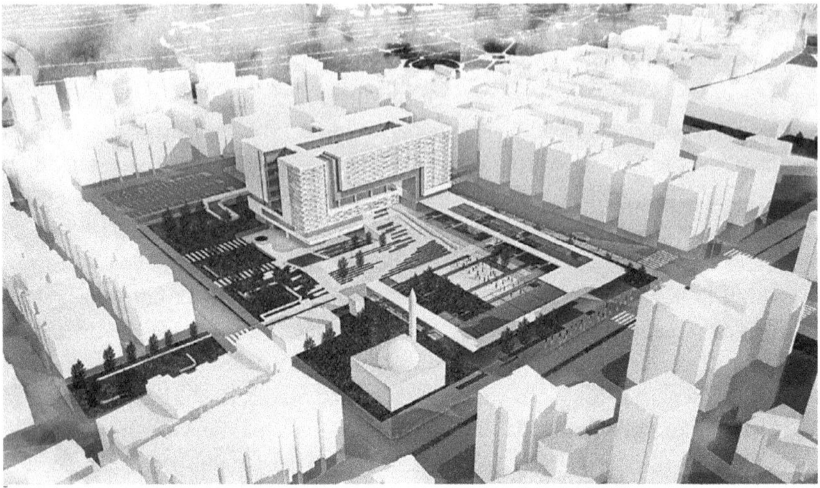

Open Space Setup *Publicity*

The way the area in which the building takes place is handled has been embodied with an urban entablature work which will provide solutions to the many problems included in the construction parcel. The formation of a number of small commercial units (bookstores, boutiques, stores, etc.) under the canopy which have a strong relationship with the municipality, as well as open pedestrian spaces, is handled in a way to boost the active usage of the area during the day and at night in Bornova, where the open public space and its surroundings are structured with a low level of commercial understanding. The idea that the historical Bornova Bazaar might merge here with a new bazaar, and that one end of the 555th street reaching the historical bazaar might have developed the proposal of an overhanging road from central Bornova to the municipal square. Another solution provided by the urban canopy was the contribution in terms of softening the scale of the gap that was revealed at the beginning. It is possible that people will have a relationship here with a gap that they never see, thanks to this entablature. The canopy, which can be defined as a threshold, is, in fact, a permeable border that protects the square from climatic conditions, contains car parks underground and kiosks.

Planimetric Setup

Undoubtedly, ground floors are the most intensively used places of public buildings and require higher sensitivity in the relation with their surroundings. At this point, getting maximum efficiency from ground floor use is a very important point for a municipal service building. The ground floor of the Municipal Service Building should be of a quality that absorbs pedestrian movements around it, feeds its entrances from these spaces and guides its users with the intermediate spaces created. In this sense, the axis coming from the 555th street on the ground floor and the axis coming from the open activity area have been taken into an intermediate space with a cultural street, and the main entrance, exhibition hall, multipurpose hall, and cafeteria entrances are provided from this place. Access to the financial services' pay desks, which is the busiest unit of the municipality, is also provided in a detached fashion from here, and waiting areas on the street are envisioned. A second ground was built at + 4.50 elevation. Access to the Assembly Hall is provided with the help of amphitheatres and ramps that face the activity area. This solution provides an advantage for municipal councils where public participation is very intense, and encourages the design of parliament halls as city councils. The differentiation of circulation and a good analysis which depends on the intense usage of the public and the municipal staff in the structure (the document circulation between archives and units, protocol, parliament hall, the differentiation of exhibition hall and citizen entrances, the freight elevator required for services, etc.) are the important points in the planimetric setup of the structure. Designing a structure suitable for disabled use is also one of the essential topics.

Section Setup

Sections of the structure compose the design character which describes the structure best. The section setup of the structure is based on the intensity of the public use of the program parts and climatic data. At this point, the gap created by the axis coming from the 555th street is designed to allow the program parts to be spread accross the floors, according to their usage intensity and, in the approach from the main square, the vaulted space and the public usage areas such as the assembly hall, the multipurpose hall and the amphitheatre which are connected to this vaulted space. Products of this section design include the street formation which is set up in the continuation of the pedestrian axis starting from the historical Bornova Bazaar, the formation movement of the canopy surrounding the urban green area which gives the scale of the main road and forms the building mass with continuity, amphitheatres which describe the use of public space, the social equipment and the city council that has the characteristics of an urban parliament. When the sections of the design are examined, we can see how fully empty mass formations are organized in the light of urban continuity; and clues are given that a structure can be set up not only by occupancy but also by the design of space.

Fig. 7: The section setup of Bornova Municipal Services Building

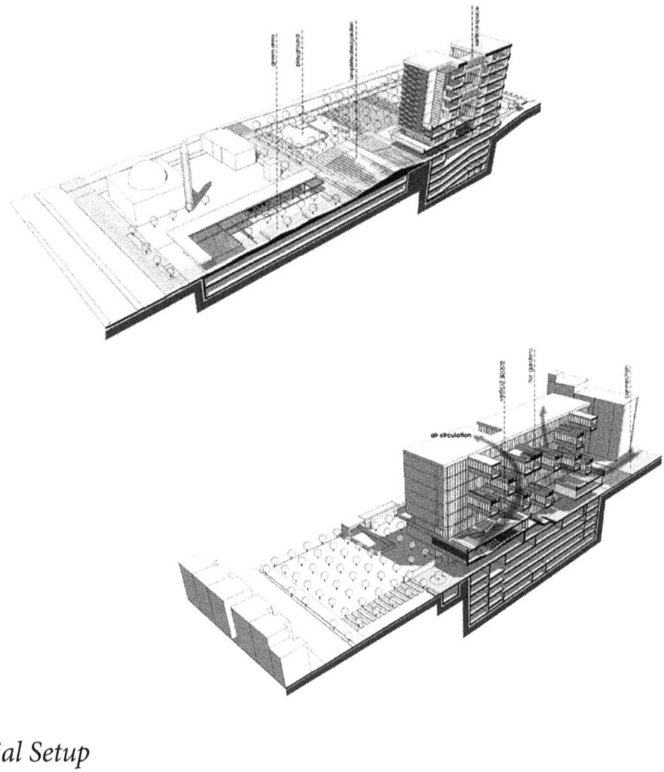

Material Setup

Economical and easily accessible materials were prioritized and climatic data were taken into the account in the selection of material. The right choice of material to be used in a building is of great importance in terms of cost, future heating, cooling, ventilation, etc., based on the continuity of construction in the building. The selection of materials in the Bornova Municipal Service building design consists of a reinforced concrete skeleton system, which is easily obtained and practiced in the Turkish construction sector, and ready-mixed concrete covering materials. The canopy, which describes the public space setup of the design, is set up together with a steel conveyor system, ready-mixed concrete and wooden elements.

Fig. 8: The open space setup of Bornova Municipal Services Building

Climatic Setup *Direction / Sun / Wind / Northern Façade*

The main decisions relating to the building are handling the structural mass, which is the main point of the design and is divided into two from edge to edge in a way that enables air circulation instead of it being a closed volume, as the parts are connected to each other by green-roofed steel bridges, saving up in terms of the climatization of this intermediate space instead of a seven-floor- high gallery. The more major handling of the northern façade depends on its orientation; the preference for horizontal sun refractors on the southern façade and the setup of the shaded sub-spaces created by the vaulted space and street have strengthened the ecological decisions of the building. Façade design forms the most prominent and most effective point in the perceptual context and gives identity to the structure in the structural design. When each façade's solar radiation values are considered to be different, the settlement of the structure in terms of north-south orientation which are the two opposite directions in the context of where the climatic data such as the wind, the light, etc. that the structure receives leads to, determines important criteria in façade decisions. The fact that the southern direction receives continuous sunlight, whereas the northern direction is exposed to the wind effect and does not receive direct light, has had an influence in the differentiation of the southern and northern façades of the structure. But when it is considered that this difference will be reflected in the general plastic mass of the structure, an arrangement with the design decisions has been made. According to this, there have been attempts to bring into being the creation of a refractive design located on the southern façade and relieving the interior with different refractions of light; the establishment of a barrier against the northern winds which addresses the refractive design used on the southern façade vertically on the northern façade; and the differentiation of the southern and northern façades of the structure within the same design unity and in the light of climatic data.. Because of the fact that the eastern and western façades receive very little sunlight in winter and a lot of sunlight in summer, The permeable and transparent façade logic, which emphasizes thepublicness of the ground elevations of the structure, has created an inviting groundwork for the structure and the aim is especially that the big frame, which leads people in and directs them towards the street level, gathers people with the influence of a vaulted space. The proposed canopy design for the urban open space surrounds the green space that offers a breath of fresh air to the city with its shaded areas, framing it, and captures the large urban green space that the structure faces. This architectural formation setup, which is used to increase the intensity of public use, also allows data such as the natural weather, light and wind, in the light of the climatic data of the design, to become a part of the design. [11]

CAFERAĞA SPORTS HALL AND CULTURE CENTER

The design area, which offers a different identity and life on the Anatolian side of Istanbul, located in an intensively structured area in the centre of the town of Kadıköy, is on a corner parcel created by the pedestrianized paths in a central location. The gymnasium and the cultural centre, which exist in the design area, have not been able to establish efficient relations with these pedestrian paths and the slope of the land. The competition program not only expects to enhance these relations but also to lead to a more intensive building program on the construction site. The functions to be analyzed are the functions that people use in a massively intensive way (sports halls, cultural centres) and also are indoor-oriented (swimming pools). Apart from the relationships of these function groups with each other being a design problem, their relationship with outdoor and public open space is discussed as a fundamentally problematic issue.

Fig. 9: The perception of Caferağa Sports Hall and Culture Centre from the main approach

Tectonic / Formation / Form Setup *Urban context / City / Texture / Environment*

Taking into account the structural intensity of the environment, a design which is sensitive to height and mass has been set up; for that purpose, the gym and swimming pool functions were resolved under an inclined shell set up in the direction of Chios Street. Leaving Caferağa Sports Hall's tribunes as single tribunes, which is the most important datum in the name of urban use, and handling the area in a way that will give it reference to the park before its construction as a sports hall, are the data considered in the design in the name of urban memory.

Open Space Setup *Publicity*

This design, through which public outdoor spaces are interlinked, in which elevations are used fluently and the usages as a city square are emphasized, brings itself into existence through maximum use of public open space. This design has transformed the Chios Street elevation to a 6% sloping platform in an effort to bring a refreshing new square to the city and has created an inner street that invites people from other pedestrian arteries. This inner street also serves as a common foyer for cultural and sporting activities as anticipated in the program. This level amphitheatre connected to the open foyer creates new spaces for the use of cultural concerts, shows, etc. and also hosts informal artistic events. The setup of open / semi-open public spaces with a flexible usage that are able to directly relate to the "street" at all the possible elevations, but primarily on the ground, is the aim of this inner street.

Fig. 10: The location of Caferağa Sports Hall and Culture centre in the environmental texture

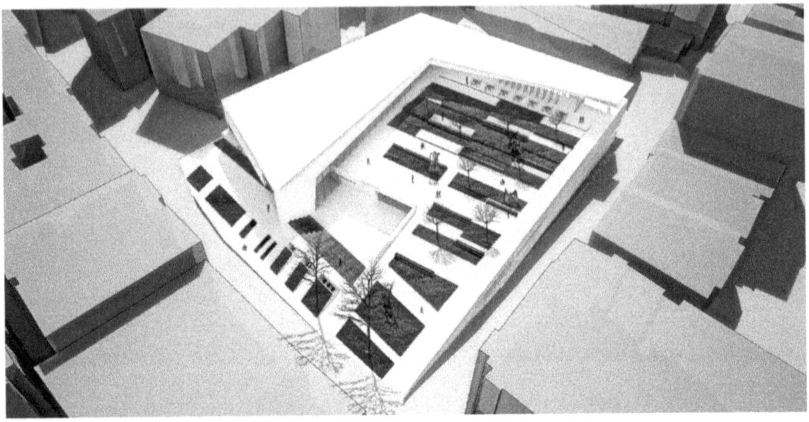

Planimetric Setup

The planimetric setup of the structure is about feeding the building programs from the different open spaces created. In this direction, while the centre of culture takes its main approach over the street of Chios, it stands out with its foyer connections from different elevations and forms new focal points with the amphitheatre that connects the upper foyer elevation with the lower foyer elevation. The platform connected to the street of Chios ends with a cafeteria, and entrances to the cultural centre from this elevation are anticipated. While the sports hall and the swimming pool spectators' entrances are provided from the open foyer on the inner street, sportsmen entrances and outdoor and indoor parking entrances are enabled from the intensively structured area, through the vehicle road.

Section Setup

The point of origin of the design is related to the section setup. The fragmentation of the urban shell, which is formed by raising a 6% slope from the main street, to form open and semi-open spaces, and the existence of the program parts together with these gaps both constitute the language of the building and bring the sectional life into existence through the urban context.

Fig. 11: The section setups of Caferağa Sports Hall and Culture Centre

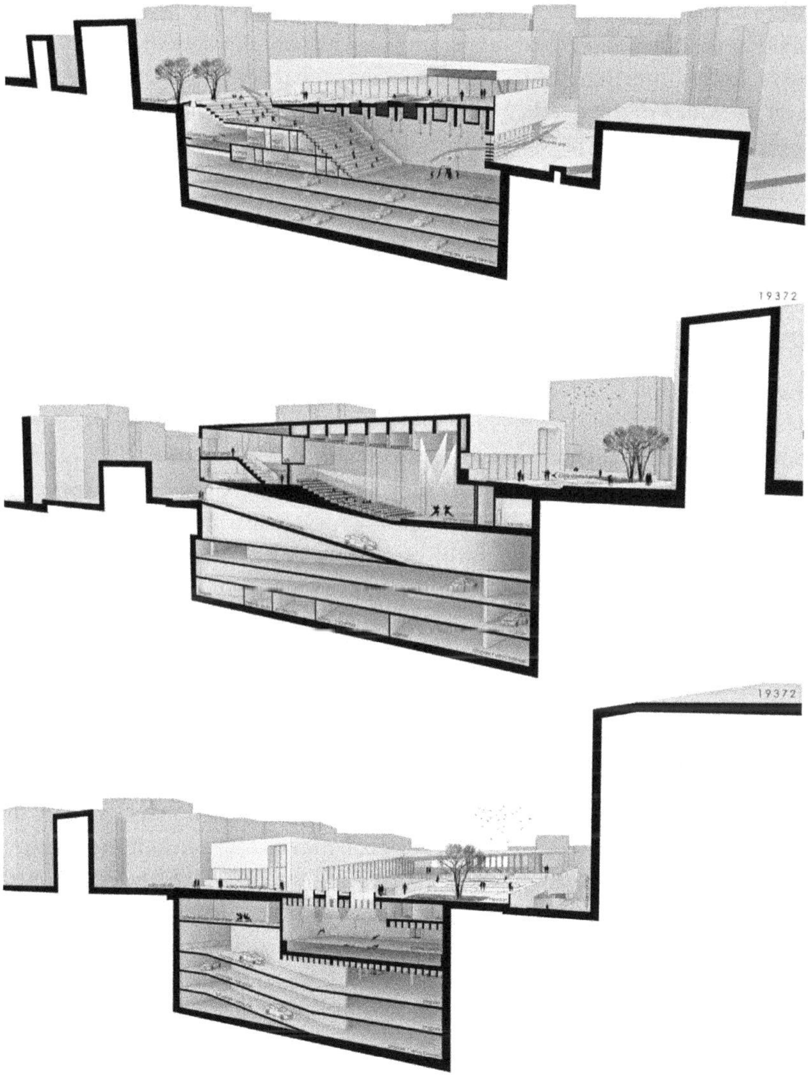

Material Setup

The material setup of the design consists of the elements that will represent the building tectonics and create simple surfaces that will not ruin its sculptural attitude. In this direction, the structure handled with a reinforced concrete system is set up with a dry concrete system which will represent the whole appearance instead of fragmented façade elements, and there has been an attempt to reduce the effect of concrete with thermal wooden elements in places. The green roof setup that forms the landscape of the new public space which has been brought to Kadıköy and the wooden elements that are articulated within this setup intend to create an alternative to the concrete surfaces in the area.

***Climatic Setup** Direction / Sun / Wind / Northern Façade*

The structure that is based on the need for urban open spaces in Istanbul, which are constantly restructured, defines the urban space created by itself with its green roof. The determinants of the climatic setup are: increasing the surface of green spaces by using the structure of the roof; the attitude that the shell movement adopts towards the heat movements of the sports hall with this formation; the unification of structural tectonics with characters such as orientation; and natural light taken in a controlled manner to the sports field.

3. Conclusion

The concept of sustainability is regarded as a holistic phenomenon that needs to be understood in an environmental, social, cultural and economic sense. It is clear that this holistic approach must be fundamental to the disciplines of architecture and design. When the approaches towards a concept and the differentiation of the disciplines in which the concept is discussed are taken into account, especially in the field of architecture, the concept of place and the entirety of the relations established between place and design gain importance. In this sense, it is possible to say that sustainability can start in the form of an approach that minds, senses and evaluates environmental values in the design phase. In this sense, harmony with topography, integrating design with the consideration of local values, protection from basic environmental factors such as solar radiation and wind movements according to climatic characteristics or providing the utmost benefit from these data become important as the forms of existence of the notion in the design process. Therefore; **in order to achieve a green, sustainable and smart building, it is necessary to acquire this quality, not only through automation systems but also through architectural design**. A good design provides an opportunity for the use of alternative clean energy sources by comprehending the urban area with the main decisions it makes within the framework of sustainable building concepts such as settlement, building envelope opacity/permeability, natural lighting, heating/cooling, acoustic and interior comfort. The relationships established with the environmental values and the space reduce the consumption values in the building use process by also providing sustainability in the economic sense. Another point that is ignored when the consumption values are taken as a basis is the functional setup of the structure. The resolution of the pedestrian circulation intensity of the structure on the ground and first floors, by handling these with the right operations, and providing as little circulation as possible on the upper floors gain value in terms of the usage criteria of building automation systems. Also, the flexible usages of spaces and the fact that building cores provide criteria such as all wet areas, warehouses, archives, etc., can lead the spaces to be transformed with minimum intervention and can increase the life of the building. This flexibility and transformability enables natural and sustainable resource utilization of the structure. On the other hand, sustainability of the structure is also assessed in terms of social and cultural values. Especially the usage possibilities of public buildings through public open/closed areas and the links established with the cultural and social values of the building environment gain importance in this respect. Criteria such as developing and maintaining the spatial characteristics of an existing usage habit, presenting new usage possibilities, accessibility and ease of use are all values that should be taken into account when considering sustainability as a whole. The main ideas to be expressed when the projects presented in the scope of the study are evaluated in the concept of sustainability are: **the coexistence of the notions of design and sustainability and the emphasis on the relationship between the structure and the environment, or "place", in which it is located.** Moreover, the potential of the factors and main ideas that shape the design of each project is to be directly related to the notion of sustainability. While the environmental values and climatic data determine the mass setup and settlement decisions in the Kayseri Chamber of Commerce and Bornova Municipality Service Building, topography, the use of open spaces in the closely-spaced city centre, the social-cultural values and the features of "space" determine the design of the Caferağa Sports and Culture Centre.

Notes

1. W. McDONOUGH: *The Hannover Principles: Design for Sustainability. William McDonough Architects, New York,* (1992).
2. D. OKTAY: *Sürdürülebilirlik Bağlamında Planlama ve Tasarım: Kuzey Kıbrıs'ta Yöresel Mimarinin Geleneklerinden Çağdaş ve Duyarlı Çevrelere. Mimarist Dergisi,* (6), 67–71 (2002).
3. K. YEANG: *Kean Yeang, Green Architecture. Architectural Design,* 71 (4), 60 (2001).
4. N. FOSTER: *Lord Foster of Thames Bank, Green Architecture. Architectural Design,* 71 (4), 32 (2001).
5. A. SEV: *Sürdürülebilir Mimarlık. Yapı-Endüstri Merkezi, Yem Yayın, (2009).*
6. D. EKİM: *Sürdürülebilirlik Kavramı ve Mimari Form Üzerindeki Etkisi. İstanbul Teknik Üniversitesi Fen Bilimleri Enstitüsü Yüksek Lisans Tezi* (2004).
7. *Deniz Dokgöz, Ferhat Hacıalibeyoğlu ve Orhan Ersan'a ait olan tasarımlardan Bornova Belediye Hizmet Binası ve Caferağa Spor ve Kültür Merkezi 1. Ödül, Kayseri Ticaret Odası eşdeğer ödül kazanmıştır.*
8. *Kayseri Ticaret Odası Hizmet Binası Ulusal Mimari Fikir Yarışması Şartnamesi*
9. *Bornova Belediye Binası ve Çevresi Mimari Proje Yarışması Şartnamesi*
10 . *Caferağa Spor ve Kültür Merkezi Mimari Proje Yarışması Şartnamesi*
11. F.HACIALİBEYOĞLU, D. DOKGÖZ, O.ERSAN: *Investigation of Some Design Decisions on Energy Efficiency: A Case Study of Bornova Municipality Service Building. SBE16 ISTANBUL, International Conference on Sustainable Built Environment,* 117–126 (2016).

Umut Pekin Timur, Özgür Burhan Timur, Ferhat Özden

Importance of Greenways for Sustainable Cities

1. Introduction

The urban areas inhabited by the majority of the world population are the most appealing settlement areas, which offer various opportunities for urban people. But today they are also face to face with many environmental problems due to rapid urbanization, population growth and developments in industrialization. Urban issues not only stand out for the environmental dimension but also on economic and social levels [1]. According to estimates made by the United Nations and many other sources, it has clearly been stated that this tendency will continue throughout this century and by 2050 70% of the world population will be living in cities [2]. Parallel to this, the existing problems in cities will certainly increase.

Environmental issues were brought to the United Nations Conference on the Human Environment in Stockholm and every dimension of the global agenda in 1972 for the first time. Following this conference focused on the human and development, the Brundtland Report was prepared by the World Commission on Environment and Development (WCED) in 1987, and the concept of sustainable development gained importance and triggered a wide range of environmental movements in many countries of the world [3, 4].

Thinking about sustainability in relation to cities is a relatively new phenomenon. According to Yazar (2006), with the 'Habitat-2 Urban Summit' held in 1996, sustainability was moved to the dimension of urban and life environment [5]. Thus, the concept of the sustainable city gained importance and 'sustainable cities' were mostly raised as one of the alternatives for a better environment and social life in cities [6].

At this point, the use of greenways in urban areas supports three dimensions of sustainable development, with potential benefits. This is because they are excellent bodies for protecting the environment, balancing the needs of urban growth, providing recreation and alternative transportation to the public of a city [7].

The origin of greenways was based on Olmsted (1865). They are used for solving the environmental problems we see increasing in cities, decreasing open areas and meeting the increasing need of people for recreation, all issues which has never been so important in any period as they are today [8, 9, 10]. It is possible to see many successful examples of greenway plans and practices in the world, especially in the USA.

This chapter discusses the greenway concept, its definition, basic features and types, a brief history, and its benefits, and puts forth a relationship between urban sustainability and greenways.

2. The definition and brief history

Greenways serve many purposes, for this reason many different definitions of them can be found in the literature.

A greenway is a linear open space established along either a natural corridor, such as a riverfront, stream valley, or ridgeline, or overland along a railroad right-of-way converted to serve recreational uses; a canal, scenic road, or other route. This linear open space is a connector linking parks, nature reserves and cultural heritage sites, including classified urban settlements, with each other and with populated areas [9].

Fabos (1995) defined greenways as "corridors of various widths, linked together in a network in much the same way as our networks of highways and railroads have been linked". The author added that greenway systems represent "nature's superstructure" [11].

Turner (1998) explained the difference between greenways and historic public parks. He said that they can provide for movement while the historic public park is essentially a static concept [12].

The term greenway is named by combining *green* from greenbelt (term with a British origin) and *way* from parkway (term with an US origin) [9]. Although the term *greenway* is relatively new, having emerged only in the late 1950s, the design of linear greenspace in North America predates the name by nearly a century [13].

The first generation of ancestral greenways are boulevards (in Europe) and parkways (in the USA), and go back the 1700s [14], but the greenways movement started with Frederik Law Olmsted in 1865 in the United States. His proposal project, which may be the root of the greenways, is the California Campus Project in Berkeley. Olmsted suggests two greenway elements here. The first is the area covering the Strawberry Creek valley located at the top of the campus park area. The second is the way that recommends connecting the campus to the Oakland in order to provide a visual landscaping experience [9]. Olmsted's first complete greenway was the 'Emerald Necklace' for the Boston Park System in the 1880s. Today, this system is still being used for recreation, transportation, flood control, scenic beauty and a home to wild species [8].

With the mass production of the car, greenways were transformed into scenic roads that were used for recreational purposes in the 1890s. In the 1960s, the damage caused by automobiles to the environment increased, and bicycle paths and pedestrian paths returned. [1, 9, 10, 14].

In the 1980s, the loss of open space and an increased need for recreation in urban and metropolitan areas focused attention on greenways. The President's Commission on Americans Outdoors (1987) found strong support for greenways, to address the need for additional open space and recreational land and propose a national system of greenways [8]. Since 1995, the greenway concept has evolved into a comprehensive greenway network concept, in order to meet the needs of an increasingly urban society. Greenways are embraced as a framework for urbanisation and intelligent land use planning [1].

3. The basic features and types of greenway

According to [15], greenways are networks of land including linear components that are planned, designed and managed for multiple purposes including ecological, recreational, cultural, aesthetic, or other purposes compatible with the concept of sustainable land use. By the same author, the basic qualities of greenways are listed below:

- Their spatial forms are linear; this feature means they contribute to recreation and ecology
- They make connections between life spaces by forming associations with landscape forms of all scales with their binding qualities
- They can be multifunctional, such as being ecological, recreational and cultural
- They are consistent with sustainable development
- They have a form that completes landscape planning.

Although there are different greenway categories made by different scholars, the most common classification is that made by [9] into five categories (Figure 1). These are:

1. **Urban riverside greenways**, usually created as part of a redevelopment program along neglected, often run-down city waterfronts
2. **Recreational greenways**, featuring paths and trails of various kinds, often of relatively long distance, based on natural corridors as well as canals, abandoned railbeds and other public rights-of-way.
3. **Ecologically significant natural corridors**, usually along rivers and streams and ridgelines, to provide for wildlife migration and an interchange of species, nature study, and hiking.

4. **Scenic and historic routes**, usually along a road or highway, the most representative of which make an effort to create pedestrian access along the route or at least places to alight from the car.
5. **Comprehensive greenway systems or networks**, usually based on natural landforms such as valleys and ridges. This system sometimes presents simple opportunities with assemblage of greenways and open spaces of various kinds in terms of creating green infrastructure on an urban or regional scale.

Figure 1. Examples of greenways type [9].

4. Relationship Greenways and Urban Sustainability/ Sustainable Cities

Many definitions have been made regarding the term sustainability, which has been widely used in many platforms since the 1990s [3]. UNESCO (1996) defines sustainability as the 'viability of socially-shaped relationships between society and nature over long periods of time' [16]. According to Childers et al. (2014), most interpretations of this concept focus on human needs and values and place an emphasis on the future [17].

The pioneer of the concept was Hans Carl von Carlowitz in the 17th century, with his book on the protection of forest areas. But the rise of the concept was found in the second half of the 20th century. The foundations of the sustainability concept were laid down by the Stockholm Conference in 1972. With the World Conservation Strategy (WCS) document, published in 1980 by the International Union for the Conservation of Nature and Natural Resources (IUCN), sustainability and development concepts were brought together. With the 1987 Brundtland Report, the concept of sustainable development became popular and its definition became clear [3].

According to Yazar (2006), with the 'Habitat-2 Urban Summit' held in 1996, sustainability was moved to the dimension of urban and life environment [5]. This summit's main goal was to have healthy, safe, equal, and sustainable cities, towns and villages, now and in the future [18].

A common characteristic in sustainability literature is that the concept of sustainability, and in particular of urban sustainability, involves environmental, social and economic components [1]. Çahantimur (2007) explained that environmental, economic and social sustainability are independent components of each other but these three different components should be taken within a whole in order to achieve sustainable development goals [19].

According to Shearer et al. (2006) and Jenny and Ericson (2006); it is in no doubt that the cities have problems in terms of environmental dimensions just as much as economic and social dimensions today. Urban sustainability is becoming increasingly important as a key objective to reduce or eliminate these problems [20]. Also, Altuntaş (2012) points out that the most emphasized alternative is 'sustainable cities', for a better environment and social life in cities [6].

Keskin (2012) defines sustainable cities as places where socio-economic interests are harmonized with environmental concerns to ensure a continuity of change and development [21].

According to Huang (1998), the sustainable city is not the only city where the environment is protected; it also requires economic and social changes to reduce the need for protection of the environment [20]. Sustainable urban components can be explained as shown in Figure 2.

Greenways are an important planning tool for sustainable cities. [22] stated that sustainability has emerged as a major theme in planning. He/she added that planners often suggest greenways as ways to make places greener, healthier, and more in urban planning processes.

Figure 2. Anonymous (1996): Sustainable urban components [4].

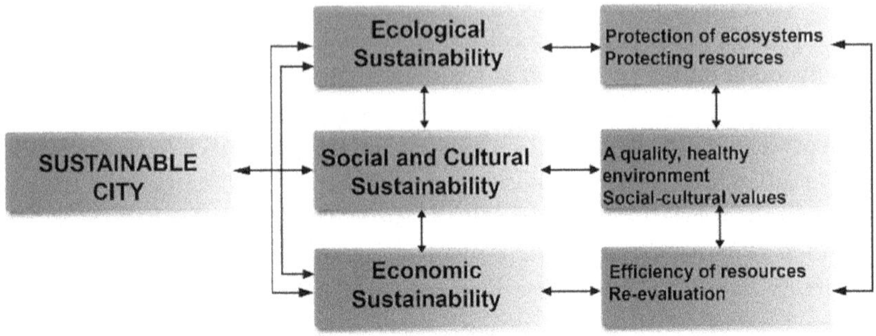

Greenways are not created only for the protection of nature and the environment, but also they enable people to use them. This feature of greenways harmonizes with the principle of sustainable development, which is that of building a balance between resource use and protection [15].

They provide environmental, economic and social benefits to urban areas through their basic features. Their benefits are compatible with the components of sustainability. These benefits were summarized in Table 1 [23].

Jongman and Pungetti (2004) and Kappler and Miller (2009) stated that the environmental benefits of greenways are numerous and varied depending on the design and scale of the project [24].

These benefits vary from the protection of water resources and biodiversity, the limitation of urban development, the provision of a life environment for plants and animals, carbon sequestration, an enhancment of environmental quality, and the provision of more accessible urban green spaces against erosion and for flood control [1, 22].

Greenways, whose features are linearity and connectivity, offer both environmental benefit and social benefit. While linearity promotes recreational activities such as walking and bicycle use, on the other hand it makes an ecological contribution by promoting the circulation of species, various materials and nutrients in terms of wildlife [15, 10]. The linking feature of greenways is that they associate landscapes as well as strengthening social ties by linking communities, and at the same time they may be responsive to the notion of regionalism [22]. This situation provides a significant social benefit for the cities.

Table 1. Adapted from [23] greenway benefits

Environmental Benefits	Economical Benefits	Social Benefits
✓ Help to protect and restore natural environments ✓ Connect between living environments ✓ Protect and supports biodiversity ✓ Provide protection of water resources ✓ Reduce flood hazards ✓ Reduce problems of soil erosion and sedimentation ✓ Provide control of urban development	✓ Increase property value in neighboring areas ✓ Provide development of various business and commercial facilities ✓ Promote high efficiency ✓ Provide tourism attraction ✓ Reduce public investment and private spending ✓ Provide a cost effective strategy for outdoor recreation ✓ Reduce the need for second vehicle	✓ Provide training for people of all ages ✓ Ensure that historic and cultural resources are protected and linked with each other ✓ Offer recreational opportunities ✓ İnfluence on public health and quality of life ✓ Help society to raise awareness of responsibility for nature and the environment ✓ Strengthen social ties ✓ Increase aesthetic values ✓ Provide an alternative transport route

Three of the social benefits of greenways are closely related. According to Anonymous (2007), while they provide alternative transportation facilities with recreational uses such as climbing, cycling and hiking trails, at the same time they help keep the community healthy by providing physical activity. With these benefits they prevent air pollution in cities [25].

Some environmental and social benefits of greenways also bring economic benefits. The most important of these is perhaps the reduction of private spending and some public investments. It is possible to give many examples of economic benefits provided by greenways.

According to estimates of benefit analysis made in Houston, Bayou Greenway, which is in the implementation phase, will provide approximately $13.9 million a year in health benefits for increasing physical activity. In addition, the annual benefit of reduced air pollution, vehicle operating savings, and reduced crashes is estimated at $4.2 million/year [26].

There are many similar examples. Anonymous (2010b), New York City, was told by the EPA to construct a $8 billion water filtration facility in order to improve the quality of drinking water supplies. Instead, officials decided to protect 80 000 acres of natural land (including greenways) in their water-supply watershed and achieve the same water quality result for $1.5 billion. For example, according to Anonymous (2007b), a greenway was designed along Mingo Creek in the flood-plain in Tulsa, Oklahoma for flood control. As a result of this and other measures, flood insurance rates in Tulsa dropped by 25 percent [27].

The greenways and trails, in particular the plants, are, according to Novak (1994), the air filtering feature of trees provided in Chicago, Illinois, avoiding the annual $1 million cost of air pollution removal [28].

[23] stated that together with the benefits they provide, greenways are a means of planning which can help make progress towards greater urban sustainability. In addition, they have taken main topics for urban sustainability such as quality of life, land use planning, good connections, sense of place, identity and belonging, development of a sustainable transport system, greener urban areas and open spaces (Figure 3). In particular, they have added that greenway networks offer a wider range of benefits. Also they pointed out that greenways support these topics providing sustainability in urban areas through their benefits, and that they are a solution which will relieve urban areas.

Figure 3. Greenway network and urban sustainability [23].

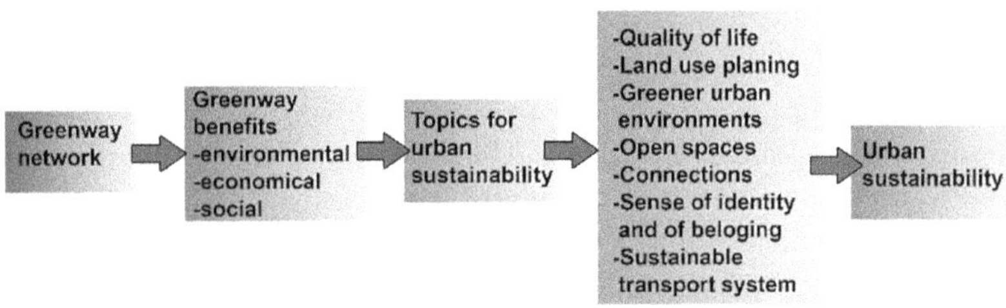

5. Conclusion

Cities have confronted many environmental problems today because of fast urbanization, a population increase and developments in industrialization.

According to Spirn (1994), although the problems arising from urbanization may vary from city to city and from country to country, they are many common features, and for this reason the solutions can have similar qualities too [29].

In this context, the 'sustainable city' approach has become very important in the recent period. Sustainability components need to be addressed together in order to achieve sustainable cities. At this point, greenways which are based on a sustainable approach with environmental, social and economic benefits are important for ensuring urban sustainability. According to [30], it is possible to see greenway examples of more than 600 on the state scale in the USA.

Greenways enable many resources to be saved within the least area [8]. They are based on natural or artificial corridors which form open green areas in cities. According to Riaz et al. (2002), the amount and distribution of these areas in urban landscapes are important in terms of sustainability and livability [24].

Greenway projects are composed of very complex and comprehensive processes. Also, multi-disciplinary work and public participation is required.

Greenways can be planned at the local level as well as at the regional level [15]. Today, it is possible to see greenways that connect cities and even countries. When greenways are planned, especially in the form of greenway networks, their benefits in cities will be increased and sustainability can be ensured.

References

1. P. VASCONCELOS SILVA: Greenways, a Path towards Urban Sustainability. Master's Thesis. Faculdade de Ciências e Tecnologia da Universidade Nova de Lisboa, Portugal. Date of Access:02.01.2011, Available from: http://www.ece.auckland.ac.nz/~sinnen/VasconcelosSilva2006.pdf (2006).
2. J. AHERN: From *fail-safe* to *safe-to-fail*: Sustainability and Resilience in the New Urban World. Landscape and Urban Planning Volume 100, Issue 4, (341–343) (2011).
3. SKD TÜRKİYE: İş Dünyası ve Sürdürülebilir Kalkınma Derneği. 100 Madede Sürdürülebilirlik Rehberi. 59 s., İstanbul (2017).
4. F. AKLANOĞLU: Geleneksel Yerleşmelerin Sürdürülebilirliği ve Ekolojik Tasarım. Ankara Üniversitesi Fen Bilimleri Enstitüsü Peyzaj Mimarlığı Anabilim Dalı, Doktora Tezi, 209 s., Ankara (2009).
5. E. KARAKURT TOSUN: Sürdürülebilirlik Olgusu ve Kentsel Yapıya Etkileri. PARADOKS, Ekonomi, Sosyoloji ve Politika Dergisi (e-dergi), ISSN 1305–7979, 5 (2), Retrieved February 22, 2015, from PARADOKS: http://www.paradoks.org (2009).
6. A. ALTUNTAŞ: Sürdürülebilir Toplumlar ve Metropollerin Baskılarından Kurtulmak İçin Alternatif Bir Yol: Sürdürülebilir Kentler. Mustafa Kemal Üniversitesi Sosyal Bilimler Enstitüsü Dergisi 9 (17), 135–148 (2012).
7. A.CONINE, W. XIANG, J. YOUNG and D. WHITLEY: Planning for Multipurpose Greenways in Concord, North Carolina. Landscape and Urban Planning, 68, 271–287 (2004).

8. J. AHERN: Greenways in the USA: theory, trends and prospects. Ecological Networks and Greenways, Cambridge University Press, ISBN 0 521535026, 34–53 p., USA (2004).
9. C. E. LITTLE: Greenways for America. The Johns Hopkins Press Ltd., 237 p., London (1995).
10. M. ARSLAN, E. BARIŞ, E. ERDOĞAN, Z. DİLAVER: Yeşil Yol Planlaması Ankara Örneği. Ankara Üniversitesi Bilimsel Araştırma Projesi Kesin Raporu, 139 s., Ankara (2007).
11. J. G. FABOS: Introduction and Overview: the Greenway Movement, Uses and Potential of Greenways. Landscape and Urban Planning, 33, 1–13 (1995).
12. T.TURNER: Greenway Planning in Britain: Recent Work and Future Plans. Landscape and Urban Planning, 76, 240–251 (2006).
13. P.C. HELLMUND and D.S. SMITH: Designing Greenways Island Press, 270 p. , London (2006).
14. M. R. SEARN: The Evolution of Greenways as an Adaptive Urban Landscape Form. Landscape and Urban Planning, 33, 65–80. (1995).
15. J. AHERN: Greenways as a Planning Strategy. Landscape and Urban Planning, 33 (131–155) (1995).
16. A. ÇAHANTİMUR, T. H. YILDIZ: Sürdürülebilir Kentsel Gelişmeye Sosyokültürel Bir Yaklaşım: Bursa Örneği. İTÜ Dergisi/a 7 (2), 3–13 (2008).
17. L.D. CHILDERS, S.T.A. PICKETT, J.M. GROVE, L. OGDEN and A. WHITMER: Advancing Urban Sustainability Theory and Action: Challenges and opportunities. Landscape Urban and Planning, http://dx.doi.org/10.1016/j.landurbplan.2014.01.022 (2014).
18. A. ÖZCAN: Ekolojik Temele Dayalı Sürdürülebilir Kentsel Gelişme: Malatya Kent Örneği Üzerinden Bir Değerlendirme, 38. ICANAS Uluslararası Asya ve Kuzey Afrika Çalışmaları Kongresi, 10–15 Eylül 2007, Bildiriler Kitabı, s. 689–710, Ankara (2007).
19. E. KARAKURT TOSUN: Sürdürülebilir Kentsel Gelişim Sürecinde Kompakt Kent Modelinin Analizi. Dokuz Eylül Üniversitesi Sosyal Bilimler Enstitüsü Dergisi 15 (1), 103–120 (2013).
20. Z. A. GÜREL ÜÇER: Kentsel Politikaların Belirlenmesinde Bir Araç: Sürdürülebilirlik Göstergeleri. Çağdaş Yerel Yönetimler, Cilt 26 Sayı 1, 103–124 (2017).
21. E. B. KESKİN: Sürdürülebilir Kent Kavramına Farklı Bir Bakış: Yavaş Şehirler (Cittaslow). PARADOKS Ekonomi, Sosyoloji ve Politika Dergisi 8 (1), 81–99 (2012).

22. G. LINDSEY: Sustainability and Urban Greenways: Indicators in Indianapolis. Journal of the Amerikan Planning Association, 69, 2 (165–179) (2003).
23. P. VASCONCELOS and M. PRITCHARD: A Greenway Network for a more Sustainable Auckland. Proceedings of the 2nd international conference on sustainability engineering and science, Auckland, New Zealand, Feb 21–23 (2007).
24. M.E. BARIS, E. ERDOĞAN, Z. DİLAVER and M. ARSLAN: Greenways and the Urban Form: City of Ankara, Turkey. Biotechnol. & Biotechnol. Eq. 24/1, 1658–1664 (2010).
25. U. PEKİN: Kentsel Akarsu Koridorlarının Geliştirilmesi Ve Ankara Çayı Kavramsal Yeşil Yol Planı. Doktora Tezi. Ankara Üniversitesi Fen Bilimleri Enstitüsü, 283 s., Ankara (2007).
26. URL1:http://www.bayougreenways.org/benefits (04.09.2015).
27. U. PEKİN TİMUR, M.E BARIŞ, Ö.B TİMUR: Greenway for Protection of Water Resources. International Soil Sciences congress on "Management of Natural Resources to Sustain Soil Health and Quality" Proceedings Book, 294–302, Samsun (2010).
28. URL2:https://www.railstotrails.org/resourcehandler.ashx?id=3062 (05.09.2016)
29. M. ARSLAN: Yeşil Yol Planlaması: Ankara örneği. Çevre Planlama ve Tasarımına Bütüncül Yaklaşım Sempozyumu, s. 277–285, Ankara (1996).
30. J. A. BUENO, V. A. TSIHRINTZIS and L. ALVAREZ: South Florida Greenways: A Conceptual Framework for the Ecological Reconnectivity of the Region. Landscape and Urban Planning. 33, 247–266 (1995).

Hülya Öztürk Tel[1], Elmas Erdoğan[2]

The Evaluation of Sanliurfa Traditional City Center in the Framework of Ecological Design

[1]*Harran University, Technical Sciences Vocational School Department of Architectural Restoration*

[2]*Ankara University, Faculty of Agriculture, Department of Landscape Architecture E-mail:hulyaozturktel@gmail.com*

1. Introduction

The 'ecological crisis' created by the rapid depletion of natural resources, the decline of living diversity, the inevitable rise of environmental pollution and global warming has made ecological practices in architecture a necessity. Since the 1980s, the concept of "ecological design" has become an important phenomenon in the field of architecture.

Ecological design is an important phenomenon in terms of increasing the quality of urban life and having healthy cities with an identity. A design approach that strengthens the ecology-economy-energy relationship must be based on a healthy life that is in harmony with nature. The main objective in ecological design is to create a living construct of the city.

At this stage, the technology which has been developed is guided by experiences and applications that have been experienced in the past. For this reason, the ecological priorities that exist in the traditional urban texture, but which are now disappearing, contain important data in terms of design. Understanding these approaches in the formation of traditional cities necessitates an ecological approach which will bring these results to the new settlements.

As a result of the ecological design approach, cities and urban areas gain identity as the spatial quality improves. Historical cities (examples of civil architecture, monumental structures, roads, streets, squares and green spaces, etc.) should be assessed by adapting their framework for urbanisation and intelligent land use, planning [1] current uses to meet current needs, rather than maintaining them in traditional form. Thus, while benefiting economically from available sources, it may be possible to transfer the values of past civilizations and cultures onto the future.

Şanlıurfa is a settlement that can be evaluated within the scope of ecological design, with traditional urban texture, open green areas, traditional structure and life culture.

The aim of this study is to determine the ecological landscape design approach that will enable the traditional urban texture to be a sustainable settlement, and to determine the implementation methods and tools. It is clear that the ecological principles determined for Şanlıurfa, the design proposals created by an ecological approach and the implementation methods will be guided in new studies for Şanlıurfa. Şanlıurfa is an example of ecological design studies on the regional and local scale, to be carried out in the current and future processes.

2. Material and Method

Material

The main material of the study is the Şanlıurfa urban conservation area. This area has been selected as a research area due to its historical identity as well as the intensive concentration of central activities, the use of (registered) cultural property in nature, and the potential for ecological design. The maps, documents and reports used in the inventory and the analysis and evaluation stages of the research constitute the other materials of the study.

Figure 1 Research area; Şanlıurfa urban site area (Original 2013)

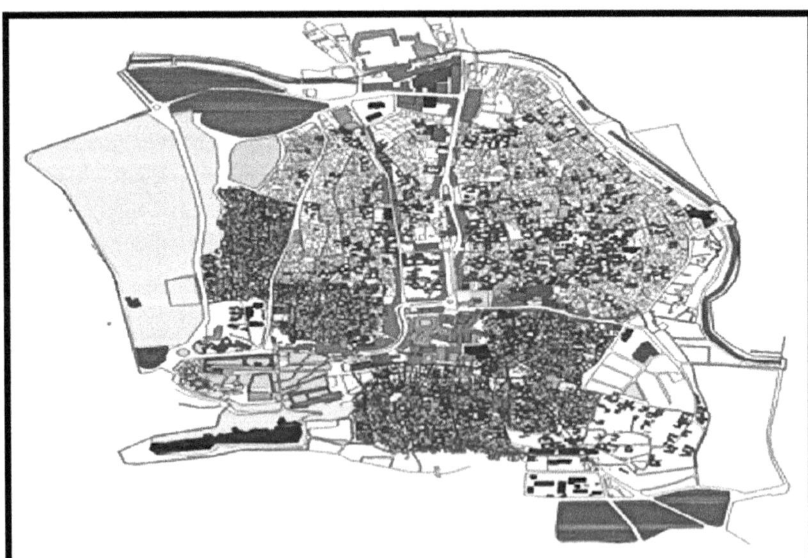

Within the scope of generating the basic data base for research, a 1/1000 scale reconstruction plan with the purpose of preservation, which was taken as 10 layouts in the image format of 1992 Sanliurfa Metropolitan Municipality Conservation Development Directorate, was merged and updated and used as a base map by drawing in AutoCAD 2010 program. The residential buildings, religious buildings, mansions, military buildings, water facilities, educational buildings, baths, commercial areas, accommodation areas, traditional houses, cemeteries, parking lots, constructed and under construction projects, health projects and green areas are indicated (Figure 1).

Method

The objective of the study was to evaluate the ecological landscape characteristics of Şanlıurfa traditional urban texture selected as a research area and to examine the national and international approaches in order to ensure the sustainability of the research area as an ecological settlement. In order to achieve this, an ecological-based design approach has been adopted. Then, analyses and evaluations were carried out in order to create an ecological design suitable for the area in terms of landscape architecture. In this context, the limits of the study area were determined, then the research area study-analysis studies were made; conservation development plans and the ecological characteristics of the research area were evaluated. In the final phase, ecological design and suggestions were developed with conclusions resulting from the evaluation of the research area.

Studies conducted in relation to the aim and scope of this research [1, 2, 3, 4] are used. As a method in this study, the ecological design approach in Aklanoglu, [2]' s study has been adopted. In terms of the ecological point of view and design approach and targets; survey and analysis studies of the research area were carried out, assessment schedules were prepared according to determined ecological design criteria, ecological design potential was scored and an ecological design proposal was developed.

4. Research Findings

In this chapter the location of Şanlıurfa city is specified, and abiotic and biotic factors and cultural characteristics and data about the research area are evaluated.

4.1 Location

The Şanlıurfa province is located in the Southeastern Anatolia Region of Turkey between 36° 40' and 38° 02' north latitudes, 37° 50' and 40° 12' east longitudes [5]. Şanlıurfa is surrounded by Birecik and Suruç in the southwest, Harran and Ceylanpınar in the southeast, Akçakale in the south, Bozova and Halfeti in the northwest, Viransehir in the east, Hilvan in the north and Siverek in the northeast (Figure 2).
The Şanlıurfa province, which has been inhabited by various civilizations throughout history, has been a settlement for various religious and cultural communities. The Şanlıurfa province is one of the important historical cities in the Middle Euphrates section of Southeastern Anatolia Region [6].

Figure 2 Şanlıurfa City Center [7].

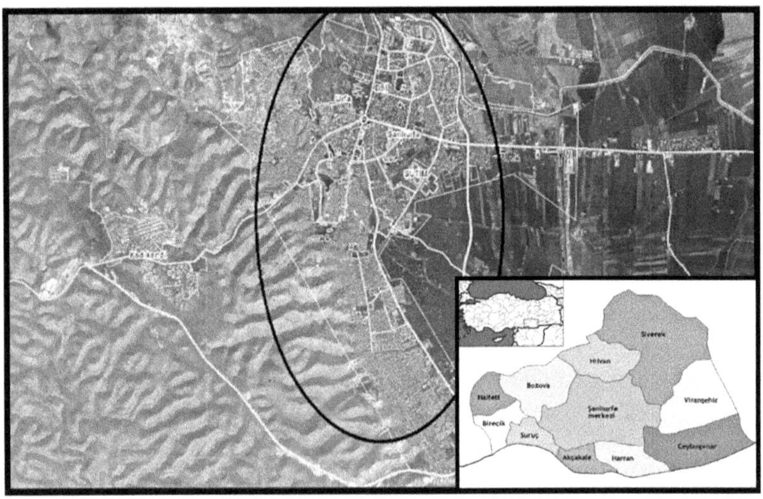

4.2 Abiotic Factors

4.2.1 Climate

There is a continental climate in Şanlıurfa. Summer months are long and very warm, whereas winter is very cold. The continental climate characteristic is dominant in Şanlıurfa. There are huge differences between the average temperatures for day and night and winter and summer. The annual temperature difference sometimes exceeds 40° C [8]. Şanlıurfa is the city with the most sunny days in Turkey. The daily average insolation is 8.30 h/m [9].

4.2.2 *Topography*

Şanlıurfa is on the southern outskirts of the middle part of the Southeast Taurus Mountains. The mountains and high hills in the north of the province generally descend towards the south. The large plains are located to the south of Şanlıurfa. The Suruç, Harran and Viransehir plains are lined up from west to east. The city center of Şanlıurfa does not have a sloping structure but is close to being flat (Figure 3) [10].

Figure 3 Height groups map (Original 2013)

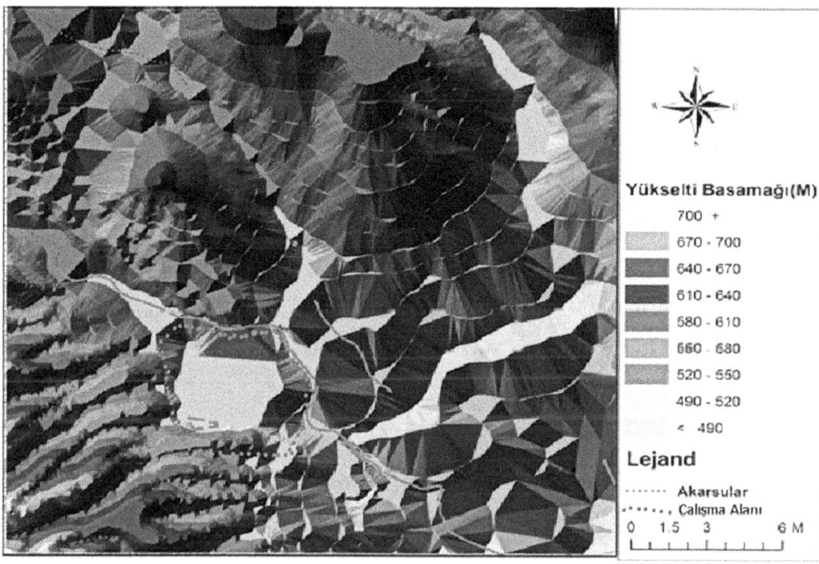

4.2.3 *Soil*

Due to climate, topography and bedrock differences, various large soil groups are formed in Şanlıurfa. These are colluvial soils, basaltic soils, river flood beds and brown forest soils.

4.2.4 *Geology*

The study area and its immediate surroundings are rich in limestone and basalt rocks. The presence of hard (Eocene age) and soft (Oligo-Miocene age) limestone rocks have been used to make shelter throughout history since they are easily processable. In particular, the fact that the Neogene limestones are softer, resulting in the use of cut limestone stone (Urfa stone) for the construction of houses. Urfa stone was used in the creation of the unique architecture of the city due to its easy processing. Basalt sites near the study area provide great convenience in providing this material [6].

4.3 Biotic Factors

4.3.1 Flora

The dominant vegetation cover in Şanlıurfa is the steppe. Forest vegetation is usually present on the river bed (Populus L. (Poplar), Salix L. (willow). In addition, Quercus L. (Oak), Crataegus L. (Hawthorn) are also some forest remains. There is Pistacia khinjuk Stocks (wild nut tree) over a vast area. Due to the effect of the Ataturk Dam Lake, which looks like an inland sea, it is expected that dry air will be replaced by humid air depending on the increase of humidity in the air. Therefore, it is possible to say that it will provide an enrichment in vegetation cover [11]. Şanlıurfa has seasonal and perennial flowers (geophytes) which are unique to the region.

4.4 Urban Characteristics

In this section, the characteristics of Şanlıurfa City such as its traditional urban texture, traditional houses, open green spaces, squares and transportation are given in detail.

4.4.1 Şanlıurfa Traditional Urban Texture

Şanlıurfa's old city is within the walls, with a large area surrounded by the Karakoyun Stream. There are many immovable cultural assets in this area. After the Islamization of Şanlıurfa, monumental buildings, cemeteries and traditional houses around Balıklıgöl created a dense texture.

The traditional urban texture of Şanlıurfa has developed into a flat topography. Urfa Castle, which overlooks the city, is built on the southern slope.

As people caused destruction before the city was put under preservation as an archaeological site, the traditional Şanlıurfa urban texture has been destroyed. Despite this, there are many local residences that could protect their identity (Figure 4–5).

Figure 4 View from the castle: in 1923 (Akkoyunlu 1989)

Figure 5 Traditional city center of Şanlıurfa in 2013 (Original 2014)

Culture and climatic conditions of the region play a big role in the formation of Şanlıurfa's urban texture.

The organization of space suitable for settlement has been implemented. The position of the structures and the organization of the street directions in the direction of the dominant wind are of great importance in terms of natural ventilation. The fact that traditional residences come with courtyards and are adjacent to each other has created an urban texture in harmony with nature. A compact form is observed in the settlement and a balance is established between the green texture and the structural density in the courtyards of traditional houses.

Figure 6 Şanlıurfa streets (Original 2014)

At first glance this organic and complicated street texture was thought to have emerged spontaneously, but the main streets were shaped in the west-east direction because the dominant wind was in the westerly direction. To provide solar radiance control, the streets are built narrow and delimited by courtyard walls. Thus, on hot summer days, the refreshing effect of the wind provides natural ventilation to the streets. The elements of the street are boulders, overhangs, terraces, squares and basalt stone (blackstone) roads (Figures 6–7).

Figure 7 Street Texture (Original 2014)

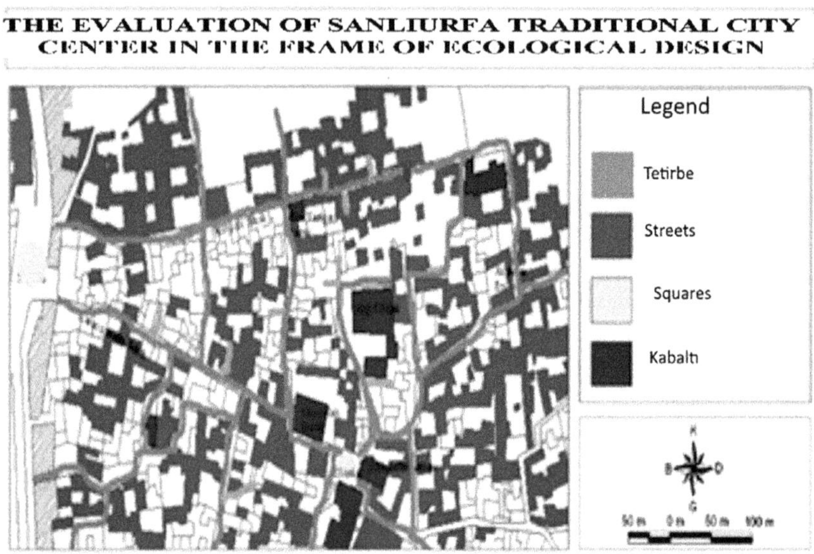

4.4.2 *Traditional Urfa Houses*

The economic and social structure, building materials and climate features have been influential in the formation of the unique housing in Şanlıurfa. Traditional houses combine with the characteristic of narrow streets.

Traditional Urfa houses have a maximum of two floors and a maximum floor height of 7.5 meters. Traditional Urfa houses consist of spaces lined around an inner courtyard. On the ground floor there are daily living spaces and service areas, and there are living spaces on the first floor. It is also possible to see rooms planned to be half a floor below or half a floor above the ground. These spaces are used as pantries (zerzembe).

Some of the Urfa houses have been constructed with places where women and men sit separately, and they have been hidden by the streets with high walls without windows, facilitating the privacy of family life [5].

The absence of wood materials in the surrounding areas led to the use of stone as the dominant material in Şanlıurfa architecture due to the easy processing of Urfa stone. Regarding the interior decoration of the rooms in the houses, there is a large amount of richly decorated wood on the window and door wings. It is possible to examine the traditional Şanlıurfa houses as the main rooms, iwan and separate sections where women and men sit: rooms, pantries, floor furnace, bath, dam and courtyard sections (Figure 8).

Figure 8 Views of traditional Şanlıurfa houses (Original 2013)

Structural elements in the courtyards of traditional houses are pools, wells, cisterns, water basins, flower beds and bird houses.

The flower beds are inside the courtyard and are mostly left as soil. Some of them are surrounded by safety rails. In the flower beds, there are *Platanus ori-*

entalis (Oriental plane trees), *Thuja orientalis* (gallnut) *Juniperus communis* vile juniper*Olea europea* (olive), *Punica granatum* (pomegranate), *Citrus aurantium* (bitter orange), *Juglans regia* (walnut), *Morus sp.* (white mulberry) *Ficus carica* (Fig.), *Washingtonia filifera* (palm tree), *Vitis vinifera* (grapevine), *Hedera helix* (rock ivy), *Ampelopsis veitchii* (American ivy), *Nerium oleander* (oleander), *Campsis radicans* (trumpet vine), *Rosa sp.* (rose), *Buxus sempervirens* (boxwood), and *Lonicera caprifolium* (honeysuckle).

4.4.3 Open green areas

The open green areas of the city can be considered as Balıklıgöl and its surroundings, City Park, Haleplibahçe Archeo-park, Karakoyun Stream Landscape Area, Kızılkoyun Ridge, and squares and courtyards.

The open green areas of the traditional city, which is on top of a flat topographical landscape, are very limited. The Şanlıurfa Conservation Development Plan has included green areas, but due to the rapid increase of Şanlıurfa population, the green areas have become insufficient.

The Balıklıgöl City Park was constructed in 1995 and the buildings are in conformity with Şanlıurfa architecture which is completely harmonious with the environment, not oppressing the old texture. Exterior façades are covered with traditional Urfa stone. Wooden material, metal material and Urfa stone were used for the landscape of Balıklıgöl. Mostly exotic plants are used (Figure 9–10).

Figure 9 Views from Balıklıgöl.

Figure 10 Balıklıgöl Plan (Original 2014)

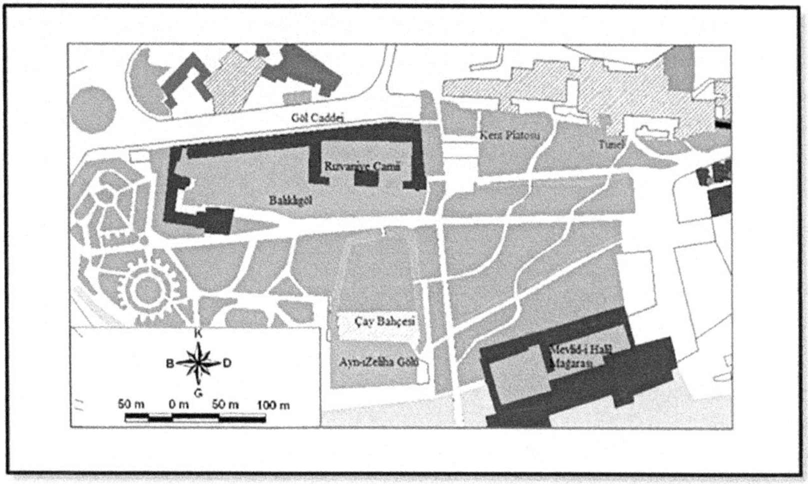

4.4.4 Squares

There are 3 squares in the city center. These squares, on the same axis, are Hasimiye Square, Yıldız Square and Kara Square. Due to the parking of the vehicles around the squares and the narrow space, traffic complications arise (Figure 11).

Figure 11 Hasimiye Square (Original 2004)

4.4.5 Transportation

Vehicle traffic is very dense and limited in the settlement because the roads are narrow and in an organic order within the city. Narrow streets are paved by cobblestones and there are no sidewalks. The width of the streets is of such a distance that a vehicle can pass, and it does not pose a danger even if it causes the pedestrians and vehicles to mix. The use of vehicles in the streets is for service purposes. Traffic is mostly dense at Göl, Demokrasi and Sarayonu Streets. There are commercial buildings around these streets and they have a busy pedestrian and vehicle traffic circulation.

4.5. General Evaluation

Şanlıurfa's 1/1000 scaled conservation zoning plan decisions of 1993 were evaluated according to the ecological settlement criteria; ecological deficiencies were defined and then the findings were evaluated according to ecological design criteria (effective resource use, effective energy use, effective water use and ecosystem conservation) .

When the conservation zoning plan decisions of 1993 were examined, it was seen that these decisions did not provide an adequate contribution in the context of ecological criteria, because the plan was not detailed enough, gaps existed in practice, and different implementation occurred.

The general evaluation and scoring of Şanlıurfa Traditional Urban Texture in terms of Ecological Design Criteria were made. (Table 1)

Table 1 Evaluation of Şanlıurfa Traditional Urban Texture in terms of Ecological Design Criteria.

EVALUATION OF ŞANLIURFA TRADITIONAL URBAN TEXTURE IN TERMS OF ECOLOGICAL DESIGN CRITERIA				
	CRITERIA	**EVALUATION**	**SCORE**	**TOTAL**
EFFECTIVE ENERGY USE	Use of renewable energy sources	The availability of winter and summer rooms, the natural ventilation of the chimneys called pherik in summer iwans, the ventilation of the summer rooms with higher ceilings, construction of pantry in the basement and keeping cool	10	
	Ensuring compact planning in the settlement	The layout of the constructions is suitable for topography and a compact settlement is visible.	10	
	Proper orientation and positioning of buildings	The houses were oriented in the north–south direction and winter and summer rooms were formed.	10	

EVALUATION OF ŞANLIURFA TRADITIONAL URBAN TEXTURE IN TERMS OF ECOLOGICAL DESIGN CRITERIA				
	Buildings consist of a compact form	Buildings have courtyards, in the form of compact structure with a flat roof. Building form is in low energy consumption forms, simple plan type and compact building form.	10	
	Providing air conditioning with natural methods	The traditional houses are oriented according to the prevailing winds, and the positioning of the winds The positioning of rooms in north–south direction and establishment of summer and winter rooms Buildings are with courtyards and they are positioned in a grift way The courtyard walls are kept high and shade is provided Use of Urfa stone on outer walls of buildings and natural isolation of Urfa stone Placing a pantry and establishment of a natural cool place The pools placed in courtyards provide the increase of micro–climatic comfort.	10	
Chart 4.18				
Eva-luati-on		**The total score is considered as 50 and 50 points are obtained on the full score. If 50 is 50 and 100 is set as x, then x = 100**	**100% of the energy efficiency was de-termined.**	**50**
EFFECTIVE WATER USE	Application of rainwater collection systems	None.	0	
	Use of permeable surfaces and protection of groundwater level	There are no excavation and impermeable building materials affecting underground water around the structures. Partly permeable surfaces were used in the streets of the study area	5	
	Water efficient and local landscape application.	Drip irrigation system and spring systems are used. Vast grass areas are present. Deep excavations or impermeable floor coatings are not carried out.	5	
	Waste water system	There is no waste water system in the study area.	0	

EVALUATION OF ŞANLIURFA TRADITIONAL URBAN TEXTURE IN TERMS OF ECOLOGICAL DESIGN CRITERIA				
Evaluation		**Full score = 40 was accepted and 10 points were obtained from the full score. When 40 is 10 and 100 is set as x, x = 25**	**25% of the water efficiency was determined.**	10
EFFECTIVE RESOURCE USE	Use of durable materials	Use of durable building materials such as Urfa stone	10	
	Recycled material usage	Use of natural materials such as wood, iron, Urfa stone	10	
	Use of local resources	Local materials have been evaluated by using Urfa stone in traditional houses and floor slabs.	10	
Evaluation		**The total score is considered as 30 and 30 points are obtained on the full score. If 30 is 30 and 100 is set as x, then x = 100**	**100% rate is determined in environmental pollution control.**	30
PROTECTION OF ECO-SYSTEM	Achieving proper green texture ratio and correct landscape application in the city	The natural plant cover is provided by plant seedlings in flower beds, and the presence of bird crates can be used to protect the fauna.	10	
	Waste reduction and control	There is no system for recycling solid wastes or recycling wastewater.	0	
Evaluation		**The total score is considered as 20 and 10 points are obtained on the full score. If 20 is 10 and 100 is set as x, then x = 50**	**50% was determined for the protection of eco-system.**	10

5. Ecological Landscape Design Recommendations

- Within the context of ecological landscape design, the areas to be re-evaluated are: Göl Caddesi, Hasimiye Square, Sarayönü Street, Kazancı Bedih Street and Ellisekiz Square.

It is suggested that the Gol Street road is designed as a one-way passage going from the Balıklıgöl to the Hasmiye square and supported by a green area and a pedestrian strip on the side bands. A reduction in the number of lanes on the street has been foreseen in order that they cover less surface.

Hasimiye Square is a square which is embedded between the Sarayonu Street, Gol Street and the surrounding traditional bazaar region. This place should be decorated with ecological design by disposing of the traditional texture of the waste. In the arrangement of Hasimiye Square, harmony of design along with the traditional texture should be given importance, and thus the use of Urfa stone and basalt stone in the ground as durable material on the hard floor, by giving importance to ecological design and increasing the amount of green texture. **Permeable Pavement**

The groundwater is preserved and sustainability is ensured by providing a controlled flow of rainwater. Considering Sarayönü Street to the north of Hasimiye Square: a solar-powered light rail system is proposed, which can be transformed into a promenade that feeds Hasimiye Square and produces its own energy in a pedestrian and ecological context. On the streets of Sarayönü, ecological equipment needs to be developed. Operating lighting elements, such as a power plant, can be provided as an example.

Care has been taken to select the natural species (*Berberis crataegina, Pistacia khinjuk, Centaurea obtusifolia, Scilla mesopotamica*) in Hashimiye Square where the garbage cans that separate paper, metal, and plastics are placed along with the placement of a kiosk.

Cooling is provided by water element design; natural cooling is provided by reducing unwanted heat load. The bird pool was designed to protect the fauna (Figure 12).

Figure 12 Ecological landscape design project of 1/200 scale Lake Street, Hasimiye Square and Sarayönü Street

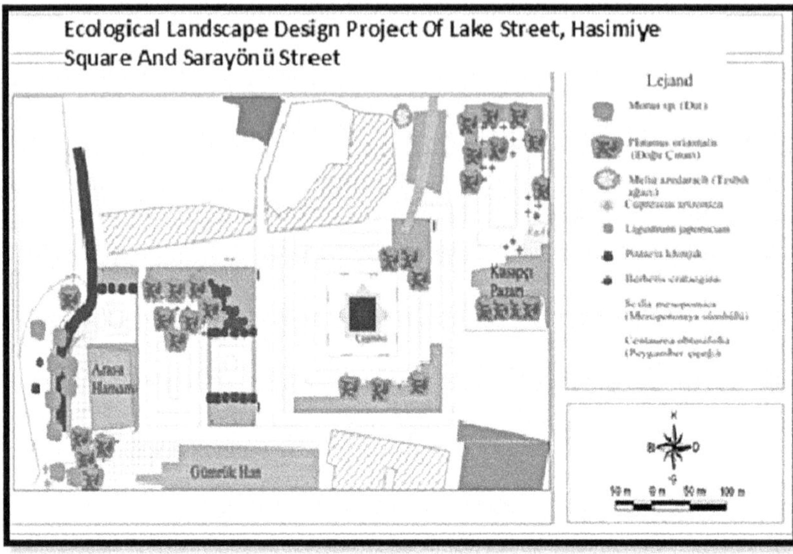

- Recommendations about Transport

Ecological design suggests pedestrian priority design. Balıklıgöl surroundings, the Karakoyun Deresi landscape area and the Haleplibahçe Archeopark Area are the three most important components of open green areas in the Şanlıurfa traditional city center (Figure 13). These components are supported by pedestrian, bicycle and transportation roads.

Enriching the ecological system by promoting pedestrian access should be considered as a shared road (light rail system + pedestrian zone) starting from Sarayönü Street and going up to Hasimiye Square.

The narrow streets, even for a single vehicle, should be reserved for pedestrian use, and street reclamation work should be completed.

It should be considered as t part of Haleplıbahçe Street (bicycle + pedestrian + vehicle).

Figure 13 Recommended transport plan (Original 2014)

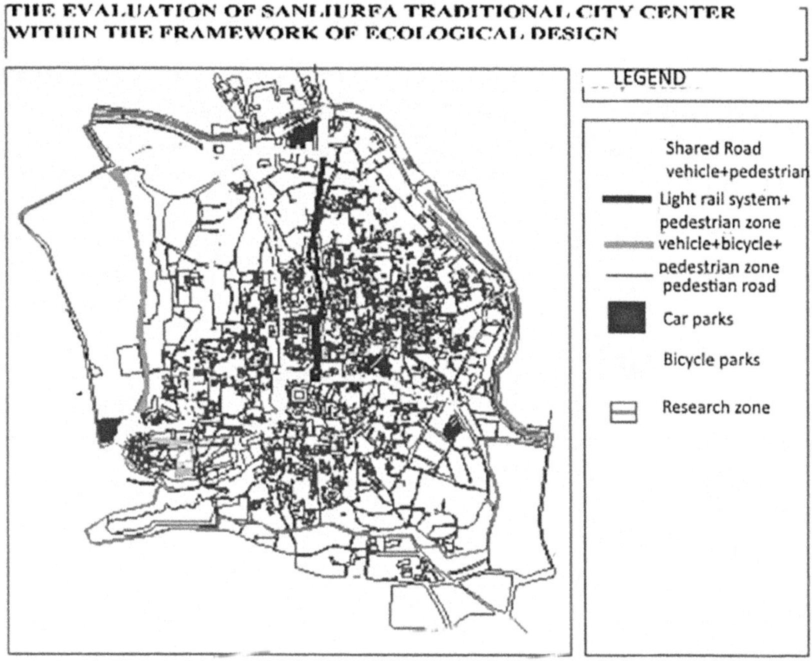

6. Discussion

In ecological design, the physical, social, cultural and economic characteristics of the area as well as ecosystems and ecological factors, planning and design, are important in the first instance. Success in landscape planning and design work depends on preserving and improving the ecosystems and ecological factors within the area.

In the study, it was seen that Şanlıurfa had no ecological basis for the conservation zoning plan and that it was inadequate in terms of directing the landscape design. The fact is that the Şanlıurfa Conservation Development Plan does not totally cover the buildings (traditional houses, baths, fountains and churches) but also does not fully take into account the other parts related to the city (buildings' surroundings, gardens and courtyards, roads, equipment). There was an attempt to close this gap by the recommendations in the field.

When the traditional urban texture of Şanlıurfa is evaluated in terms of effective resource utilization, it is seen that the use of durable materials, recyclable materials and local resources have been provided in large amounts. When the research area is evaluated in terms of effective energy use, it is considered that buildings are made up of compact forms and that the natural climate, effective orientation and positioning in the buildings, compact planning and easy accessibility for pedestrians, and sustainable transportation systems are provided on a large scale.

The research area seems to show that the permeable surface and underground water conservation provided a significant proportion of effective water use, water efficiency and local landscape use. Rainwater use, gray water usage and wastewater control are not applied.

In this research, the ecological characteristics of Şanlıurfa's traditional urban texture were evaluated on the criteria of effective use of resources, effective use of water, effective energy use and protection of the ecosystem, as well as preserving existing properties. Ecologically based planning and design proposals were developed.

In consequence:

- It is possible to create settlements with a higher quality of living standards that respect the environment and human beings by protecting the ecological approaches in traditional urban culture.
- The designs in Turkey are implemented without taking environmental impacts into consideration. Environmental awareness should be established by training various professional sectors in ecological design.
- All actors and users involved in the structural activity process should feel responsible for the environmental impacts of the settlements that have the least impact on environmental values, and should act in this direction. For this reason, using the advantage of still existing traditional life and collaborating with the local people is considered as a solution to enable the transformation of both sides.
- The discipline of landscape architecture, which is knowledgeable about ecosystems and ecological cycles, should be guided by other design disciplines in the design of solutions to ecological problems.
- The adoption of ecological principles and design proposals for the traditional city texture in Şanlıurfa will serve as a guide in making the right decisions in the new development areas to be made by local authorities, practitioners, planners, designers and local people.

References

1. T. KIPER: *Safranbolu Yörükköyü Peyzaj Potansiyelinin Kırsal Turizm Açısından Degerlendirilmesi.* Ankara Üniversitesi Fen Bilimleri Enstitüsü. Ankara, (2006).
2. F. AKLANOĞLU: *Geleneksel yerleşmelerin sürdürülebilirliği ve ekolojik tasarım: Konya-Sille örneği.* Ankara Üniversitesi, Fen Bilimleri Enstitüsü, Ankara, (2009).
3. İ. YÜKSEK: *Geleneksel Anadolu mimarlığında ekolojik uygulamalar üzerine bir araştırma (Kırklareli kırsal alan örneği). Trakya Üniversitesi Fen bilimleri Enstitüsü.* Edirne, (2008).
4. N. MANISALI : *Ekolojik Yerleşimler Üzerine Bir Değerlendirme, İstanbul'dan Örnekler.* T.C. Gebze Yüksek Teknoloji Enstitüsü Mühendislik ve Fen Bilimleri Enstitüsü. Gebze, (2011).
5. Z. AKKOYUNLU: *Geleneksel Şanlıurfa evlerinin mimari özellikleri,* Kültür Bakanlığı Yayınları, Ankara, (1989).
6. M.S. ŞAHINALP: *Şanlıurfa şehri'nin kuruluş ve gelişmesi.* Ankara Üniversitesi Sosyal Bilimler Enstitüsü. Ankara, (2005).
7. Web site: https://www.google.com/earth/ Erişim tarihi: (02.10.2014).
8. Şanlıurfa Hava ve İklim Durumu Web site:http://kongre.hasuder.org.tr/public/conferences/3/schedConfs/1/accommodation-1.pdf). Erişim tarihi: (13.08. 2014).
9. Trc2 Diyarbakır-Şanlıurfa bölgesi bölge planı (2011–2013) mevcut durum analizi. Erişim Tarihi: (13.08.2013).
10. G. TURAN: *Yeni işlev verilen geleneksel Urfa evlerinde koruma sorunları* Dicle ün. Fen Bilimleri Enstitüsü. Diyarbakır, (2009).
11. M.N. MIRDESLIOĞLU: *Şanlıurfa ilinin leguminosea florası.* Harran Üniversitesi Fen Bilimleri Enstitüsü. Şanlıurfa, (2008).

www.ingramcontent.com/pod-product-compliance
Ingram Content Group UK Ltd.
Pitfield, Milton Keynes, MK11 3LW, UK
UKHW020122290926
15471UKWH00003B/221

9 783631 734391